Challenging climate change

Competition and cooperation among pastoralists and agriculturalists in northern Mesopotamia (c. 3000-1600 BC)

Arne Wossink

Sidestone Press

This publication is a result of the project *Settling the steppe. The archaeology of changing societies in Syro-Palestinian drylands during the Bronze and Iron Ages.*

© 2009 A. Wossink

Published by Sidestone Press, Leiden
    www.sidestone.com

ISBN 978-90-8890-031-0

Cover Illustration: The Wadi Jaghjagh as seen from Tell Hamidiyeh.
    Photo by the author (mirrored for esthetical purposes)
Cover design: K. Wentink, Sidestone Press
Lay-out: P.C. van Woerdekom, Sidestone Press

Challenging climate change

Competition and cooperation among pastoralists and agriculturalists in northern Mesopotamia (c. 3000-1600 BC)

Proefschrift

Ter verkrijging van
de graad van Doctor aan de Universiteit van Leiden,
op gezag van Rector Magnificus prof. mr. P.F. van der Heijden,
volgens het besluit van het College voor Promoties
te verdedigen op woensdag 28 oktober 2009
klokke 15:00 uur

door

Arne Wossink

Geboren te Hilversum
in 1978

Promotiecommissie

Promotor:	Prof. dr. J.L. Bintliff
Co-promotor:	Dr. D.J.W. Meijer
Overige leden:	Prof. dr. P.M.M.G. Akkermans
	Prof. dr. R.T.J. Cappers
	Dr. G. van der Kooij
	Prof. dr. E. Peltenburg
	Prof. dr. W.H. van Soldt

Contents

List of figures

List of tables

1 Introduction

1.1 Setting the stage: the context of this study

This study explores human social responses to environmental change in northern Mesopotamia during the late third and early second millennium BC, insofar as these can be reconstructed from the archaeological record. This research developed in two different yet related contexts. The first context is that of the research project of which this study was a part. The second context is that of the current interest in future environmental change and its effects on modern society. These contexts will be discussed below.

This study was part of the research project *Settling the steppe. The archaeology of changing societies in Syro-Palestinian drylands during the Bronze and Iron Ages*.[1] The *Settling the steppe* project was initiated to address common interests of by two earlier Leiden University projects: the excavations of Tell Deir 'Alla (Jordan), and Tell Hammam et-Turkman (Syria). A relatively rapid succession of occupation (re-)establishments and abandonments was observed at Iron Age Tell Deir 'Alla in the Jordan Valley, leading to questions about the stability of these occupations and the nature of past human-environment interaction in the region (van der Kooij 2001). Evidence from the Early and Middle Bronze Age layers at Tell Hammam et-Turkman in the western Syrian Jezirah suggested that the site may have played a key role in the relations between urban and rural populations, which were argued to have been a vital part of life in such marginal environments (Meijer 2000, 2007, forthcoming). The fact that today both the Jordan Valley and the Syrian Jezirah are important agricultural production areas made questions about the sustainability and nature of past occupations even more relevant. There was thus a shared interest in human-environment relations in marginal Near Eastern environments, and the different research lines were brought together in the *Settling the steppe* project. Three central research questions emerged from the common interests of the Deir 'Alla and Hammam et-Turkman projects emerged: (*a*) what was the reason for and nature of human occupation in the steppe zone, (*b*) how stable was this occupation, and which variables determined this level of stability, and (*c*) how did human occupation in the steppe zone relate to more favourable zones? These general questions were to be applied to the project's two regional clusters, focusing on Tell Deir 'Alla and Tell Hammam et-Turkman respectively, and were to be specified as necessitated by local environmental, socio-cultural, and historical conditions. The present study deals with the so-called Hammam et-Turkman cluster. Understanding how human societies in marginal environments coped with environmental change is critical to answering these questions, especially in areas where societies had to maintain a careful balance with their environment in order to survive.[2]

The present study furthermore emerged from the recent concerns with climate change, as elaborated in the reports of the Intergovernmental Panel on Climate Change (IPCC). It has now been established beyond doubt that the world is experiencing global climate change that is at least partly anthropogenic in nature (IPCC 2007). As a result, modern society faces the tremendous challenge

---

1   The *Settling the steppe* project was funded by the Netherlands Organisation for Scientific Research (NWO; project number 360-62-020) and based at Leiden University. The project team consisted of Dr. Gerrit van der Kooij (director), Dr. Diederik Meijer (co-director), Dr. Lucas Petit (post-doctoral researcher, Deir 'Alla cluster), Eva Kaptijn (PhD-student, Deir 'Alla cluster), Dr. Fouad Hourani (geomorphologist, Deir 'Alla cluster), Ellis Grootveld (archaeobotanist, Deir 'Alla cluster), and the present author (PhD-student, Hammam et-Turkman cluster). See Kaptijn et al. 2005 for a general introduction to the Deir 'Alla cluster, and Kaptijn 2009 and Petit forthcoming for detailed studies focusing on that region.
2   Steppe is not by definition a marginal environment, but only under specific socio-cultural and economic conditions, because marginality is an expression of the ability of human societies to survive in a given environment. The term steppe should therefore be strictly used in a descriptive sense and not as a synonym for marginality.

to adapt to a rapidly changing environment. How modern society should mitigate, and adapt to, the effects of climate change is subject to worldwide scientific discussion. Archaeological and historical studies documenting human responses to environmental change can play a role in this discussion, because decisions about the future are made with knowledge about the past. For that reason, many researchers, policy makers and authors have turned to the disciplines of archaeology and history to find models of successful and failed adaptations to environmental change, and to draw lessons from these cases (e.g. deMenocal 2001; Weiss and Bradley 2001; Diamond 2005; Costanza et al. 2007).

There are numerous archaeological case-studies documenting environmentally triggered social disruptions across the globe. Many of these studies fit remarkably well into what McIntosh et al. (2000a: 6) called the 'macrosocietal technological-economic approach', adopted among others by the Intergovernmental Panel on Climate Change (IPCC) in their reports on current climate change. This approach is characterized by biases manifested 'in issues of scaling, in overreliance on economic theory, in little consideration for history, and in misplaced concreteness' that 'limit understanding of the range of human responses to climate' (McIntosh et al. 2000a: 4). Archaeological case-studies that adopt this approach are therefore of limited usefulness to the discussion on current environmental change. This observation justifies a study of past human responses to environmental change that aims to break away from the 'macrosocietal technological-economic approach' characterized above. The present study hopes to achieve this break by explicitly championing a small-scale social and diachronic perspective, paying close attention to regional social, cultural, and environmental diversity (cf. Brooks 2006: 44).

1.2 The emergence of the late third millennium BC climate change hypothesis

Ever since it was first recognised that the transition from the Early to the Middle Bronze Age represented a period of significant social and cultural change in many parts of the Near East, there have been attempts to explain the changes through a single overarching model (cf. Marro 2007a: 14). The archaeological record of this transitional period was characterized by changes in settlement patterns, the emergence of new cultural traits, and apparently abrupt settlement destruction and abandonment. Explanations have ranged from foreign invasions or penetration of new (ethnic) groups (Kenyon 1966) to system collapse (Butzer 1997; Peltenburg 2000). Based on ancient Egyptian texts, Bell (1971) proposed that an interruption of the annual Nile floods may have been responsible for the social disruption of the contemporary First Intermediate Period in Egypt. However, this hypothesis could not be substantiated by palaeoenvironmental data at the time and was not taken up for other parts of the Near East until the early 1990s.[3]

In the early 1990s, climate change was put forward as an explanation for the changes observed in settlement patterns and in the political constellation of the ancient Near East and beyond (Weiss et al. 1993). The study by Weiss et al. held that (a) soil micromorphological studies at and around Tell Leilan indicated the occurrence of an abrupt dry phase from 2200–1900 BC, possibly caused by an otherwise unidentified volcanic eruption, (b) this evidence was supported by indications for drought elsewhere, notably Egypt, (c) this drought led to (i) the abandonment of urban settlements in the area around Tell Leilan including Tell Leilan itself, (ii) an overall decline of settled population in the same area, (iii) an increase in pastoral subsistence strategies at the cost of agricultural production, (iv) the collapse of the Akkadian state due to grain shortages, and (v) the subsequent collapse of the Ur III state due to pressure from refugees coming from the north and grain shortages as reported in the royal correspondence from Ur, and (d) this collapse was synchronous with collapses elsewhere in the Old World, notably the Levant, Egypt, the Aegean, and the Indus Valley (Weiss et al. 1993). In the present study, this model of climate-induced societal change will be referred to as the Leilan Climate Change Model (LCCM) (see Section 2.3.5 for further discussion).

---

3   Advocating climate change as an explanation for socio-cultural change goes back much further in (Near Eastern) archaeology however (see Section 1.3, and summaries of previous research in Issar and Zohar 2004; Gronenborn 2005; Rosen 2007).

The LCCM has been critically received. Butzer succinctly summarized the criticisms as follows: '(1) Are the empirical climatic data valid? (2) If so, was the climatic anomaly effective and synchronous across the area in question? and (3) Did sociopolitical unrest or collapse everywhere follow upon climatic change, in a form consonant with greater aridity?' (1997: 249). Indeed, many discussions of the LCCM focused on exactly these points. Firstly, environmental data did not seem to uniformly support abrupt late third millennium drought (e.g. Bottema 1997). Secondly, historical sources were uncritically used, and it has not been proven that the Upper Khabur was of critical importance for grain production in the Akkadian state (Zettler 2003). Thirdly, urban collapse in northern Mesopotamia was probably not as wide-spread as argued by Weiss et al. (Dohmann-Pfälzner and Pfälzner 2001; Oates and Oates 2001; Wossink forthcoming). Fourthly, the cause, date and nature of the abrupt arid phase are still contested (e.g. Courty 2001; Kuzucuoğlu 2007a). Fifthly, the role of late third millennium BC environmental stress in other so-called collapses across Asia is still debated (e.g. Madella and Fuller 2006). Still others argued that the urban collapses did not result from abrupt climate change, but from frequently recurring strong oscillations in rainfall. In this scenario, climate change may have taken place but it only marginally affected processes that were already in operation (Wilkinson 1994, 1997). Finally, inconsistencies in the geomorphological data from Tell Leilan and a problematic correlation with other palaeoclimate proxy records may have contributed to the continuation of the debate (see Section 2.3.5).

Nevertheless, the LCCM has continued to be propagated in a relatively unmodified form since its earliest conception (e.g. Courty and Weiss 1997; Weiss 2000; Ristvet and Weiss 2005; Staubwasser and Weiss 2006). There is now a stronger focus on globally recognized climate mechanisms, rather than volcanism, as the cause of the so-called 4.2 ka BP event. Furthermore, the causal link between abrupt drought and the end of the Akkadian state has been downplayed in favour of an emphasis on the effects of drought observed at and around Tell Leilan itself. These possibilities were, however, already presented in the 1993 publication. Therefore, they do not represent significant changes to the LCCM (see Section 2.3.5).

At the same time, many researchers are aware that the basic argument of the LCCM, namely that there was some form of aridification, has stood the test of time. The shift from a relatively wet climate during the early third millennium BC to a drier climate similar to that of today during the early second millennium BC is now relatively well documented as a result of the increased number, reliability, and resolution of Near Eastern palaeoclimate proxy records. Although there is still conflicting evidence, it seems beyond doubt that, at least in northern Mesopotamia, the third millennium climate change involved a shift toward more arid conditions (Kuzucuoğlu 2007a). Even though specific details may still be a matter of debate, the LCCM has thus succeeded in (re-)introducing climate change as an important variable of socio-cultural change (Akkermans and Schwartz 2003: 283).

1.3 Studying human responses to climate change: methodological considerations

Before methodology can be discussed, important terms must be defined and clarified. The online *Oxford English Dictionary* defines the *environment* as 'The conditions under which any person or thing lives or is developed; the sum-total of influences which modify and determine the development of life or character'. Halstead and O'Shea (1989a: 2) give a similar definition in which environment is defined as all external factors acting upon the social unit that is studied. Climate, then, is one of these external factors, and changes therein are at the same time environmental changes. Humans outside the studied social unit are strictly speaking also part of that unit's environment; this is a matter of the scale at which the study is carried out. *Climate* 'in a narrow sense is usually defined as the average weather, or more rigorously, as the statistical description in terms of the mean and variability of relevant quantities over a period of time ranging from months to thousands or millions of years' (IPCC 2007: 78). This definition allows the distinction to be made between climate variability and climate change. *Climate variability* represents 'variations in the mean state and other statistics (such as standard deviations, the occurrence of extremes, etc.) of the climate on all spatial and temporal scales beyond that of individual weather events' (IPCC 2007: 79), whereas *climate change* is a 'change in the state of the climate that can be identified (e.g. by using statistical tests) by changes in the mean and/or variability of its properties, and that persists for an extended period, typically decades or longer' (IPCC 2007: 78). Derived from this definition, *abrupt climate change* can be understood as climate change where the involved 'time scales [are] faster than the typical time scale of the responsible forcing [mechanism]' (IPCC 2007: 76).[4] Weather is nothing more than day to day variations in climatic variables within the long-term mean. The difference between the terms climate change, climate variability, and weather as used here, then, is largely a matter of scale (Butzer 1982: 23). A further difference between variability and change is that variability is predictable because it is to a certain degree cyclical in nature, whereas change can be conceptualised as directional, by moving away from previously established conditions toward new conditions.[5]

Archaeological and historical records provide a unique and effective means of following changes through time in the demography and social organization of human societies. At the same time, the past two decades have seen a continuous refinement of palaeoclimate proxy records, and palaeoclimate models. Synchronization of material culture chronologies with palaeoclimate histories is of prime importance to researchers who wish to evaluate the impact of climate on socio-cultural change. It is only through correct synchronization that archaeologists can perceive potential linkages between climate and socio-cultural changes.

However, synchronization of palaeoclimate developments and socio-cultural histories, as reflected in material culture change, is only the first step that must be taken in order to understand the relation between these two mechanisms. After this crucial first step, it must be demonstrated that climate was the key factor in determining socio-cultural change. In order to achieve this goal, models of human-environment interaction must be more carefully considered (Minnis 1996: 57; Coombes and Barber 2005: 305). Failure to consider the complex interactions between humans and their environment will lead either to an environmentally deterministic view of socio-cultural change or to a complete neglect of possible environmental impact on human action and history. Either of these two approaches has its own drawbacks, as will be discussed below.

Environmental determinism, or 'the idea that a society's physical environment can control its cultural development' (Coombes and Barber 2005: 303) has long been a strong paradigm in anthropology and archaeology alike. During the second half of the twentieth century, support for environmental determinism waned. The 1990s saw a revival in environmental determinism that may be associated with concerns about the present anthropogenic climate change, and the realization that past Holocene climate was less stable than had hitherto been thought. These recent studies often constructed their argument for climate induced socio-cultural change along three steps: (*a*) presenting well-dated high-resolution evidence for climate change in a given region, (*b*) presenting archaeological evidence for synchronic change in one or more socio-cultural systems, and (*c*) pointing out that causality between the observed climatic and socio-cultural change was very likely, given the synchronicity of these changes. However, insufficient attention was paid to the socio-cultural mechanisms leading to a given response, and it was not shown that the socio-cultural change was a response to environmental change at all. In other words, these studies failed to show that climate change was the critical factor in the observed socio-cultural change (Coombes and Barber 2005: 304–5; Gronenborn 2005: 5). Coombes and Barber called this approach a 'basic "black box" determinism, correlating climatic and cultural changes and largely neglecting the processes involved' (2005: 304), of which they argued that 'little can be said about what forms environmentally triggered transitions might take, why certain cultures should be susceptible to them, or – most pertinently – if the proposed relationships can indeed be corroborated' (2005: 305). Winterhalder (1994: 36) similarly argued that there is a general tendency to foreground environmental data, whereas much more attention should be given to the formulation of models of environmental

---

4   In light of this definition, it seems premature to call the 4.2 ka BP event an *abrupt* climate change, given the fact that the forcing mechanism of the event is still not understood, and there is still not sufficient agreement on its timing and nature (see Section 2.3.7).

5   Even random events that fall within climate variability can to a certain degree be considered cyclical because they occur at statistically regular intervals (Rosen and Rosen 2001: 536), for instance a spring tide that statistically occurs once every hundred years.

variability that can also be applied in the social sciences (cf. Rappaport 1979: 59–60; Rosen and Rosen 2001: 539). These problems with current approaches are fundamental issues that any new study of environmentally triggered social change must tackle.

The reverse of environmental determinism, namely the belief that environmental change cannot cause socio-cultural change, or only in very specific cases, is based on a flawed perception of human socio-cultural evolution. Early human evolution and prehistory, that is, the emergence of early modern humans and the adoption or invention of new tool types and technologies including agriculture, are usually studied with reference to the environmental context within which these developments occurred. This perspective partly results from a relative lack of material relating to socio-cultural complexity, when compared with later periods. This evolutionary and somewhat environmentally deterministic perspective increasingly tends to make way for a socio-cultural theoretical framework as social complexity and the scale apparent in the material record of the studied societies increase. Once socio-cultural complexity and diversity become apparent in the archaeological record, explanation of this diversity is given priority, and human-environment relations are no longer studied. In fact, for later periods these relations are assumed to be rather one-sided, according humans increasing control over their environment as their technology becomes better and socio-cultural complexity increases (Crumley 1994: 2–3).

This dichotomy between evolutionary and culturally determined histories clarifies the need for a framework for the study of human-environment relations that allows the recognition and inclusion of multiple causes and causalities. Within this framework, human-environment relations must be understood as the respective impacts, modifications, adaptations, and influences of humans and their environments on each other. This definition of human-environment relations supports the recognition that neither nature, or the environment, nor culture is the determining factor in these relations. There is a recursive relationship between humans and their environment, and whether human individuals or groups adapt to or impact on their environment, or do both, must be considered on a case to case basis.

A further important consideration is that human-environment interactions are shaped by past experiences (Winterhalder 1980: 147). Human individuals and groups base their decisions and actions on what they learned or experienced in the past, and apply this knowledge to the situations they are currently facing. Thus, these decisions are based on knowledge collected by individuals and groups (McIntosh et al. 2000a: 24). It can in fact be argued that the environment itself results from an extremely long history of human-environment interaction, especially for the period and region that are the subject of this study (Köhler-Rollefson and Rollefson 1990; McCorriston 1992). More abstractly, it can be argued that 'Humans as components of the landscape are not merely exogenous system disruptors, but one of many contributors to the various environmental constraints in operation' (Gragson 1998: 218). In order to fully justify this historicity, any study of human-environment relations must be explicitly diachronic in nature (Butzer 1982: 279; Crumley 1994: 6; McIntosh et al. 2000a: 8–9; Costanza et al. 2007: 522).

These observations on historicity and the role of humans in the shaping of their environment lead to the conclusion that the susceptibility of human populations to environmental change is primarily a *social* condition. Humans are not inherently susceptible to climate change, but only as a result of very specific local environmental, socio-cultural, and historical circumstances. As McIntosh et al. put it: 'To a human population, an environmental crisis is primarily a matter of the social realm, implying a failure of adaptation. The truly interesting questions in such cases involve not just meeting the biophysical challenge but also understanding why human problem solving seems on critical occasions to become ineffective' (2000a: 7).

It is therefore suggested that providing evidence for climate change and contemporaneous socio-cultural change is insufficient to prove causality between these two processes. Rather, it must be demonstrated that climate change affected human communities in such a way that they could no longer continue their traditional way of living and had to adapt. In other words, it is not sufficient to prove that the climate became drier during the late third to early second millennium BC, but it must also be shown that this change resulted in drought, that is, insufficient water to sustain the biomass of a given region (Barrow 1987: 36). Because drought is a function of environmental as well as biotic factors, it is in fact a social condition when it is applied to human communities.

As such, drought cannot be postulated from decreasing rainfall levels alone, as drought may also be triggered by ineffective subsistence strategies or by increasing population levels (Barrow 1987: 36–7).[6] In this sense, northern Mesopotamian palaeoclimate remains a relatively poorly known factor in the present discussion (see Section 2.3.6). It may therefore be in fact much more fruitful to look not at how societies responded to drought, but to look at which responses could be expected, had there been drought (see Sections 1.4 and 3.5).

Finally, the scale at which responses to environmental change are studied must be explicitly considered. Many previous studies found a society or culture to be a convenient or appropriate unit of analysis (e.g. deMenocal 2001; Weiss and Bradley 2001; Diamond 2005). However, there are serious objections to the adoption of culture or society as a unit of analysis in the study of responses to environmental change. Cultures are not ecological units in the sense of human individuals, or populations. If human populations are studied in an ecosystemic context, culture is an important property of those populations in the maintenance of human-environment relationships, rather than a distinct unit in and of itself (Rappaport 1979: 62). Societies on the other hand are complex entities consisting of multiple components, each of which may or may not respond in a certain way to environmental change. Decisions on how to respond to environmental change are made not by societies as a whole, but by groups such as communities, households, and ultimately individuals (McIntosh et al. 2000a: 4). Nevertheless, at each of these levels, decision-making may be influenced by higher-order groups or individuals, such as elites (Rosen 2007: 147). Recognizing this diversity, or explaining its absence, is necessary in order to improve our understanding of past societies. It should therefore be of prime importance to anyone studying the effects of environmental change on human societies to reconstruct the decision-making processes at these diverse scales, and to identify an appropriate scale for responses to environmental change (see Section 3.2).

1.4 The central research question

As has been outlined above, the present study is explicitly concerned with social responses to environmental change. Rather than focusing on changes in subsistence strategies, the present study aims to explore changes resulting from environmental change that occurred in the social relations between groups such as communities and households. Because humans are social beings, relations with other groups and individuals are fundamental to the maintenance of health and well-being. When these properties are compromised, changes in these relations may be expected as part of the adaptation process.

Within the context of the *Settling the steppe* project, the present study is concerned with the appearance, stability, and abandonment of settlements in the marginal areas of northern Mesopotamia during the Early and Middle Bronze Ages (*c.* 3000–1600 BC). As the late third millennium BC was a period of significant environmental change (see Section 2.3.7), changing social relations may be expected for these communities. However, these sedentary communities were only part of the total society living in this region, and they not only had to maintain relations with other sedentary communities, but also with more mobile groups alluded to by historical records from at least the late third millennium onward (see Chapters 6 and 7). The central research question thus becomes: how did social relations develop between sedentary groups, and between sedentary and (semi-) nomadic groups in response to natural environmental stress in northern Mesopotamia during the late third to early second millennium BC?

Security maximization against resource stress or shortage, whether resulting from adverse environmental conditions, enemies, or otherwise, is a strong motivation for human behaviour (Halstead and O'Shea 1989a: 1; Binford 2001: 41). In times of scarcity, much attention and energy will be

---

6   Note a similar comment by Lewis on the transitional zone between the desert and the sown: 'The limits of the intermediate zone were not fixed; they depended on politics more than on geography, particularly on the eastern border, because the desert "is an economic rather than a geographic expression". The desert was not necessarily arid and uncultivable, but rather the area in which the nomads wandered and which was devastated by their flocks and herds' (1987: 23).

devoted to the maintenance of the individual's and group's well-being, including the development and/or modification of social relations with other individuals and groups. These social relations can be classified under one of two general behavioural strategies to ensure survival for the individual or group: cooperation and competition. The social expressions of these mechanisms are very diverse and depend on local socio-cultural and ecological/environmental conditions. Thus, trade and the formation of shared identities are cooperative strategies, whereas territoriality and warfare are competitive strategies. Whether groups develop cooperative or competitive strategies, depends on the environmental constraints operating on the involved groups, and on the socio-cultural context within which they operate (Section 3.5).

Detailed discussion of these concepts will be deferred to Chapter 3 but the resulting hypotheses will be stated in advance here. Application of the concepts of competition and cooperation leads to the formulation of two hypotheses regarding the development of social relations in northern Mesopotamia during the late third-early second millennium BC: (*a*) it is to be expected that, as a result of environmental stress, social relations between sedentary agricultural communities will develop along competitive lines in order to improve each communities' chance of survival and access to scarce and therefore contested resources, and (*b*) sedentary and (semi-)nomadic groups will develop cooperative mechanisms ensuring mutual survival and maximization of their respective subsistence strategies.

1.5 About this book

This book is divided in eight chapters, including this introduction. Chapter 2 provides the geographical, environmental and historical background of the study area and discusses the evidence for climate change during the third and early second millennium BC. Additionally, it considers the archaeological and historical framework, and the political history of the region.

Chapter 3 explores the theoretical foundations of the previously advanced hypotheses. It furthermore aims to highlight archaeological correlates for the various responses. Rather than rely on the patchy historical record to infer conflicts between settlements and (city-)states, the development and organization of settlement patterns is presented as a suitable method to follow changes in relationships between sedentary communities. The development of supra-regional social identities is presented as a social mechanism that can strengthen ties between sedentary and (semi-)nomadic communities. It is argued that changes in these two mechanisms correspond to changes that would be expected in the face of environmental stress.

Chapters 4 and 5 are concerned with the development of social relations among sedentary communities as can be gleaned from the record of excavated and surveyed sites. Because survey data do not allow periods of settlement use and abandonment to be precisely established, it is necessary to determine the degree to which settlements dating to the same chronological period were actually contemporaneous. Chapter 4 presents a method for estimating the number of contemporaneous settlements, and discusses approaches to determining the number of persons living at a site in order to establish community sizes and changes therein. In Chapter 5, three regions are selected for which sufficient data are available from surveys and/or excavations. After discussing these case-studies and detailing the steps that are taken in constructing the databases, the statistical analyses outlined in Chapter 4 are performed on each dataset, and the results are compared.

Chapter 6 explores the organization of pastoralism in northern Mesopotamia during the third and early second millennium BC. Historical and archaeozoological data are the primary sources for this analysis. Although these records are unevenly distributed over the period and region under discussion, it is nevertheless possible to postulate the emergence of a specialized (semi-)nomadic pastoralism during the late third to early second millennium BC.

Chapter 7 is concerned with changes in the social relations between sedentary and (semi-)nomadic communities. The reconstruction of these relations draws primarily on historical sources, since other archaeological evidence pertaining to third and second millennium BC (semi-)nomadic groups is scarce, if not absent. Even so, it must be realized that the historical sources also tend to present a one-sided view, since these texts were primarily written by city-dwellers. However, despite this bias, it will be argued that these sources may be used in a reconstruction.

Chapter 8 brings together the evidence presented in Chapters 4 through 7. This chapter evaluates the research question and hypotheses presented above and in Chapter 3, and discusses them in light of the data presented in the previous chapters. Here, the degree to which the data conform to the expected patterns is discussed, and explanations for any deviations are offered. Central to this chapter, and to this study as a whole, is an overview of the social responses to late third to early second millennium BC climate change in northern Mesopotamia. Some final comments and directions for further research conclude this chapter and this study.

2 Northern Mesopotamia from 3000 to 1600 BC

2.1 Introduction

This chapter aims to reconstruct the environment and history of the northern Mesopotamian world during the third and early second millennia BC.[7] It starts with an outline of the geography of northern Mesopotamia of this study (Section 2.2) in which the current climatic and environmental conditions and the potential vegetation are discussed. Furthermore, the effects of recent land-use patterns on the northern Mesopotamian landscape will be evaluated. Section 2.3 presents the available data on past climate changes in northern Mesopotamia. Section 2.4 outlines the archaeological and historical chronologies that will be employed here. Finally, Section 2.5 outlines the political history of the region in so much as it can be reconstructed from written records.[8]

2.2 The geographical background

Geographically as well as geologically, modern Syria can be divided into zones that are roughly aligned along an east-west axis (Wirth 1971: 17). Bordering the Mediterranean Sea is a narrow coastal strip that until recent times was insalubrious and very sparsely settled. To the east, the land rises up to the mountain ranges of the Amanus, Jebel Ansariya and Libanon. Continuing to the east, the land descends into the valleys of the Orontes and the Jordan Rivers, and rises up again into the lower mountain ranges of the Jebel Samane, Zaouiye, Anti-Libanon and Hermon. Finally, the large flat steppes of inner Syria are reached, which stretch eastwards to the Zagros Mountains and are bordered by the Taurus in the north. These flat steppes are intersected by several rivers, of which the Euphrates and Tigris are the most important ones. Between these two rivers lies the area that is today known as the Jezirah (Fig. 2.1).

The Jezirah is part of the Arabian plate that slopes down towards the east. The northern Jezirah consists of flat plains, with few striking relief features. Of these, the Jebel Abd al-Aziz and the Jebel Sinjar are the most marked, with heights of 920 and 1480 m respectively. Using both rainfall and geological data, Reifenberg (1952: 86–7) described the soils of the Jezirah for those regions receiving less than 200 mm of precipitation annually as desert soils, and as Mediterranean steppe soils and alluvial Mediterranean steppe soils those areas receiving more than 200 mm. The northern half of the Jezirah is characterized by calcareous soils, whereas to the south gypsiferous soils dominate. In the northeast corner of the Jezirah, reddish-brown loams on igneous rock occur (Wolfart 1967).

Rivers and wadis in the northern Jezirah have cut into the plains, resulting in relatively shallow valleys. Most of the rivers and wadis drain into the Euphrates. The Khabur and Balikh are the most prominent of these rivers. The Khabur is fed by the karstic springs of 'Ras al-Ain, and is joined by multiple wadis including the Wadi Jaghjagh and Wadi Radd, creating the Khabur Triangle. Today, the Wadi Jaghjagh is an intermittent stream but it carried water permanently until quite recently.

---

7   The term northern Mesopotamia is used very loosely here to designate the upper part of the Euphrates and Tigris basins and the land in between, including the foothills of the Taurus Mountains. As such, it is largely synonymous with the Jezirah, the term currently used to designate the land between the Euphrates and Tigris rivers north of modern Baghdad and south of the Taurus. The term greater Mesopotamia is cursorily used to describe the Jezirah and the Tigris–Euphrates alluvium together (cf. Wilkinson 2000a: 222).

8   Throughout this chapter and those that follow, historical dates will be expressed as normal calendar years (BC). Where necessary, radiocarbon ages were calibrated using the calibration programme OxCal 3.10 and are expressed as calibrated calendar years (cal BC).

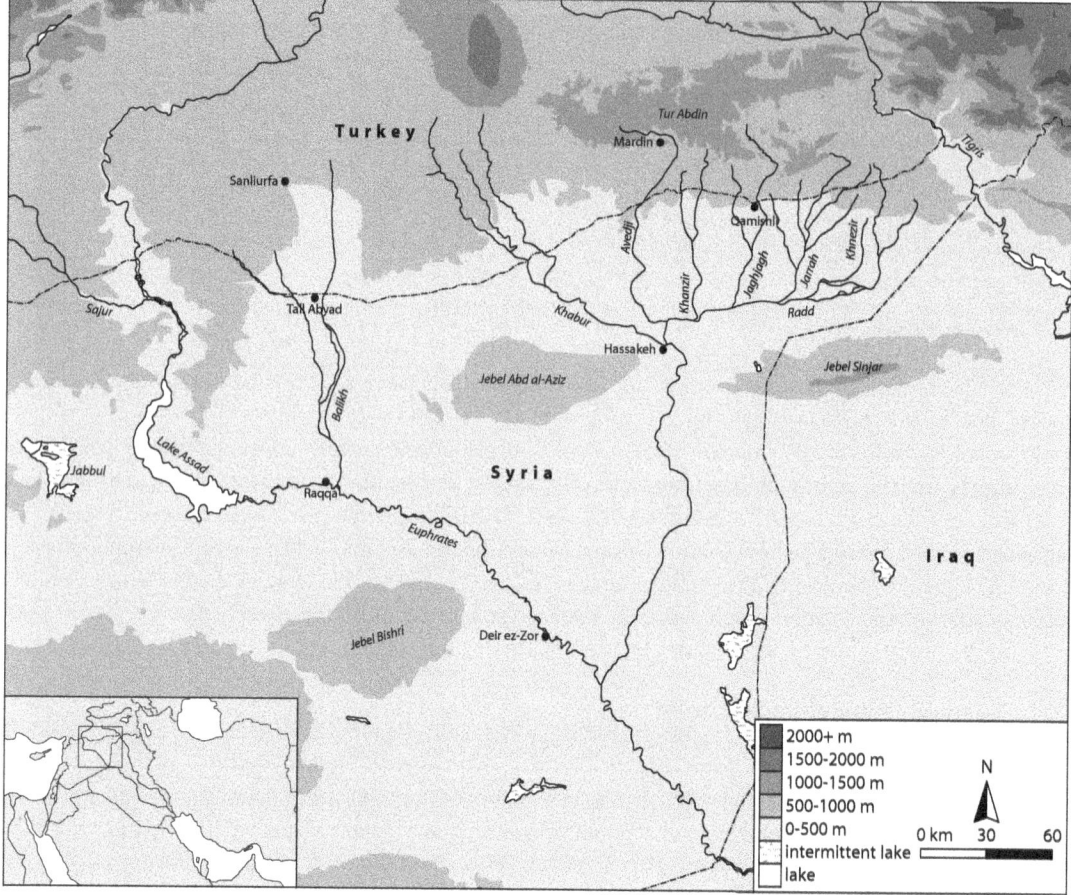

Fig. 2.1: Map of northern Mesopotamia showing major rivers, mountain ranges, modern places, and political borders. Inset map shows the research area in its wider Near Eastern context.

Mean discharge of the Khabur was 50 m³/s with a peak of 300 m³/s and a minimum of 35 m³/s. The Balikh received most of its water from the spring at 'Ain al-Arus, just south of the present-day Turkish–Syrian border, and had an average discharge of 6 m³/s, with a minimum flow of 5 m³/s and a maximum flow of 12 m³/s. By comparison, Euphrates discharge in Syria ranged between 5000 m³/s in April-May, and 250 m³/s in August–December, with a mean of 840 m³/s (Wirth 1971: 109–10). Average discharge of the Tigris near Baghdad was 1236 m³/s, with a maximum of 2909 m³/s in April, and a minimum of 352 m³/s in October (Buringh 1960: 51). Note that most of these figures describe the situation before the introduction of large-scale irrigation works and the construction of hydroelectric dams in the Euphrates, Khabur, and Tigris Rivers in Iraq, Syria, and Turkey.

2.2.1 Modern climate

Climate and geography create distinct local weather patterns. Today, the Jezirah has a dry continental climate characterized by hot, dry summers and cool, wet winters. Winter climate is heavily influenced by moisture-carrying cyclones from the Atlantic Ocean and the Mediterranean Sea. The lion's share of this moisture is deposited in western Syria, where the Jebel Ansariya and Lebanon Mountains create a rain shadow for the interior. Moisture is partly carried further east and deposited in northeast Syria, where the Taurus–Zagros Mountains act as a barrier (Fig. 2.2). During summer, winds come from Anatolia or Iran and cause dry, stable weather. Rainfall is limited to fall, winter, and spring; the summer months being almost completely dry. Temperature and potential evapotranspiration reach their maxima during summer. In western Syria, December and January are the wettest months, whereas to the east, including the Jezirah, the highest precipitation is recorded for January and February (Fig. 2.3). From west to east, reliability of rainfall generally decreases as

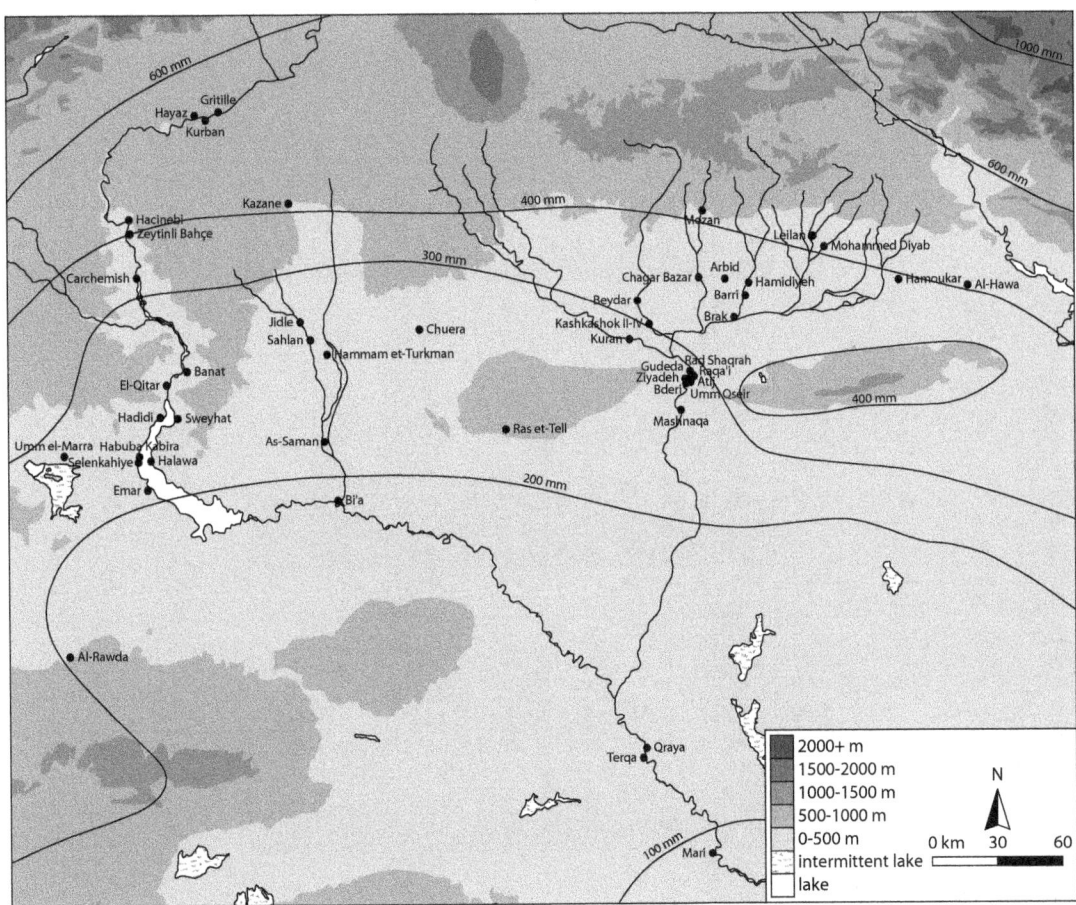

Fig. 2.2: Map of northern Mesopotamia showing mean annual rainfall based on long-term measurements (based on Wirth 1971: Karte 3), and archaeological sites mentioned in this study.

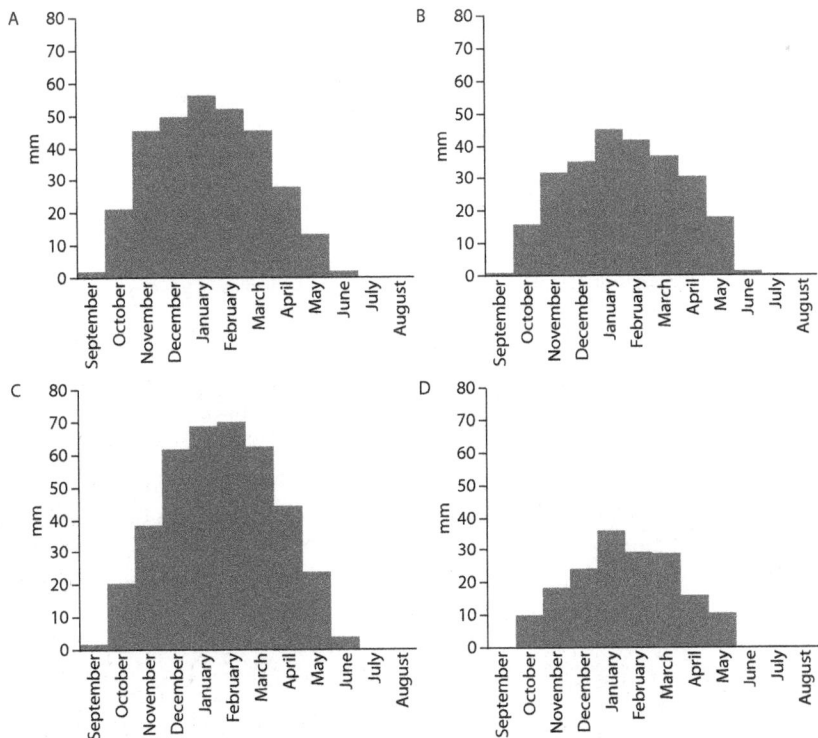

Fig. 2.3: Average monthly rainfall based on a 20-year record (1986–2005) for selected weather stations in northern Mesopotamia (data from the Syrian Meteorological Department). (A) Aleppo. (B) Hassakeh. (C) Qamishli. (D) Raqqa.

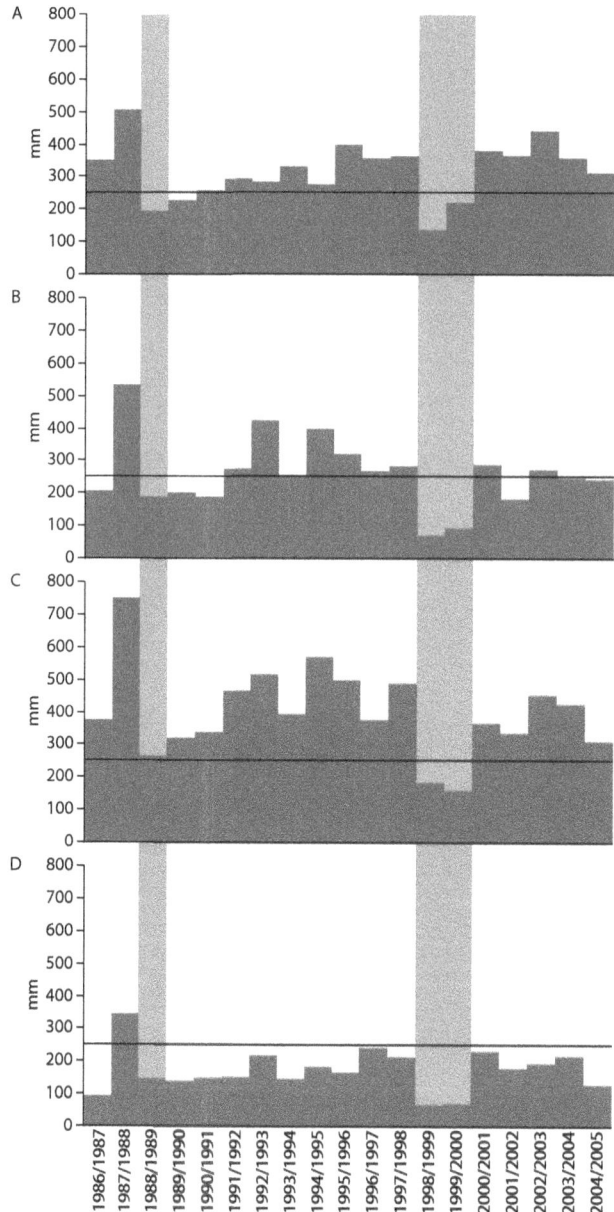

Fig. 2.4: Annual rainfall in hydrological years at selected weather stations in northern Mesopotamia from 1986 to 2005 (data from the Syrian Meteorological Department). Horizontal line indicates 250 mm. Shading indicates years with less than 250 mm at all stations. (A) Aleppo. (B) Hassakeh. (C) Qamishli. (D) Raqqa.

does the amount of rainfall. Comparison of annual precipitation from selected weather stations in and around the Jezirah suggests a high correspondence between these locations. Fig. 2.4 indicates that in three out of twenty years, rainfall at all stations drops below 250 mm. In other words, low rainfall in one area is likely to be matched by low rainfall elsewhere (Wilkinson 2004: 44). Long-term rainfall and temperature observations seem to indicate that dry years occur in groups, rather than individually dispersed among average years. Wirth (1971: 92) has calculated that during such dry years, the 250 mm isohyet coincides with the long-term average 400 mm isohyet. Conversely, wetter than average years seem to cluster as well (Wirth 1971: 97).

2.2.2 Current vegetation and land-use

The potential vegetation of the Jezirah, that is the vegetation as it would exist in the absence of human activities such as agriculture, pastoralism, and deforestation, has been reconstructed by Moore et al. (2000). For this reconstruction, they used a combination of climatic, botanical, geographical, and historical evidence, and the vegetation as it survives in areas that have been relatively shielded from grazing and human intervention. Eight vegetation zones are distinguished, ranging from true forests in the mountain ranges along the Mediterranean coast to true desert in Syria's interior. For the northern Jezirah, the number of vegetation zones encountered is more limited. The vegetation zones are roughly aligned on a north-south axis, except for vegetation zones related to river valleys and mountain ranges (Fig. 2.5). For the northern parts of the Jezirah, dense forest-type vegetation was reconstructed, dominated by various types of oak, including the Turkey oak (*Quercus cerris*) and Lebanon oak (*Q. libani*). Other important plants include various almond (e.g. *Amygdalus communis* subsp. *microphylla*) and terebinth species (e.g. *Pistacia atlantica, P. palaestina*). This vegetation type can potentially be supported in areas with annual rainfall exceeding 400 mm. In slightly drier areas, this oak-dominated forest gives way to park-woodland vegetation. Oak is still present, but its distribution is less continuous and interspersed with grassland areas. Multiple shrub species occupy the border zones of the oak patches in these park-woodlands. Still further the south, as the Jezirah becomes progressively drier, forest trees can no longer be supported, except for some drought resistant species. Overall tree density is low, and steppe species such as *Artemisia* spp., dominate. True steppe appears further south, well below the 200 mm isohyet. This area is today dominated by *Artemisia* and Chenopodiaceae, but would support a large variety of grasses in the absence of heavy grazing. The Euphrates and Tigris valleys, as well as parts of the Balikh and Khabur, support riverine forest vegetation, including poplar (*Populus euphratica*), ash (*Fraxinus rotundifolius*), and elm (*Ulmus* spp.). Swamp areas, as can potentially be encountered along the Euphrates, Balikh, and the Wadi Radd, would support wetland species. The borders between zones are generally correlated to rainfall levels but plant species could extend well beyond their hy-

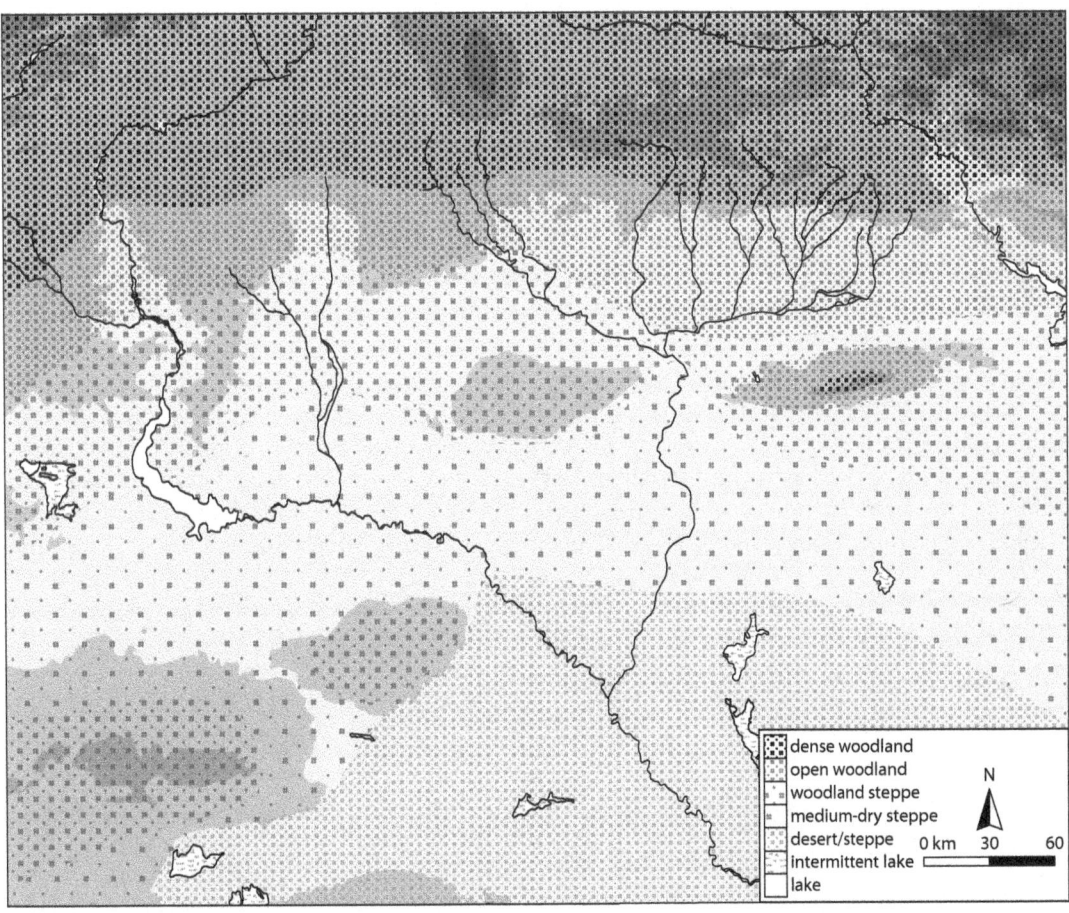

Fig. 2.5: Map of northern Mesopotamia showing simplified potential vegetation zones under modern climatic regime (based on Moore et al. 2000: fig. 3.7).

pothesized zones due to the moisture carrying capacities of specific soils. Especially in and around wadis, and in coarse-grained soils, plant species may be encountered that need more water than rainfall alone would offer (Moore et al. 2000).

However, the current vegetation in the northern Jezirah, and the Near East in general, is characterized by millennia-long human modification and use and far removed from the potential vegetation described above (Köhler-Rollefson and Rollefson 1990; McCorriston 1992). Like natural vegetation, cultivated crops are subject to environmental constraints, particularly rainfall, in the absence of artificial irrigation and other agricultural production-enhancing strategies. Thus, as rainfall diminishes, drought-tolerant crops will probably increasingly dominate the crop spectrum. The number of production strategies aimed at reducing the effects of water shortage, such as fallowing, will also increase. In addition, the overall number of crops is likely to decrease with lower precipitation. Under relatively recent production strategies, as rainfall diminishes from over 300 mm per year to less than 200, wheat is gradually replaced by barley as the main cereal crop, and the variety of other crops also diminishes significantly (Wilkinson 2004: fig. 3.3).[9] Below 200 mm, agriculture as a subsistence strategy becomes altogether unimportant. The 200–300 mm isohyet clearly represents a border zone in terms of the dominant subsistence system, with agriculture being gradually replaced by pastoralism. Even so, specific soil properties in the relatively dry areas can allow farmers to grow crops with high moisture requirements there (Wilkinson 2004: 43).

Today, the greater part of the northern Jezirah is cultivated, with cereals being the main crop, followed by cotton (Lévêque n.d.: 11–13). Much of this cultivation is sustained through irrigation, practised both on a local and regional scale. Artificial wells tap the groundwater table, which as a result has dropped considerably in recent decades. Extensive canal networks provide water from the major reservoirs in the Euphrates and Khabur rivers, or are projected to do so in the near future (Hole and Zaitchik 2007: fig. 4). Mechanical pumps distribute water from perennial as well as intermittent streams to the fields. As a result of the irrigation possibilities, summer cropping in the northern Jezirah, requiring large amounts of water and previously virtually non-existent, increased dramatically in recent years. Additionally, there has been a shift in the location of agricultural plots away from the river valleys towards off-river plots (Hole and Zaitchik 2007: 142). The current agricultural production regime with its heavy reliance on irrigation water is causing problems of soil salinization and groundwater quality degradation, and agricultural plots and farms in unfavourable locations have been abandoned in recent years (Hole and Zaitchik 2007: 144).

The currently established practice of mechanized cultivation can have disastrous effects on local environmental conditions if agricultural fields are abandoned, as has been observed in Iraq (Thalen 1979). Cultivation requires that natural vegetation is cleared, large stones are removed, and fields are deeply ploughed. In marginal areas, the most fertile fields will be cultivated first, for instance wadi bottoms or areas with slightly higher rainfall. These areas are also the most productive grazing lands for pastoralists. A few dry years may lead to abandonment of these fields, and relocation of cultivation to more fertile lands. On the marginal fields, exposed and loosened topsoil remains, which is easily affected by wind and water erosion. Surface sealing, whereby a hard crust is formed, or the formation of a desert pavement may inhibit the re-colonization of the area by the original vegetation. Abandoned fields are likely to be colonized by species with low grazing value.

	300> mm	300–200 mm	<200 mm
barley/fallow	+/-	+	-
wheat	+	+/-	-
fodder crops	+/-	+/-	-
grain, legumes, summer crops, tree crops	+/-	-	-

Table 2.1: Ethnohistorical correlation between various agricultural crops and annual rainfall (+: significant; +/-: present; -: insignificant) (based on Wilkinson 2004: fig. 3.3).

---

9   One reason why barley is more suitable as subsistence crop in drier areas than wheat is that barley is more salt-tolerant. According to Barrow, wheat can take water with a maximum salinity of 15000 parts per million (ppm), whereas barley has a salinity tolerance of 35000 ppm (1987: table 7.4).

Vegetation regeneration would require between five and fifteen years without grazing and changes in soil condition. This means that short-term unsuccessful cultivation of marginal lands will lead not only to the destruction of potential farming lands, but also to the degradation of land that is normally used for grazing flocks of sheep and goat by pastoralists (Thalen 1979: 295–6, 310).

Sheep/goat pastoralism remains important in the Jezirah and the more arid regions south of the Euphrates (Lewis 1988; d'Hont 2004; Lévêque n.d.: 15, 24–5). Three types of pastoral populations can be distinguished in the ethnohistorical past in Syria: Bedouin who relied on nomadic camel-rearing, full-time nomadic pastoralists relying on sheep/goat, and semi-nomadic sheep/goat pastoralists. Nomadic sheep/goat pastoralists traversed between 50 and 200 km and spent the summer above the 200 mm isohyet. There, the sheep grazed on fallowing fields or wastelands. During the colder winter months, the pastoralists camped more to the east and south of the summer pastures, in areas that during the summer were often used by Bedouin. Nomadic sheep/goat pastoralists left the summer pastures in November and stayed in their winter camps until May. The largest Syrian tribes had winter pastures northwest of the line Damascus–Palmyra–Raqqa. Semi-nomadic pastoralists (*Halbseßhaften*) were former fully nomadic pastoralists who had settled during the French mandate period and undertook transhumance-like activities for only part of the year (Wirth 1971: 256–8). Hole (1991) described a migration pattern that was dependent on the available precipitation, that is the wetter the summer season, the farther nomadic pastoralists could move away from the permanent water sources (see Section 6.2 for a full discussion of the terms pastoralism, (semi-)nomadism, and transhumance).

2.3 Past environment of northern Mesopotamia

This section discusses relevant climate proxy records, the combined results of which will be used to reconstruct the palaeoclimate of the Jezirah. Focus will be on high-resolution records outside northern Mesopotamia, as well as low-resolution local records (Table 2.2, Figs. 2.6, 2.7).[10] Furthermore, discussion will be limited to the centuries framing the late third millennium climate change. In other words, focus will be on the period from 3000 to 1500 cal BC. Before local palaeoclimate proxy records are discussed, the evidence from high latitude ice and deep sea cores

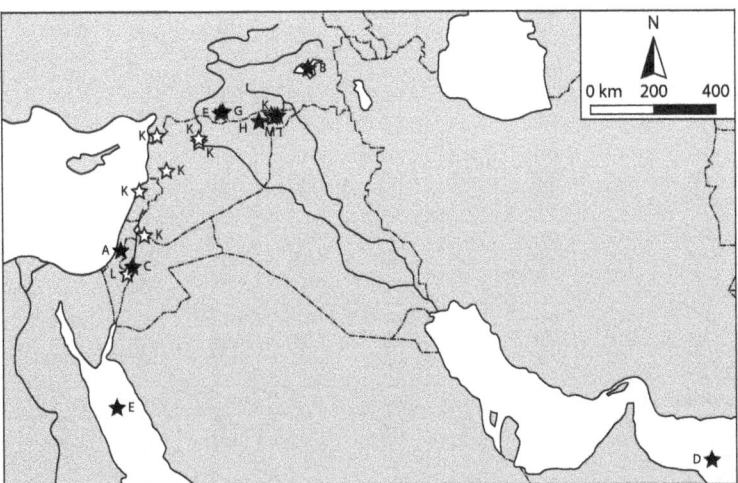

Fig. 2.6: Map of the Near East showing palaeoclimate proxy records used in this study (black stars), as well as sites with evidence for drought in plants and humans (white stars). (A) Soreq Cave. (B) Lake Van. (C) Dead Sea. (D) Gulf of Oman. (E) Red Sea. (F) Kazane Höyük. (G) Göbekli Tepe. (H) Wadi Avedji. (I) Wadi Jaghjagh. (J) Tell Leilan. (K) Sites with botanical samples. (L) Mount Sedom. (M) Tell Barri.

---

10 Palaeoclimate proxy records documenting late third millennium climate change elsewhere in the Near East are discussed in Rosen 1995 (Palestine), Bell 1971, Said 1993, Krom et al. 2002 (Nile), van Zeist and Bottema 1977, Stevens et al. 2001, 2006 (Iran), and Eastwood et al. 1999 (Turkey). See Arz et al. 2006 for a discussion of late third millennium climate change in the monsoon system. See Issar and Zohar 2004, Mayewski et al. 2004, Staubwasser et al. 2006, and Rosen 2007 for extensive treatments of Near Eastern climate during the entire Holocene.

map	record	proxy	resolution	range	dating
A	Soreq Cave	$\delta^{18}O$ / $\delta^{13}C$ analysis	~50 yrs?	60 ka BP–present	TIMS ^{230}Th/^{234}U dates
B	Lake Van	$\delta^{18}O$ / trace element analysis	~1 yr	14.7 ka BP–present	varve counting
C	Dead Sea	lake-level curve	10-100 yrs	8 ka BP–present	^{14}C-dates
D	Gulf of Oman	$\delta^{18}O$ / $CaCO_3$ / dolomite analysis	~100 yrs	200 ka BP–present	AMS ^{14}C-dates
E	Red Sea	$\delta^{18}O$ analysis	~200 yrs	5.9 ka BP–3.9 ka BP	AMS ^{14}C-dates
F	Kazane Höyük	geomorphological analysis	low, discontinuous	12 ka BP–present	cultural chronology
G	Göbekli Tepe	$\delta^{18}O$ / $\delta^{13}C$ analysis	low, discontinuous	10 ka BP–4 ka BP	^{14}C-dates
H	Wadi Avedji	geomorphological analysis	low, discontinuous	8 ka BP–present	cultural chronology
I	Wadi Jaghjagh	geomorphological analysis	low, discontinuous	3.5 ka BC–present	TL dates
J	Tell Leilan	geomorphological analysis	low, discontinuous	2.6 ka BC–1.9 ka BC	^{14}C-dates, cultural chronology
K	Sites with botanical samples	$\Delta^{13}C$ analysis	low, discontinuous	3.0 ka BC–1.6 ka BC	^{14}C-dates, cultural chronology
L	Mount Sedom	$\delta^{13}C$ / $\delta^{15}N$ analysis	low, discontinuous	2.3 ka BC–1.9 ka BC	^{14}C-dates
M	Tell Barri	dental analysis	low, discontinuous	EB II–Neo–Assyrian	cultural chronology

Table 2.2: Comparison of various properties of Near Eastern palaeoclimate proxy records used in this study (A–J), as well as sites with evidence for drought (K–M). Letters in the left column refer to Figs. 2.6 and 2.7. See main text for references.

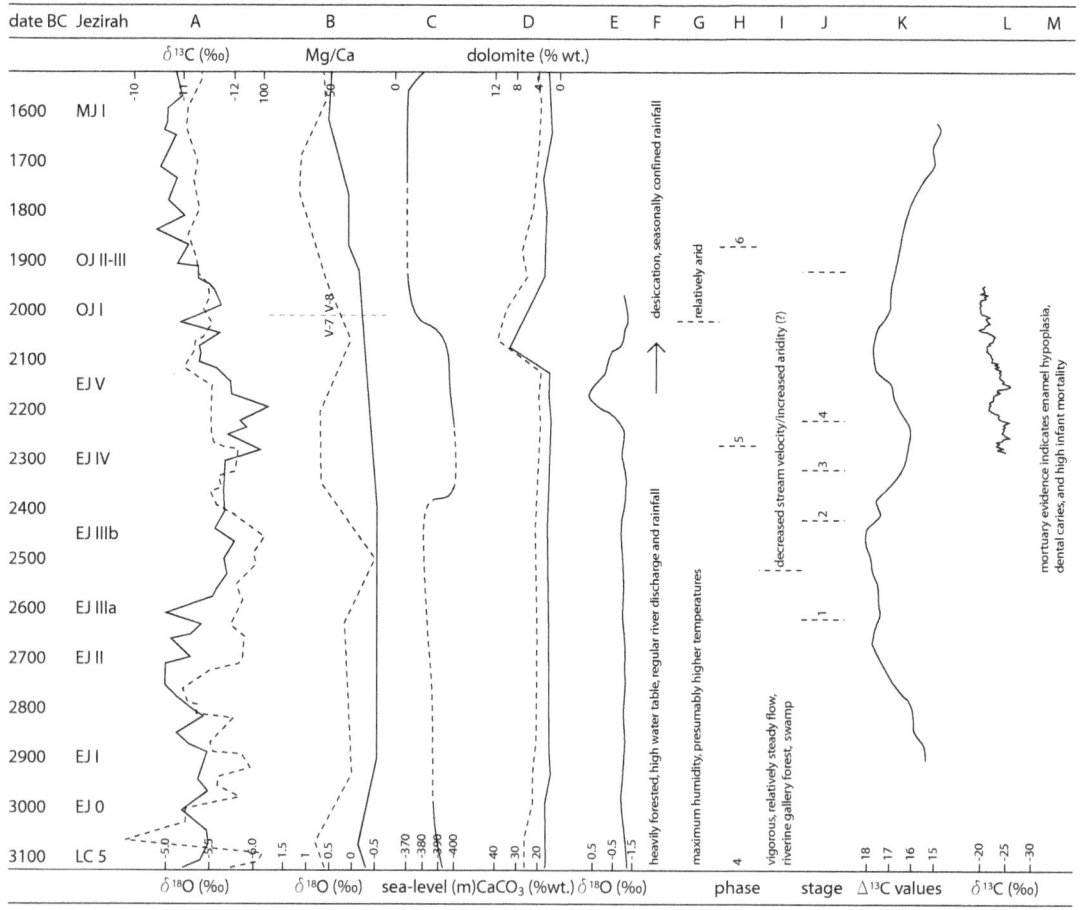

Fig. 2.7: Near Eastern palaeoclimate proxy records used in this study for the period 3100–1500 BC. See main text for references. (A) $\delta^{13}C$ (solid line) and $\delta^{18}O$ (dashed line) values at Soreq Cave. (B) Mg/Ca (solid line) and $\delta^{18}O$ (dashed line) values, and PAZ (grey dashed line represents transition) at Lake Van. (C) Dead Sea water level changes. Uncertainties are dashed. (D) Dolomite (solid line) and $CaCO_3$ (dashed line) values at the Gulf of Oman. (E) $\delta^{18}O$ value at the Red Sea. (F) Environmental history around Kazane Höyük. The transition is marked by an arrow. (G) $\delta^{18}O$ and $\delta^{13}C$ values at Göbekli Tepe. (H) Sedimentation history in the Wadi Avedji. (I) Sedimentation history in the Wadi Jaghjagh. (J) Stages of the LCCM at Tell Leilan. (K) Averaged $\Delta^{13}C$ values of barley grains from sites in the Levant and northern Mesopotamia. (L) $\delta^{13}C$ values of a subfossil tree from Mount Sedom. (M) Mortuary evidence from Tell Barri.

should be mentioned. It has long been recognized that cold spells in high latitudes may indicate increased aridity in lower latitudes, as more water is stored in glaciers and ice sheets. For the third millennium BC, the Greenland Ice Sheet project 2 (GISP2) ice core indicates cold spells, which are correlated to arid events at lower latitudes, in 2800, 2650, 2450, 2350, 2050, and 1950 BC. Deep sea cores from the North Atlantic document the occurrence of drift ice, and these cores suggest cold periods around 2600 BC and 2200–2050 BC (Kuzucuoğlu 2007a: 463–4). It should be noted, however, that there was considerable variety in intensity of these events, and that numerous smaller changes in temperature occurred during the third and second millennium. Recently, a correlation has been proposed between these events and magnetic reversals documented in mudbricks from Mari, suggesting that these high-latitude events indeed affected the Near East in some way (Gallet et al. 2008).

2.3.1 Soreq Cave

The proxy record from Soreq Cave, Israel, is among the most important of the Near East. It is located at the intersection of climate systems from Europe, Africa, and Asia, and hence susceptible to changes in any of these. Today, Soreq Cave experiences a climate with an average annual rainfall of 500 mm, mainly deposited during winter. Water entering the cave deposits annually laminated calcite speleothems that have been used to construct a continuous record of climate change for the last 185000 years. Palaeoclimate proxies that are analyzed include $\delta^{18}O$ and $\delta^{13}C$ values and uranium and strontium isotope ratios in the speleothems. Chronological control is achieved through multiple thermal ionization mass spectrometry (TIMS) $^{230}Th/^{234}U$ dates, taken at regular intervals (Bar-Matthews et al. 1997; Bar-Matthews et al. 2003).

The oxygen isotope and carbon isotope values are important in the present context. The $\delta^{18}O$ value indicates the ratio of the ^{18}O isotope to the ^{16}O isotope. Variations in the $^{18}O/^{16}O$ ratio indicate variations in temperature and/or precipitation, depending on the proxy record that is studied. For Soreq Cave, it has been shown that $\delta^{18}O$ variations can be correlated to changes in precipitation, whereby low $\delta^{18}O$ values indicate moist periods, and high $\delta^{18}O$ values indicate dry periods (Bar-Matthews et al. 1997: 161). By applying the current correlation between annual amount of rainfall, $\delta^{18}O$ values of rainfall, and $\delta^{18}O$ values of the corresponding speleothems to the entire speleothem record, it is possible to reconstruct changes in rainfall through time (Bar-Matthews et al. 1998). The carbon isotope composition of speleothems is influenced by numerous factors, including the picking up of plant-derived CO_2 as rainwater passes through the ground towards the cave. This process allows $\delta^{13}C$ to be used as an indication of the ratio of C3 to C4 plants, which grow in moderate and drier climatic conditions, respectively. Thus, high $\delta^{13}C$ values generally indicate a stronger presence of C4 plants and by inference drier conditions, whereas low $\delta^{13}C$ values indicate fewer C4 plants and moister conditions. It should be noted, however, that due to the complexities of speleothem formation, these correlations are very rough approximations. Under specific environmental conditions, the described processes may have operated in slightly different ways, and human actions may have played a role as well (Goodfriend 1999: 510).

The $\delta^{18}O$ and $\delta^{13}C$ values recorded at Soreq Cave suggest the following reconstruction of climate trends (Fig. 2.7A). For the Holocene, the combined $\delta^{18}O$ and $\delta^{13}C$ records clearly document the major climatic shifts of the Late Glacial Maximum (LGM), the Younger Dryas, and the Holocene dry event at 6100 cal BC. After this event, there followed a wet period during which $\delta^{18}O$ and $\delta^{13}C$ values were generally lower than today. The period from c. 3000 to 1500 cal BC is characterized by several extreme fluctuations in $\delta^{18}O$ and $\delta^{13}C$. Whereas the period from c. 2600 cal BC to 2200 cal BC is relatively wet, the period from 2200 to 1800 cal BC is characterized by a strong rise in especially $\delta^{18}O$, indicating a strong trend toward aridity. Between 1800 cal BC and 1500 cal BC, $\delta^{18}O$ and $\delta^{13}C$ values remain relatively stable, and the variation remains within a range that is on average higher than during the preceding 1000 years. Thus, the Soreq Cave speleothems show that the third and early second millennium BC climate is characterized by severe fluctuations, and eventually settled into a drier and more stable climate around 2000 cal BC.

2.3.2 Lake Van

The sediments of Lake Van, southeast Turkey, have provided important insights into Near Eastern palaeoclimate. Their main importance in their having been dated by the radiocarbon-independent method of varve counting. The Lake Van region is at the intersection of three different climatic regimes and should therefore be sensitive to climate changes in each of these regimes. Lake Van is a terminal, soda-rich lake with poor animal and plant life. Sediment cores were obtained and studied in the 1970s (van Zeist and Woldring 1978) and 1990s (Landmann et al. 1996; Lemcke and Sturm 1997; Wick et al. 2003). The 1990s investigations have considerably revised the 1970s palaeoclimate reconstructions. The varves at Lake Van consist of annual carbonate and organic layers that can be almost continuously counted for the last 14000 years. The sediment cores from Lake Van provided a number of different palaeoclimate records, including pollen, charcoal, $CaCO_3$, $\delta^{18}O$, and Mg/Ca records corresponding to various aspects of the past environment.

The Mg/Ca record has been shown to indicate changes in the salinity of the lake water with respect to the current Mg/Ca ratio (Lemcke and Sturm 1997: 669). A high Mg/Ca ratio indicates high salinity, whereas a low ratio indicates low salinity. The $\delta^{18}O$ value depends on temperature, humidity, and water influx into the lake. In order to detect the correlation between these variables and $\delta^{18}O$, a model was developed based on data from the last 60 years. The modelled $\delta^{18}O$ values were compared to the $\delta^{18}O$ values recorded for that period, and a good correlation was found. As in Soreq Cave, a high $\delta^{18}O$ ratio corresponds to increased aridity. The vegetation development as reconstructed from the 1990s sediment cores corresponds to the reconstruction by van Zeist and Woldring (1978), but shows that their chronology should be pushed backwards by several thousand years. Pollen Assemblage Zones (PAZ) V-1 to V-9 broadly indicate the vegetation responses to climate change and human impact in the region.

Together, these records indicate the following climate trends. A late glacial dry and arid climate was replaced by more humid conditions at the onset of the Holocene. At *c.* 6000 cal BC, relative humidity at Lake Van increased, in contrast to the globally documented trend toward a drier climate. Between 4000 and 2000 cal BC, relatively optimal climatic conditions persisted in the Lake Van region, with Mg/Ca and $\delta^{18}O$ values reaching minima around 2500 cal BC (Fig. 2.7B. The expansion of oak and minimal presence of Chenopodiaceae indicate forest-steppe vegetation (PAZ V-7). The $\delta^{18}O$ ratio between *c.* 2500 and *c.* 2200 cal BC indicates a drier climate, followed by a brief reversal around 2100 cal BC. After 2100 cal BC, both the $\delta^{18}O$ and the Mg/Ca ratios increase, pointing to a more arid climate, lasting for several centuries. At the same time, the importance of oak diminished in favour of a more open landscape (PAZ V-8). Wick et al. argue that it is unlikely that this reduction resulted from human activity, and was probably caused by summer drought. Slightly later, around 1800 cal BC, human activity in the area became apparent through evidence for disturbance by livestock. Around 0 cal BC, modern climatic conditions were established, in tandem with a further decline of oak (Lemcke and Sturm 1997; Wick et al. 2003).

2.3.3 Dead Sea

Several Near Eastern palaeoclimate records measure variations in the water level of the Dead Sea (Frumkin et al. 2001; Enzel et al. 2003; Migowski et al. 2006). Most proxies measure variations in the water level or in salt deposition of the Dead Sea. These variables are directly correlated with the amount of water the Dead Sea receives, as well as local evapotranspiration, which in turn are correlated with rainfall, humidity, and temperature. As the Dead Sea itself receives relatively little rainfall, lake level changes directly reflect precipitation in the wetter parts of its drainage basin, especially the northern part. It has been demonstrated that Dead Sea water level changes are in step with precipitation changes in and around the Jordan Valley. Precipitation is largely determined by cyclones coming from the Mediterranean, and for dry years it can be shown that cyclones affect areas to the north and west of the Levant, rather than the Levant itself. If this correlation holds for the past as well, changes in storm tracks may explain some of the variety observed in the Dead Sea water level (Enzel et al. 2003).

Reconstructed Dead Sea water level changes are shown in Fig. 2.7C. General trends are as follows, the caveat being that several temporal gaps in the reconstruction remain. The Younger Dryas and the 6000 cal BC dry event are clearly detected by the long sequence recovered by Migowski et al. (2006). Wet phases are reconstructed for 8000–6600 cal BC and 5600–3500 cal BC. These periods are, however, interspersed with very abrupt and short-lived reversals in climate. Major dry events occurred at 8200, 6600, 2200, and 1500 cal BC. The drop around 2200 cal BC is corroborated by Enzel et al. (2003), although they do not detect the return to wet conditions reconstructed by Migowski et al. Finally, Frumkin et al. (2001) also reconstruct several wet phases, dated to 8300-6000 cal BC, 4000–2900 cal BC, and 1400–800 cal BC. These phases, again, only partly overlap with the data presented above, but they do suggest that the late third–early second millennium experienced a relatively dry climate, resulting in a Dead Sea water level lower than 400 m below sea-level, compared to -390 m before the introduction of modern large-scale mechanized irrigation schemes.

2.3.4 Gulf of Oman and Red Sea

Winds transport significant amounts of dust from Mesopotamia and the Zagros toward the Persian Gulf and the Gulf of Oman. The mineral composition of this dust has been analysed to determine the provenience of this dust, resulting in the recognition of a mineral signature of Mesopotamian dust. Hence, sediment cores from the Persian Gulf and the Gulf of Oman provide information on the amount of Mesopotamian dust present. Changes in dust deposition correlate directly with past aridity changes in Mesopotamia, even though climate in the Gulf of Oman itself is not governed by the same mechanisms as Mesopotamia itself. To investigate these changes, Cullen et al. (2000) sampled a sediment core from the Gulf of Oman at regular intervals. Analyzed sample variables included oxygen isotopes, calcium carbonate and dolomite content, and strontium (Sr) and neodymium (Nd) isotope composition. Calcium carbonate and dolomite indicate the amount of dust in the core. The Sr/Nd isotope composition analysis confirmed that most of the dust deposited here derived from Mesopotamia. Several accelerator mass spectrometry (AMS) radiocarbon dates provided an age control for the reconstructed sedimentation rate. The Gulf of Oman core documents dust deposition for the last 25000 years. The Younger Dryas cold period is evidenced by a high content of wind-borne dust. A second dust peak is recorded for *c.* 3200 cal BC. Finally, a major dust peak appears at *c.* 2100 cal BC, levelling off until *c.* 1700 cal BC, dated by four AMS ^{14}C-dates (Fig. 2.7D).

A similar climatic anomaly is recorded in a core from the Shaban Deep in the northern Red Sea, which is located between the rainfall regimes of the Mediterranean and the Afro-Asian monsoon. The Shaban Deep is a basin where brine is deposited annually. The laminated brine sediments provide a record for the climatic conditions under which deposition occurred. Brine sedimentation is influenced by climate related surface water changes. The sediments have been cored, and the upper 65 cm of the undisturbed sediments document the sedimentation sequence from 4000 to 1900 cal BC. Chronological control comes courtesy of six AMS radiocarbon dates. Various proxies, including $\delta^{18}O$ values, are analysed; all detecting a climatic anomaly between 2200 and 2000 cal BC (Fig. 2.7E). The $\delta^{18}O$ values show a distinct peak between 2200 and 2000 cal BC, indicating unfavourable climatic conditions, primarily interpreted as an arid event (Arz et al. 2006).

2.3.5 Palaeoclimate records from northern Mesopotamia

Geoarchaeological investigations at ancient river beds, pits dug for irrigation canals, and test pits around Kazane Höyük, Turkey, document past environmental change (Rosen 1997). The presence of datable artefacts in sediments afforded some crude chronological control (Fig. 2.7F). During the Late Chalcolithic and early EB periods, relatively favourable environmental conditions prevailed at Kazane Höyük. The region was more heavily forested than at present, river discharge and rainfall were more regular, and the groundwater table was higher. This favourable period was followed by a dry phase probably starting slightly before the end of the third millennium BC and continuing well into the second millennium BC. This phase was characterized by 'accelerated channel erosion and desiccated abandoned floodplains' (Rosen 1997: 405). Rosen suggests that climate change, deforestation, or diversion of local rivers for irrigation purposes could explain this erosion phase. Climate change is put forward as the most likely cause, given the fact that there was no return to previous environmental conditions when settlement in the area, and hence human action, was resumed. There is thus evidence for late third–early second millennium BC desiccation accompanied by seasonally bound rainfall at Kazane Höyük.

The Pre-Pottery Neolithic site of Göbekli Tepe, Turkey, was recently subjected to micromorphological investigations to reconstruct local climate conditions. Pustovoytov et al. (2007) investigated pedogenic carbonate coatings on stone walls, resulting in $\delta^{18}O$ and $\delta^{13}C$ curves for the period from 8000 to 2000 cal BC. that correspond to air temperature and the ratio of C3/C4 plants, respectively. During the early Holocene (8000–4000 cal BC), there was a decrease in $\delta^{13}C$ and an increase in $\delta^{18}O$ values, and both curves showed opposite peaks that were most pronounced during the earlier part of the covered time period. These curves suggest that during the period 8000–4000 cal BC, the climate was relatively humid, and that there was a development toward a more stable climate with less pronounced oscillations. For the period 4000–2000 cal BC, the carbonate coatings indicate relatively high humidity and high temperatures, with a short-lived peak slightly before 2000 cal BC toward either higher humidity, or lower temperatures. After 2000 cal BC, there was a trend towards lower humidity than recorded for the early and mid-Holocene periods (Fig. 2.7G).

Several geomorphological studies documented past climate changes in the Upper Khabur basin. Courty (1994) identified six alluviation and erosion phases in a geomorphological study of the Wadi Avedji in the Upper Khabur. These phases corresponded to different climatic conditions during the Holocene (Fig. 2.7H). During the early Holocene phase 1, there was evidence for strong alluviation. Phase 1 was characterized by a climate warmer than that of today, followed by phase 2 when a trend toward higher aridity set in. From 5800 to 3800 cal BC (phase 3), climate improved and humidity was high, followed by phases 4–5 with clear evidence for environmental change. During phase 4 (3800–2250 cal BC), the Wadi Avedji incised into the floodplain created during the previous phase, and water flow was irregular. Phase 4 (3800–2250 cal BC) also yielded evidence for human river management. Phase 5 (2250–1850 cal BC) was characterized by even more severe climatic deterioration than in the previous phase. Finally, during Phase 6 there was a restoration of climate, and present climatic conditions were established.

Further evidence for late third millennium aridity comes from recent geomorphological research by Deckers and Riehl (2007). They investigated fluvial deposits of the Wadi Jaghjagh and Wadi Khanzir in the Upper Khabur basin (Fig. 2.7I). Botanical samples from these deposits were used to support the palaeoenvironmental reconstruction. The presence of pottery sherds in these deposits allowed chronological control through thermoluminescence (TL) dating. The resulting reconstruction of the past environment stretched back to c. 5000 cal BC. During the late fourth and early third millennium BC, the Wadi Jaghjagh seems to have experienced a relatively steady flow. Botanical samples from the Wadi Jaghjagh section indicated that a riverine gallery forest with *Salix* existed during this period. This material furthermore indicated the presence of woodland vegetation. After 2500 cal BC, these favourable environmental conditions ended, and discharge of the Wadi Jaghjagh dropped. The authors suggested water extraction for irrigation purposes, increased erosion resulting from the removal of the woodland cover, or late third millennium aridity as possible causes.

Finally, a similar, though undated shift was observed for the Wadi Jarrah in the Upper Khabur basin. There, a channel system characterized by meandering channels was replaced by a braided system, which was in turn replaced by a meandering system. As meandering channel systems result from more regular flow regimes, the shift to a braided system possibly indicated drier conditions in which rainfall was restricted to fewer, but heavier rain showers (Besonen and Cremaschi n.d.). In the face of the evidence presented above on sequences in nearby wadi systems, it seems reasonable to assume that the change to a drier environment in the Wadi Jarrah also dated to the second half of the third millennium BC.

The Leilan Climate Change Model (LCCM) was based on archaeological and geomorphological evidence from Tell Leilan and Abu Hgeira (Weiss et al. 1993). The LCCM consisted of a four-stage model of state development and environmental change (Fig. 2.7J, Table 2.3). Stage 1 of this model corresponded to Leilan IIId (2600–2400 cal BC). During this period, the previously stable and relatively wet local alluvial regime was replaced by a more erratic regime. Water discharge and precipitation became more seasonal. For the second stage (Leilan IIa, 2400–2300 cal BC), no climatic data were presented. For stage 3 (Leilan IIb, 2300–2200 cal BC), geomorphological samples indicated continuation of the drying trend observed for stage 1, and a temperature that was slightly higher than that of today. Stage 4 (2200–1900 cal BC) was called 'Khabur hiatus 1', based on the apparent absence of archaeological sites for this period. Khabur hiatus 1 thus coincided with the climatic deterioration of Courty's phase 5. Based on geomorphological analysis of stratigraphic layers at Tell Leilan, Weiss et al. suggested a three-phase model for the abrupt climate change of Khabur hiatus 1. Phase 1 was characterized by two distinct layers. The lower and oldest layer was 0.5 cm thick and consisted of 'weakly altered fine silt-sized volcanic glass with a few potassic feldspars, phytoliths, and rounded calcitic sand-sized pellets' (Weiss et al. 1993: 1000). On top of this tephra layer were Aeolian sediments mixed with volcanic ash and mudbrick, suggested to represent an 'abrupt arid phase marked by intensification of wind circulation' (Weiss et al. 1993: 1001) and a higher amount of dust in the air (phase 2). Phase 3, finally, was characterized by 'the augmentation of soil moisture, a reestablishment of marked dry-wet seasonal contrast, a lessening of heavy rainstorms and a progressive soil stabilization' (Weiss et al. 1993: 1000). A volcanic eruption was invoked as a cause of the tephra deposition and the prolonged dust veil. Other palaeoclimate proxy records did not, however, detect a major volcanic eruption around 2200 cal BC. It was therefore suggested that the tephra at Tell Leilan resulted from a small-scale volcanic eruption in the 'Anatolian–Caucasian area' that would not show up in far-off proxy records such as the GISP2 ice cores.

In a recent geomorphological study of sediments from Tell Brak, Courty (2001) redefined the palaeoclimate sequence of the LCCM. According to this study, specific soil particles within a unique stratigraphic sequence recorded at Tell Brak acted as the fingerprint of a major 'dust fallout event'. This event was argued to be the same as, and therefore contemporaneous with, LCCM stage 4. This claim of contemporaneity was based on the similarity of the reconstructed event scenarios. At both Tell Brak and Tell Leilan, an event was reconstructed involving large amounts of dust being ejected into the air, followed by a period of major wind circulation. Analysis of the soil particles at Tell Brak, however, indicated that this event could not have been caused by volcanism, as originally argued. Courty considered as a working hypothesis the possibility that the dust fallout resulted from a 'violent air blast that caused multi-site ignition and widespread wild-fires' (2001: 368). Furthermore, and more importantly, it was suggested that the event should be assigned a new date. Based on the stratigraphy of Tell Brak, Courty argued that the event should not be dated to 2200 cal BC, but some 150 years earlier, to 2350 cal BC.

Several comments can be made about the palaeoenvironmental part of the LCCM, and Courty's adjustments. First, it should be noted that the sediments indicating the dust fallout event were never observed outside cultural contexts, even though wadi sections dating to this period have been investigated in the Upper Khabur area. Second, it is not clear whether the 2200–1900 cal BC tephra and Aeolian sediments were only occasionally observed in Tell Leilan trenches, or in every trench reaching the appropriate levels of the Akkadian/post-Akkadian/Ur III periods. Third, some excavations with Akkadian/post-Akkadian/Ur III or contemporary levels do not yield any evidence

stage	date BC	Leilan	social/political	climate
4: collapse	2200–1900	IIc	Leilan abandoned, regional abandonments, Khabur hiatus 1	abrupt climate change (phase 1: tephra; phase 2: marked aridity; phase 3: climate stabilization)
3: imperialization	2300–2200	IIb	Akkadian control of Khabur (Brak), water management	climate similar to the present; high seasonal contrasts; higher temperature?
2: consolidation	2400–2300	IIa	Leilan acropolis city wall, pottery mass-production	?
1: secondary state formation	2600–2400	IIId	Leilan 15>100 ha, centralization, administration	increased seasonality, higher evaporation, reduced discharge

Table 2.3: Summary of the Leilan Climate Change Model (LCCM) as described in Weiss et al. 1993.

for tephra layers, suggesting an extremely local rather than regional or even global cause and scale for the dust fallout sediments observed at Tell Leilan (e.g. Meijer 2007: n. 16, on Tell Hammam et-Turkman).[11] Fourth, the sequence of environmental change observed at Tell Leilan and Tell Brak is difficult to attribute to a single event of short duration, be it a volcanic eruption or a cosmic impact. In fact, Weiss et al. pointed to several inconsistencies between the LCCM and what might be expected for volcanic eruptions, and already suggested that ultimately some other climatic mechanism may have been responsible for the observed scenario of environmental change (1993: 1002). Finally, in a micromorphological study on sediments from Tell Brak, Matthews (2003: 385) argued that the soil particles supposedly signifying the tephra deposition event are common in sediments from multiple periods, and are thus unlikely to point to any single catastrophic event.

It might therefore be suggested that there is evidence for two distinct processes at Tell Leilan. First, there is evidence for an as yet unidentified process that led to the deposition of a thin soil layer of supposedly volcanic origin. Matthews' study on the sediments from Tell Brak has already shown that this origin may be incorrect. Furthermore, an eruption is an event of very short duration, with effects that last a few years at most (Stothers 1999). Such an event is therefore unlikely to cause a 300-year long drought or dust-veil. Second, there is evidence for prolonged wind activity and large amounts of wind-borne dust, pointing to abandonment of the site and perhaps drought. As indicated by Weiss et al., it is difficult to find a single explanation for these observations. It therefore seems most logical to disentangle both processes. In other words, the event leading to tephra deposition and the subsequent 300-year drought are unrelated (cf. Courty and Weiss 1997: 143). Summarizing, considerable doubt as to the date and nature of the dust fallout event documented at Tell Brak and Tell Leilan remains. On the other hand, the evidence for a long-term process toward higher aridity, Courty's stages 4–5, is stronger, especially because it derives from both cultural and natural deposits. This trend is commensurate with the evidence for increased aridity at Tell Leilan, if regarded separately from the tephra evidence.

2.3.6 Reconstructing climate change

It should be noted that the interpretation of palaeoclimate data and their extrapolation to other areas of the Near East remain problematic for three reasons. Firstly, data come from a variety of sources and geographically widely separated locations (Table 2.2, Figs. 2.6, 2.7). This scattering of data types and locations sometimes results in apparently contradictory palaeoclimate reconstructions. Secondly, some proxy records will fail to record short-term climate changes due to their low resolution. Such short-term climate change events may also remain undetected in high-resolution proxy records elsewhere due to the sometimes limited geographical distribution of these events. Finally, palaeoclimate reconstructions are complicated by the fact that climate-forcing mechanisms may have been different in the past, so that modern data cannot be directly extrapolated to the past. Thus, palaeoclimate reconstructions for the Near East can presently only achieve a limited precision. Gaps in local climate histories are now being filled through the modelling of palaeoclimate developments (e.g. Bryson and Bryson 1997; Riehl et al. 2008).

The geographical locations of Near Eastern climate proxy records highly influence the resolution that can be achieved for palaeoclimate reconstructions. For some areas, multiple records are available that can corroborate each others' respective results. In other regions, climate proxy records are scarce, and palaeoclimate can only be inferred from records elsewhere. The geographical distribution of climate proxy records is determined by the specific conditions under which such records are created and survive. From Table 2.2 and Fig. 2.7 it is evident that well-dated, high-resolution climate proxy records from northern Mesopotamia are lacking. This means that any reconstruction can only be an approximation, and that some details must be inferred from current climatic conditions.

---

11   It should be noted, though, that this absence could at least partly be explained by the possibility that excavators have not been actively looking for this layer, or that the geomorphological analyses that are necessary to identify it have not been carried out.

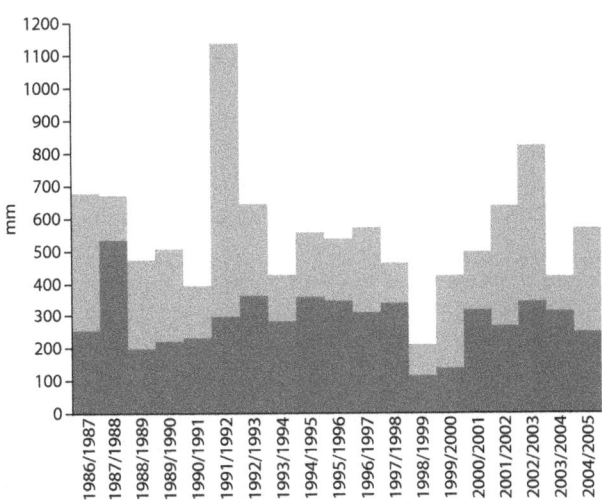

Fig. 2.8: Annual rainfall in hydrological years for the Jezirah (dark shading, average from Hasakeh, Qamishli, and Raqqa, cf. Fig. 2.4) and Jerusalem from 1986 to 2005 (light shading, Jerusalem data from the Israel Meteorological Service).

Currently, the best climate data for the Near East come from Soreq Cave and Lake Van. Climate in these locations is generally driven by the same mechanisms. Rainfall in all locations is mainly deposited during winter by cyclones from the Mediterranean, as is the case for northern Mesopotamia. For Soreq Cave, it has been shown that changes in humidity and sea-level respectively correlate well with precipitation changes in Jerusalem. Modelling of annual precipitation for the last 7000 years in both Jerusalem and Qamishli indicated similar trends on a very general level, although details and timing may be considerably off (Bryson and Bryson 1997: 579). A comparison of recent precipitation data seems to suggest that there are similarities and that dry years generally occur simultaneously. However, given the weak correlation between precipitation in Jerusalem and in the Jezirah ($r_s=0.48$), it is clear that data from Jerusalem must be cautiously extrapolated to northern Mesopotamia (Fig. 2.8).

The palaeoclimate proxy records discussed above document two types of changes. First, a number of records, especially those from higher latitudes, record the gradual replacement of relatively favourable third millennium BC climatic conditions by a more arid climate with either lower or more seasonally confined rainfall. Second, several records document an abrupt arid event lasting several hundred years at most and followed by a return to pre-event climatic conditions. Examples of the first include the Soreq Cave, Lake Van, and Harran Plain records, as well as the geomorphological evidence from the Upper Khabur Basin, except for the evidence from Tell Leilan. An abrupt arid event is recorded in the Gulf of Oman and Red Sea records, and less clearly in the Dead Sea record. Several researchers have noted this discrepancy between an abrupt arid event and a long trend toward higher aridity, but this discrepancy has not yet been fully explained (e.g. Arz et al. 2006: 438–40; Magny et al. 2009: 831).

Interestingly, some palynological studies in the Near East fail to detect the late third millennium climate change (e.g. Gremmen and Bottema 1991; Bottema 1997; Neumann et al. 2006). Both Gremmen and Bottema (1991) and Neumann et al. (2006) argued that climate remained largely unchanged for the last 6000 years, and that human impact on the environment must therefore be considered more significant. The pollen core from Lake Bouara showed some minor evidence for a dry episode around 2000 BC, but the evidence from pollen cores from Anatolia and the Zagros area was less conclusive (Bottema 1997: 509–10, 514). The vegetation reconstruction from the Golan (Neumann et al. 2006) indicated a high amount of arboreal pollen for the Early/Middle Bronze Age, indicative of Mediterranean climatic conditions. However, there are many variables involved in the response of vegetation to climate change, so that deducing regional climate change from localized pollen spectra remains problematic. Plant species react differently to climate change, with short-lived species showing the strongest reactions. Furthermore, the number of pollen spectra from northern Mesopotamia is limited. Finally, dating is relatively coarse and usually derived from a few radiocarbon dates, from which a sedimentation rate is extrapolated. These dates may be biased as a result of carbon-depletion in water-logged sites.

2.3.7 Synthesis of climate change

The combined evidence from the palaeoclimate proxy records discussed here suggests the following scenario of climate change. The fourth millennium is subject to considerable fluctuations. The Soreq Cave speleothems document notable shifts in stable oxygen isotope levels for the period 3500–3000 cal BC. This unstable climate is also reflected in the Dead Sea record. Notable drops in water-level are recorded for *c.* 3700 cal BC and around 3500 cal BC, coinciding with a major $\delta^{18}O$ peak in Soreq Cave. The Gulf of Oman core shows a minor rise in dolomite and $CaCO_3$ percentages slightly before 3000 cal BC, indicating a higher dust-content, which again is paralleled in Soreq Cave. Wadi flow regimes in the Jezirah are regular, indicating that rainfall is more evenly spread over the year than is the case today. Also, low erosion rates imply a dense vegetation cover.

Climatic conditions in the Near East became more favourable after *c.* 3000 cal BC. For the period 3000–2500 cal BC, Soreq Cave and the Dead Sea document relatively wet conditions, although at Soreq Cave short arid spikes occurred around 3000 and 2700 BC. The percentage of oak pollen in Lake Zeribar was still high, despite declining continuously from the fourth millennium onward (Stevens et al. 2001). Low pollen preservation inhibits any interpretation for this period at Lake Mirabad, but around 2500 cal BC, the percentage of oak pollen was very high (Stevens et al. 2006).

Climate records across the Near East start to attest drier conditions from 2500 cal BC, with progressively more records indicating aridity towards the end of the third millennium. The Soreq Cave speleothems signal gradually increasing $\delta^{18}O$ for the period 2500–2000 cal BC, with $\delta^{13}C$ following from *c.* 2200 cal BC. At Lake Van, pollen evidence indicates drier climatic conditions towards the end of PAZ 7. The $\delta^{18}O$ isotope level around 2250 cal BC is the highest recorded for the third millennium BC. Lake Van shows a short return to higher humidity levels at *c.* 2150 cal BC, followed by a gradual and stronger rise in both $\delta^{18}O$ and Mg/Ca. The records most prominently indicating a sudden arid event, that is the Gulf of Oman and Red Sea cores, are dated between 2200 and 2000 cal BC. Several records indicate a short-lived return to wetter conditions around 2000 cal BC, followed by more prolonged aridification. After 2000 cal BC, some records show a return to previous wetter conditions, including the Dead Sea and the Gulf of Oman. Most records indicate, however, that the late third millennium climate change is of a more permanent nature, signifying the onset of more or less modern climatic conditions. This seems for example to be the case for the Wadi Jaghjagh, Göbekli Tepe, and Kazane Höyük records. For the second millennium, virtually all records indicate drier conditions than existed previously during the fourth and early third millennium (Kuzucuoğlu 2007a: 474). Whether the combined signals should be understood as indicating one prolonged arid phase, or two smaller arid events punctuated by a short humid period as argued by Kuzucuoğlu (2007a: 476) remains unclear on the basis of the available northern Mesopotamian evidence. Modelling of northern Mesopotamian climate during this period seems to confirm that there was one prolonged dry period, although there are clear local differences (Riehl et al. 2008: fig. 2).

Summarizing, four stages of different climatic conditions can be established. The first stage lasted from 4000–3000 cal BC and was characterized by considerable climatic fluctuations. Notable dry spells occurred around 3700, 3500, 3300, and 3000 cal BC. The second stage started *c.* 3000 cal BC and was relatively wet across the Near East. Around 2500 cal BC, many proxy records start to indicate drier climatic conditions, initiating stage 3 (2500–2000 cal BC). This stage was characterized by progressively drier conditions. A possible short humid period may have occurred around 2000 cal BC, but this period has not been detected in all proxy records. Stages 1–3 fall into Rosen's Middle Holocene phase of highly erratic precipitation changes (2007: fig. 3.5). Stage 4, from 2000 cal BC onwards, was characterized by climatic signals that vary across the Near East. Some records indicate a return to wetter climatic conditions as persisted during the first half of the third millennium, whereas other records suggest that the climate remained dry, similar to the conditions existing today. The Dead Sea and Gulf of Oman records suggest a more or less rapid return to previous or even wetter conditions. At least for the Dead Sea this return is relatively short-lived, as the water level drops significantly around 1500 cal BC. The evidence from Lake Van, Soreq Cave, the Harran Plain, and the Wadi Jaghjagh on the other hand indicates that the late third millennium arid phase continues well into the second millennium.

2.3.8 Evidence for drought

The foregoing overview of northern Mesopotamian palaeoclimate has highlighted the occurrence of an arid phase at the end of the third millennium and the beginning of the second millennium BC. The establishment of an arid climate is, however, not in itself evidence for drought, which is a shortage of water for the inhabitants of a given region. It may well be that overall precipitation, humidity, and river discharges declined, but that there still was sufficient water for both humans and their crops and livestock. Although changes in the composition of the crop spectrum are regularly linked to local environmental changes, it should be noted that such changes may also come about because of changing political, socio-economic, or cultural conditions (Riehl 2009: 18–19). A change in the crop spectrum can for example result from changing food habits rather than be down to any particular environmental reason. Thus, the development of the crop spectrum from the third to the second millennium as increasingly reflecting arid conditions is likewise not in itself evidence of drought. An independent method is therefore required to test whether or not particular changes in crop selection were correlated to the occurrence of water stress or drought.

In order to address this issue, the stable carbon isotope composition has been measured from grain samples collected from numerous Early and Middle Bronze Age sites across the Levant and northern Mesopotamia (Riehl 2008; Riehl et al. 2008).[12] It has been observed that a correlation exists between the stable carbon isotope composition of C3 plants and water availability through irrigation, soil moisture, or humidity. C3 plants have lower carbon fixation in conditions of low water availability, resulting in an increase in $\delta^{13}C$ or a decrease in $\Delta^{13}C$. Even though there are other variables influencing $\Delta^{13}C$, water availability seems to be the most important environmental variable (Riehl et al. 2008: 1012). Because $\Delta^{13}C$ values can only be interpreted when data are available on the local moisture conditions, Riehl et al. used a macrophysical climate model (MCM) based on data from Lake Van and Soreq Cave to simulate ancient climate conditions. The resulting local climate conditions were compared to the stable carbon isotope compositions of barley under the assumption that this crop was generally cultivated without additional irrigation. Comparison indicated that different $\Delta^{13}C$ values were recorded for sites in different geographical settings and with different moisture conditions, that the $\Delta^{13}C$ values of other cereal crops showed similar trends to those of barley, and that a general decrease of $\Delta^{13}C$ values coincided with reconstructed increasing aridity at the end of the third millennium, and especially during the Middle Bronze Age (Fig. 2.7K) (Riehl et al. 2008: 1020). These observations, then, suggest that there was indeed a certain level of water stress or drought at various sites in northern Mesopotamia, although the severity of this water stress differed per site. This conclusion bolsters claims that crop selection was altered due to environmental change, and legitimizes the investigation of other responses to environmental change in general and drought in particular.

A similar study has recently been carried out on subfossil wood from a *Tamarix* tree from Mount Sedom, on the southern shore of the Dead Sea (Frumkin 2009). In this arid region, temperature and precipitation are the main factors constraining plant growth, and this is reflected in the ^{13}C and ^{15}N isotope composition of plants. Whereas ^{13}C enrichment can clearly be correlated to high temperatures and/or low precipitation, ^{15}N enrichment is primarily associated with low precipitation, although problems with this correlation remain. Using the correlation between ^{13}C isotopes and rainfall as it has been observed in modern *Tamarix* trees in the region, the absolute precipitation levels for the period encompassed by the ancient tree trunk could be reconstructed. Radiocarbon dates established a growth period between *c.* 2300 and 1930 cal BC, at which time

---

12 These sites were Hirbet ez-Zeraqoun (Jordan), Tell Fadous (Lebanon), Qatna, Alalakh, Tell el 'Abd, Emar, and Tell Mozan (Syria).

the tree died. During this period, the ^{13}C and ^{15}N isotope levels in the sampled tree show similar trends, which can be summarized as follows (Fig. 2.7L). Precipitation was highest at the start of this palaeoclimate record, and remained so with some minor fluctuations until 2200 cal BC. At that time, there occurred a drop in precipitation lasting some 70 years, after which rainfall briefly returned to the previous level. After 2130 cal BC, precipitation gradually dropped to a level similar to or even slightly lower than that of today, with an additional short-lived arid event taking place at 2020 BC. The analysed tree died around 1930 cal BC. This is in agreement with the current situation, in which almost no *Tamarix* trees survive in the region. This palaeoclimate record therefore strongly suggests that plants were significantly affected by aridification during the late third and early second millennium BC.

Recently, evidence was presented for subsistence stress among humans during the late third to early second millennium BC. At Tell Barri, a number of human skeletons have been studied from EB II/III to Neo-Assyrian levels, including a sample that could be specifically dated to transitional levels between the Early and Middle Bronze Ages. Although the total assemblage consisted of only 92 individuals, clear differences between each period appeared. The study of teeth allowed the reconstruction of the frequency of enamel hypoplasia and dental caries. Enamel hypoplasia is linked to nutritional stress during childhood, whereas the frequency of dental caries is an indicator of food quality. Skeletons from the EB IV/MB I and the Neo-Assyrian sub-samples showed the highest frequencies of enamel hypoplasia, pointing to a lower nutritional status during these periods. Similarly, dental caries was especially high in the EB II–III and EB IV/MB I sub-samples, again indicating subsistence stress during these periods. Finally, these sub-samples also showed a high level of child mortality that seemed to deviate statistically from the expected level (Sołtysiak forthcoming). Thus, the combined mortuary evidence suggested that the late Early Bronze Age period was marked by higher than normal infant mortality, combined with nutritional stress that was particularly evident during the EB IV/MB I transitional period (Fig. 2.7M). Despite the limited nature of the evidence, that is, the small sample and the fact that it came from a single site, it seems to accord relatively well with the indications for subsistence stress that have emerged from the palaeobotanical and isotopic evidence.

2.4 Chronology construction

Correlation of environmental developments with material culture change relies on precise absolute dating of both strands of evidence. Dating environmental developments almost invariably involves radiocarbon dating and/or similar methods from the natural sciences. Dating material culture change, on the other hand, involves a large array of dating methods including radiocarbon dating, historical dates, and typochronology. This difference in dating methods presents archaeologists with considerable difficulties (Bruins 2001: 1148). Correlation of these diverse methods has proved extremely complex, leading Reade (2001: 1) to observe that 'most scholars are averse to chronological studies'. This aversion often led to the ready acceptance of a convenient, well-established chronology, in this case the so-called Middle Chronology that dates the fall of Babylon by the Hittite king Muršili I to 1595 BC, even though this chronology is flawed in several respects.[13] There are now several studies that critically address established chronological conventions. These studies focus on the implications of the increasingly refined correlation between text passages describing astronomical observations and historical events and reigns (Gasche et al. 1998; Reade 2001).

At the basis of the Middle Chronology are Brinkman's chronological charts (1977). This chronology is based on a number of synchronisms, several copies of king lists, astronomical observations, and multiple assumptions as to calendars that were used, the methods by which regnal years were counted, and the lengths of individual reigns and periods. For example, the starting point of the reign of Sargon of Akkad largely depends on the length that is assigned to the so-called Gutian

---

13 Pruß for example observed that he 'has never met a colleague dealing with chronology, who actually believes in the correctness of the Middle Chronology' (2004, n. 94). Zettler wrote in a similar vein of the Middle Chronology that it 'has come to be cited more out of convenience than conviction by archaeologists and historians' (2003: 20).

period, variously established at between 30 and 100 years, and on whether 37 or 56 regnal years are assigned to Naram-Sin (Sallaberger 2004a).[14] Furthermore, the king lists may have sequentially ordered periods that actually overlapped or that were contemporary. The main difficulty in departing from the Middle Chronology is that it has widespread repercussions for almost all other areas and sites in the wider Near East, as well as for many other periods. This has led many researchers to the continued, albeit reluctant acceptance of the Middle Chronology (e.g. Lebeau 2000a: 170; Akkermans and Schwartz 2003: 12).

The Middle Chronology is largely based on historical sources coming from outside the Jezirah. Although there are texts from, as well as about, the Jezirah for the late third millennium, periodization for the Jezirah during this period is mainly achieved through the study of stylistic variation, stratigraphy, and radiocarbon dates. The correlation of historical events with radiocarbon dates, stratigraphic evidence and material culture is notoriously difficult. The term Akkadian may for example refer to the period during which a political entity of that name existed, or only to the period for which Akkadian influence in the north is archaeologically detectable, as at Tell Brak or Tell Mozan. Additionally, it has proven extremely difficult, if not outright impossible to define material remains as representing a typically Akkadian culture. It has recently been shown that many pottery types thought to be typically Akkadian actually had much longer life-spans, thereby further weakening the link between political and cultural designations (Zettler 2003: 27–8; Nicolle 2006: 166). Recently, however, important historical synchronisms between Mari, Ebla, and Beydar have been established, allowing a finer control of archaeological periods (Sallaberger 1998; Archi and Biga 2003).

The development of the discipline of Near Eastern archaeology has presented further problems for periodization and synchronization in the Jezirah. For a long time, the northern Jezirah in itself has not been a research area. Leaving aside Mallowan's pioneering work in the 1930s, the region has always been overlooked in favour of southern Mesopotamia and western Syria. This is reflected in the preferential use of western Syrian chronologies in the western part of the Jezirah, and traditional Mesopotamian divisions in the Khabur area. The artificial division that is thus created possibly reflects real variations between the western and eastern parts of the Jezirah, but may just as well be imposing these variations upon otherwise similar material. This division is difficult to break down (see Lebeau 2000a: 168–9).

Recent attempts at chronological refinement have taken these considerations into account, resulting in the Early Jezirah (EJ) and Old Jezirah (OJ) periodizations that rely on stratified material culture change and radiocarbon dates (Pfälzner 1998; Lebeau 2000a; Dohmann-Pfälzner and Pfälzner 2001). The EJ periodization, including its regional subdivisions, builds upon the work of multiple excavations across the Jezirah. It aims to establish a framework within which the effects of chronological and regional variation can be clearly discerned and separated (Lebeau 2000a: 170). The OJ periodization, introduced at Tell Mozan for the early second millennium BC (Dohmann-Pfälzner and Pfälzner 2001), does not seem to be as widely accepted as the Early Jezirah chronology yet. Although this new chronology has removed the ambiguous and often misleading southern Mesopotamian terminology from the archaeological discourse in the Jezirah, it should be realized that the new chronology still has many ties with the south. A new terminology masks these ties, but does not eliminate them. It rather allows some stereotypical views to be corrected.

Rigorous analysis of multiple radiocarbon dates from secure stratigraphic contexts across the Near East and the Mediterranean may ultimately lead to the revision of large parts of the conventional chronological frameworks (Bruins 2001; Bruins et al. 2003a; Manning et al. 2006). Unfortunately, third and second millennium BC Mesopotamia is only now being subjected to this

---

14 Radiocarbon dates from EJ III/IV contexts at Beydar, Brak, Mari, and Mozan are still too inconsistent to support either a 40 or 100-year length for the Gutian period. For example, the 2175 cal BC date of phase 2a of the Tupkish palace in Mozan fits only with a 40 year Gutian period using the Middle Chronology, as seems to be the case with the 2291–2200 cal BC date for the destruction of Mari *Ville II*. On the other hand, the very early EJ III radiocarbon dates from Tell Brak and Tell Beydar would favour a 100-year Gutian period, again following the Middle Chronology (Sallaberger 2007: 421).

date BC	Jezirah	Southern Mesopotamia	Beydar	Brak	Chuera	Hammam et-Turkman	Leilan	Balikh Survey	Birecik-Euphrates Dam Survey	North Jazira Survey
	Middle					VIIIa		VIIIa		
1600	Jezirah I	Kassite		Q Mitanni		VII/VIII				
1700						VIIc	0	VIIc		
1800		Old Babylonian		P Old		VIIb		VIIb		
	Old			Babylonian		lower town				
1900	Jezirah II-III					adminis- trative	I		Late MBA	
	Old	Isin/				complex				
2000	Jezirah I	Larsa				VIIa		VIIa		Khabur
2100	Early	Ur III	IVb	N post-		VI west 6				
	Jezirah V	Gutian		Akkadian		VI west 5			Late EBA/	
2200			IVa			VI west 3-4	IIc		Early MBA	
2300	Early Jezirah IV	Akkadian	III-IV	M Akkadian	IE	VI west 1-2	IIb			
2400	Early	Early				city wall VI east 5-8		VIcd		
	Jezirah IIIb	Dynastic IIIb	IIIb	L post-	ID					Late third
2500				Ninevite 5			IIa		Middle EBA	millennium
2600	Early Jezirah IIIa	Early Dynastic IIIa	IIIa	K late Ninevite 5	IC	VI east 3-4				
2700	Early Jezirah II	Early Dynastic II	II		IB		IIId			
2800			(I)			VI east 1-2	IIIc			
2900	Early Jezirah I	Early Dynastic I		J early Ninevite 5			IIIb IIIa	VIab		
3000	Early Jezirah 0			H post-Uruk						Ninevite 5
3100	Late Chalco- lithic 5	Late Uruk		G late Uruk	V		IV	V	Early EBA	Northern Uruk

Table 2.4: Comparison of third and early second millennium BC periodizations for the Jezirah, southern Mesopotamia, selected sites, and the Balikh Survey, Birecik-Euphrates Dam Survey, and North Jazira Survey (based on Brinkman 1977; Curvers 1991; Algaze et al. 1994; Wilkinson and Tucker 1995; Lebeau 2000a; Dohmann-Pfälzner and Pfälzner 2001; Oates et al. 2001; Pruß 2004; Ristvet 2005; Sallaberger 2007).

type of chronological reanalysis. The relatively subjective analysis of stylistic and textual linkages therefore currently remains the most important method to assess the implications for Mesopotamia of chronological revisions elsewhere.[15]

However, independent support for the Middle Chronology now comes from radiocarbon dated wooden beams from Anatolia. The Aegean Dendrochronology Project (AEP) has constructed and dated a 1500+ years floating tree-ring sequence from wooden beams recovered from various archaeological sites in Anatolia. This sequence has been radiocarbon dated, allowing the proposal of absolute dates for the structures from which the wood was taken, and from there the dating of historical reigns and events associated with those structures. These structures include the Sarıkaya Palace at Acemhöyük, and the Waršama Palace at Kültepe. For these buildings, construction dates of 1774 +4/-7 BC, and 1832 +4/-7 BC are proposed, respectively. In both buildings sealings and documents relating to Shamshi-Adad were found. Interpolation of these data with textual evidence suggests that the Middle Chronology must be lowered by thirteen to seventeen years. According to the authors, the High, Low, and Ultra-Low chronologies are now respectively rendered impossible, unlikely, and very unlikely (Kuniholm et al. 1996: 782; Manning et al. 2001: 2534).[16]

---

15   For example, no collection of Mesopotamian radiocarbon dates featured in a recent special issue of *Radiocarbon* devoted to Near Eastern chronology (see Bruins 2001: and references therein). The ARCANE project (Associated Regional Chronologies for the Ancient Near East and the Eastern Mediterranean) that was initiated in 2002 aims to bring together and synchronize regional culture/historical chronologies, as well as radiocarbon dates, in order to achieve a reliable absolute chronology for the third millennium BC and the preceding and following centuries (http://www.arcane.uni-tuebingen.de/index.html (Accessed 4 February, 2009)).

16   On the proposed bias towards too early dates for mid-Holocene radiocarbon determinations that would affect these chronologies, see Keenan 2002, Manning et al. 2002, and Kuzucuoğlu 2007b.

In the present study, the Middle Chronology will be followed, although accumulating evidence suggests that this chronology is in need of at least some minor revision. Nevertheless, this chronology remains the most widely employed, and thus better accommodates comparative study of multiple regions and sites than newer chronologies that are generally only applied to single sites. The debate on Middle versus Low Chronology still focuses on textual evidence, which makes extrapolation to archaeological periodizations difficult. The repercussions of the shorter historical chronologies for material culture chronologies as yet remain to be determined. A step in the right direction would be the full integration of historical data and stratified archaeological material, and a stronger focus on radiocarbon dating and other dating methods from the natural sciences. Table 2.4 shows several chronologies that are currently in use in the Jezirah and neighbouring areas.

2.5 Some notes on the political history of northern Mesopotamia

Historical information on northern Mesopotamia is available from the twenty-fourth century BC onwards in the form of the archives from Ebla palace G and Tell Beydar. During the *c.* 50 years covered by these archives, the initial political dominance of Mari was challenged by a coalition of Ebla, Kish, and Nagar (Tell Brak). Although the battle fought between these powers apparently resulted in a victory for Ebla, the city was destroyed soon after. It has recently been suggested that Mari was most likely the power responsible for this destruction, rather than Sargon of Akkad (Archi and Biga 2003: 35). Considering that Ebla and Nagar were allies, the destruction of phase L at Brak may probably also be attributed to Mari (Sallaberger 2007: 422). Again not much later, Mari was destroyed, probably by Sargon of Akkad, who recorded that he subdued Mari and Elam (Frayne 1993a: E2.1.1.1; Archi and Biga 2003: 34). The texts from Nabada (Tell Beydar) and Ebla testify to the urban nature of settlement in the Upper Khabur region. Numerous cities are men-

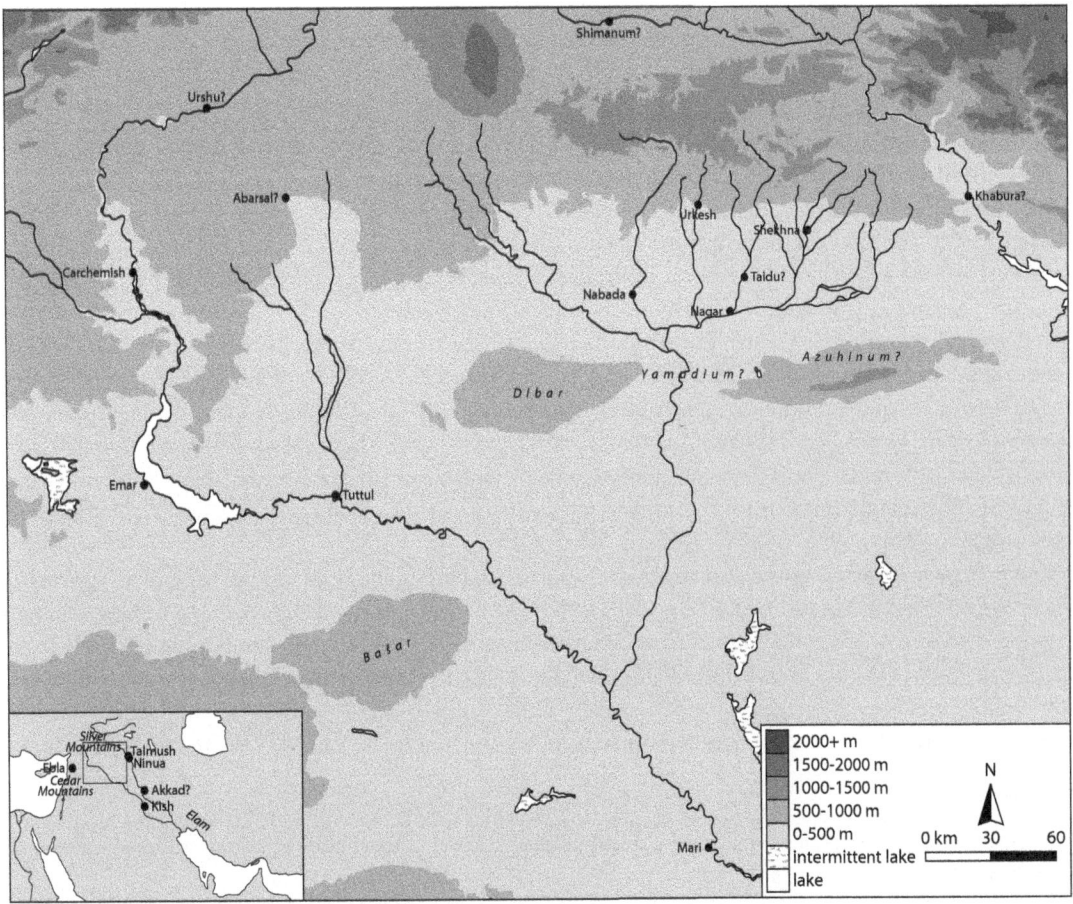

Fig. 2.9: Map of northern Mesopotamia showing late third millennium BC names of sites and regions mentioned in this study.

tioned. Among the important, and probably independent, northern Mesopotamian centres of the mid-third millennium were Nagar, Shekhna, Urkesh, Abarsal, and Azuhinum. These centres probably controlled the greater part of northern Mesopotamia (Ur 2004: 272) (Fig. 2.9).[17]

Sargon of Akkad's impact on the political situation of northern Mesopotamia is difficult to assess. He claimed to have reached Tuttul, the Cedar Mountain, and the Silver Mountains, that is Tell Bi'a at the confluence of the Balikh with the Euphrates, the Amanus, and the Taurus Mountains, respectively. However, it remains difficult to ascribe archaeologically observed destructions to his actions, as the example of Ebla, mentioned above, clearly shows. Thus, links between Sargon and archaeologically observed contemporary events or developments must be carefully considered before they are drawn. In fact, it has not yet been settled whether Sargon's western campaign was a large-scale military operation with the purpose of conquering and subduing foreign lands, or whether it was no more than a surprise military raid, taking advantage of already weakened opponents such as Mari and Ebla. The important observation must be that west of the Upper Khabur region there are no archaeological indications of lasting Akkadian influence.[18]

Possibly from the time of Sargon onwards but certainly at the time of Naram-Sin, the middle and eastern Upper Khabur region must have been under Akkadian political control. The exact nature of this control remains uncertain, however. Tell Brak probably maintained a key position in this region. Sargon's successor Rimush was mentioned in inscriptions on two stone vessels found at Tell Brak. It is difficult to see though how this relates to the position of Tell Brak at the time of Rimush, because the vessels were found in the Naram-Sin Palace, and may therefore have been brought to the site at a later time than the reign of Rimush. According to Sallaberger (2007: 424), the vessels most likely represented gifts of Rimush to a local loyal governor and were therefore heirlooms at the time of Naram-Sin.

Tell Brak was certainly an Akkadian stronghold during the reign of Naram-Sin. The Naram-Sin Palace, which should more properly be called a fortress, could be identified as such because Naram-Sin's name was stamped in mud bricks used for the construction (Frayne 1993a: E2.1.4.22). The Akkadian influence can be inferred from an administrative text from the Naram-Sin Palace listing cities and numbers of men, possibly soldiers (Eidem et al. 2001: 107). This list included, among others, Urkesh, Shekhna, and Taidum. Another list of small amounts of animals, mostly sheep, also named Shekhna (Eidem et al. 2001: 110). Urkesh is identified with Mozan (Buccellati and Kelly-Buccellati 1999: 10), Shekhna with Tell Leilan (Charpin 1987), and Taidum possibly with Tell Hamidiyeh (Eichler et al. 1985: 69; Wäfler 2001: 169). Except for Mozan/Urkesh, the relation between the Akkadian administration at Brak and the cities in this list is unclear. The relatively close proximity of Tell Hamidiyeh to Brak makes it likely that Hamidiyeh was under direct control of Brak, yet not all cities in the list were, as is clear from Urkesh.

Despite the repeated claims that Shekhna/Leilan was an important 'imperialized' centre within the Akkadian empire during Leilan period IIb (e.g. Weiss et al. 1993: 998), it is unclear what this designation means exactly, and it must be considered a possibility that Leilan was of secondary importance with respect to Tell Brak, or that it was a vassal state like Urkesh. Much of the evidence for Akkadian imperialization at and around Leilan is indirect, and the only direct evidence for a form of Akkadian administrative presence at Leilan consisted of two sealings with Old Akkadian inscriptions (Weiss 1997: 343; Weiss et al. 2002–3).[19]

At Mozan/Urkesh, sealings have been found that belonged to Tar'am Agade, a daughter of Naram-Sin (Buccellati and Kelly-Buccellati 2000: 139). She was probably the wife of an unnamed ruler or *endan* of Urkesh, known from another inscription found together with the sealing of Tar'am Agade. It is likely that Urkesh was an ally of Akkad at this time and presumably also under Tupkish, a slightly earlier ruler of Urkesh. Thus, in the time of Naram-Sin, some parts of the Upper Khabur region were controlled directly, as was probably the case at Tell Brak. In other areas,

---

17   See Sallaberger 2007 for a discussion of the problems associated with locating third-millennium toponyms. The location of Abarsal is especially problematic, and besides Kazane Höyük, a location along the Upper Euphrates has been proposed as well (Bunnens 2007: 50).

18   In fact, even Mari seems to have played a relatively marginal role up to and during the reign of Naram-Sin (Sallaberger 2007: 430).

19   See Ur (2004: 73) for a more elaborate discussion of the evidence for Akkadian control at Tell Leilan.

Akkad's influence may have been more subtle and indirect, through alliances and vassal states, as reflected in the evidence from Tell Mozan. This reconstruction would support the description of the Great Revolt against Naram-Sin, according to which the rulers of the Upper Land favoured Naram-Sin and refused to support Amar-Girid (Sallaberger 2007: 427).

The situation to the west of the Upper Khabur remains unknown during Naram-Sin's reign. Naram-Sin claimed to have reached the Cedar Mountain and also mentioned Tuttul, but it is not certain whether the western or eastern Tuttul is meant (Sallaberger 2007: 429, n. 69). The impact of Naram-Sin's western campaign is as difficult to assess as the impact of Sargon's earlier campaign.

After Naram-Sin's reign, historical evidence from southern Mesopotamia on northern Mesopotamian political affairs practically ceased until the Ur III period. The royal names preserved at Nagar/Brak and Urkesh/Mozan, however, indicate that at least these two places were the centres of local kingdoms. At Tell Brak, there is one post-Akkadian ruler, possibly contemporary with Shar-kali-Sharri or slightly later, and at Mozan the line of rulers seems to have extended almost to the end of the third millennium BC, although the evidence is admittedly meagre. It should be noted that the names of the rulers of both Nagar and Urkesh were Hurrian, possibly indicating the growing importance of this population group in northern Mesopotamia (Sallaberger 2007: 431-2) (see Section 8.3).

During the Ur III period (2112–2004 BC), northern Mesopotamia figured occasionally in historical texts. The Ur III state apparently devoted most of its attention and military power to the east, as far as can be reconstructed from royal year names. Thus, military operations to the west as are known for Sargon and Naram-Sin are unknown for any Ur III ruler. However, there were political contacts with the areas to the north and northwest of Ur as well. There were marriages between Ur III rulers and princesses from cities in the north, including Mari. The marriage with a princess from Mari could be interpreted as an alliance with that city to create a political and military buffer to the north of the Ur III state (van de Mieroop 2007: 80). The Puzrish-Dagan administrative texts indicate that messengers from cities in the north received gifts. It has been argued that the number of attestations of messengers from a particular place directly reflects the importance of that place for the Ur III state. In that case, Mari, Ebla, Shimanum (near Diyarbakır?), and Urshu (Samsat Höyük?) were the most important cities, followed by Khabura, Ninua, Talmush, Urkesh, and Yamudium. Tuttul/Tell Bi'a is also mentioned, albeit only twice, indicating that it was of even lesser political importance to the Ur III state than the cities mentioned above (Sallaberger 2007: 439). As far as can be judged from the Ur III documentation, no city in northern Mesopotamia seems to have politically controlled large territories at this time (van de Mieroop 2007: 80).

The period from c. 2000 to c. 1600 BC was characterized by competing city-states that had varying success in subduing their rivals and incorporating them in larger political entities. Many of these city-states were ruled by dynasties who claimed Amorite descent. Among the more powerful city-states of this period were Yamhad and Qatna in the west, Assur in the northeast, Mari along the Middle Euphrates, and Isin, Larsa, Eshnunna, and eventually Babylon in the south (van de Mieroop 2007). However, the transition from the third to the second millennium remains poorly understood, leading Charpin and Ziegler to describe it as 'la grande vague' (2003: 29). A clear view of northern Mesopotamian political affairs only re-emerges during the period covered by the Mari archives (Fig. 2.10).

During the Ur III period and probably until the end of the nineteenth century, Mari was ruled by *shakkanakku* or governors. Although previously a brief hiatus was reconstructed between the last *shakkanakku* and the first Amorite ruler (Durand 1985), recent epigraphical and archaeological research has provided evidence for a considerable degree of continuity (Butterlin 2007: 230-1). The first Amorite ruler was Yagid-Lim, but not much is known of his reign and he was soon succeeded by Yahdun-Lim (1810–1794 BC). Under Yahdun-Lim, Mari controlled the Euphrates northwards up to Tuttul/Tell Bi'a, and also dominated parts of the Upper Khabur region, leading to conflicts with Shamshi-Adad of Ekallatum (c. 1830–1776 BC). The border between the influence spheres of Yahdun-Lim and Shamshi-Adad may have been formed by the Wadi Jaghjagh. During his reign, Yahdun-Lim also fought against a coalition of Yaminite rulers, which he defeated at the Euphrates near Tuttul. Eventually, Shamshi-Adad defeated Mari and installed his youngest

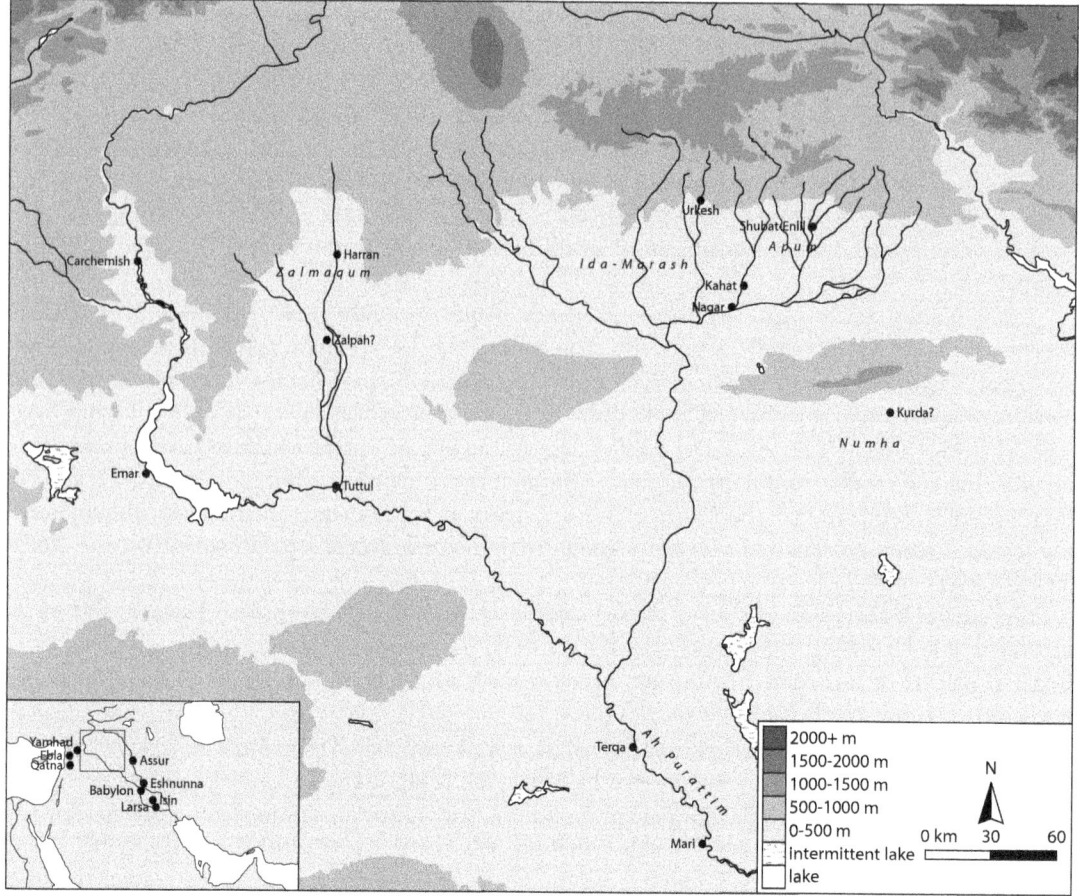

Fig. 2.10: Map of northern Mesopotamia showing early second millennium BC names of sites and regions mentioned in this study.

son Yasmah-Addu as viceroy, while establishing himself at Shubat-Enlil/Tell Leilan and leaving his eldest son Ishme-Dagan to rule Ekallatum. After Shamshi-Adad's death, Zimri-Lim replaced Yasmah-Addu as ruler of Mari. In the Upper Khabur, a kingdom of Apum appeared, centred on Tell Leilan. Zimri-Lim was in turn defeated by Hammurabi of Babylon, who destroyed Mari in 1761 BC. A new middle Euphrates state emerged at Terqa, but that kingdom of Hana failed to exert an influence over other regions comparable to the authority of Mari (Charpin 1993; Charpin and Ziegler 2003).

The political history of northern Mesopotamia during the third and early second millennium BC reflects the conflicts between local city-states and foreign powers wishing to exert influence over the region. The Akkadian presence in northern Mesopotamia, despite the fact that it remains poorly understood, signified a first attempt at creating a larger regional state, followed by the Ur III and Shamshi-Adad's northern Mesopotamian states. Each attempt proved unsuccessful in the long term, and especially the degree of administrative centralization that the Akkadian state achieved in northern Mesopotamia may have been rather low. The Ur III state was highly bureaucratized, but again relatively short-lived and it did not exert clear permanent influence on northern Mesopotamia. Zimri-Lim, on the other hand, seems to have exerted his influence by maintaining alliances with local city-states, rather than attempting to control them directly through force.

3 Theorizing social responses to environmental change

3.1 Introduction

Section 2.3.7 has shown that the late third and early second millennium BC were characterized by a climate that became progressively drier. Evidence from plants and human bone material suggests that this aridification was associated with periods of environmental and nutritional stress, probably caused by insufficient water and/or low nutritional value of food intake (see Section 2.3.8). Although the evidence for drought is currently relatively meagre, it nevertheless seems that the late third millennium was a period during which shortages in water and/or food occurred more often than before or afterwards, and that droughts must have been a regular feature of human life. The present chapter will explore how social units react to shortage and more specifically how social relations develop under conditions of environmental stress. Section 1.3 laid down the requirements that a framework for that type of study must meet. These requirements are (*a*) the ability of the framework to explicitly consider the historical development of the position of human communities in their environment, (*b*) the need for a feedback mechanism between environmental change and socio-cultural change, and (*c*) the consideration of communities, or possibly even households as the fundamental social units that respond to environmental change, rather than societies as a whole, even though these small-scale units cannot always be discerned in the archaeological record.

This chapter aims to present and discuss a framework capable of meeting these requirements. It will start with a delineation of the social scale at which social responses to environmental change will be studied (Section 3.2). The following sections discuss the organization of responses to environmental change, and the maintenance of alternative responses (Sections 3.3 and 3.4). Section 1.4 already introduced the two hypotheses regarding the interaction between sedentary agricultural and (semi-)nomadic pastoral communities: (*a*) it is to be expected that, as a result of environmental stress, social relations between sedentary agricultural communities will develop along competitive lines in order to improve each communities' chance of survival and access to scarce and therefore contested resources, and (*b*) sedentary and (semi-)nomadic groups will develop cooperative mechanisms ensuring mutual survival and maximization of their respective subsistence strategies. This chapter will explore the logic of these hypotheses, and discuss archaeological correlates to test them (Sections 3.5 and 3.6).

3.2 Defining the scale of analysis

It has been argued that a society is not an appropriate scale for the analysis of responses to environmental change, because a society rarely, if ever, acts as one single social unit (Section 1.3). Below the level of the society, there is a potentially large range of scales at which an analysis can be carried out. Following Binford, who argued that 'a researcher's choice of units and scales should be determined by the potential for knowledge growth inherent in the alternatives' (2001: 34), it is here suggested that research at the scale of human societies should be abandoned in favour of a more fine-grained perspective focusing on small-scale responses and decision-making. The present study must be understood as part of a nested scalar approach to social responses to environmental change. Within this scalar approach, there is not one legitimate approach at one appropriate scale, but rather numerous approaches at different scales that complement each other (cf. Rosen and Rosen 2001).

Examples of social units that may be taken as the basic unit of analysis include the individual actor, the household, the community, or the state. These units can be placed on a scale of increasing demographic or spatial size, and, more importantly, on a scale of increasingly larger decision-making levels. Examples of this approach can be found in Halstead and O'Shea 1989b, Tainter and Tainter 1996, and Wilkinson et al. 2007a. Alternatively, it is possible to focus on specific groups ranked along a social axis within communities of a given scale. This is the approach adopted by Rosen (1995; 2007) in her case-study on the responses to environmental stress by Early Bronze Age urban elites. These two approaches are not contradictory, but highlight different aspects of decision-making and social interaction. Rosen focused on decision-making within a social unit of a given scale, arguing that the responses of Early Bronze Age urban elites were severely constrained by social factors. The other approach focuses more strongly on interaction between independent social units, rather than on internal social processes. Thus, this approach is more strongly concerned with external social processes, and how these are influenced by environmental change. The internal and external approaches should therefore be considered complementary.

The present study takes the external approach and adopts the perspective of the settlement community. This choice is partly determined by the history of research into late third–early second millennium BC climate change, and partly by the nature of the evidence. Many studies have described the effects of late third millennium BC climate change at least partially in terms of settlement abandonment, that is, the replacement of an urban settlement system by a system characterized by fewer, and smaller sites. Although it has already been pointed out that settlement abandonment was less wide-spread than previously argued (Section 1.2), it is often still implicitly assumed that settlement abandonment logically and automatically follows upon environmental deterioration. This study hopes to break away from this implicit assumption by explicitly theorizing how settled communities can respond to environmental change. A second reason for adopting this scale of analysis can be found in the available evidence. The settlement record of northern Mesopotamia is relatively good and very diverse, offering good opportunities for detecting any social effects of environmental change. Furthermore, the settled community as unit of analysis seems to coincide relatively well with the available evidence on the political fragmentation of northern Mesopotamia. Given the fact that highly centralized states were absent in northern Mesopotamia during much of the third and early second millennium BC (see Section 2.5), it may be safely assumed that individual cities maintained a relatively large degree of independence, and that interaction between them was reasonably unrestricted by higher-order political entities.

3.3 Ordering social responses to environmental change

Every environment presents its inhabitants with risks due to unpredictability and patchiness in the availability of resources. Successful existence in any environment is not based on its inhabitants' abilities to cope with average environmental conditions, but on their ability to cope with extreme variations in these conditions. Decisions will not be based on average conditions, but on the possible occurrence of potentially harmful extreme variations. Social units will thus develop behavioural strategies to cope with unpredictable periods of resource shortage (Winterhalder 1980: 147; Halstead and O'Shea 1989a: 1). Numerous ethnographic and archaeological studies testify to the plethora of behavioural coping strategies that can be found among human groups. Observed strategies include diversification and/or intensification of subsistence strategies, group fissioning, adoption of new resources, adoption of new mobility patterns, resource storage, exchange, and the formation of social networks (e.g. Halstead and O'Shea 1989b; Tainter and Tainter 1996; Winterhalder et al. 1999: tables 3, 4; McIntosh et al. 2000b).

The variability of mechanisms to cope with environmental stress is increased by the fact that no correlation seems to exist between specific types of stress and specific responses. A review of literature on responses to stress suggests that similar conditions of environmental stress have led to a variety of responses among different groups and in different areas and periods (cf. studies cited by Winterhalder et al. 1999). Even within groups multiple responses can be adopted due to differences in, for example, wealth and availability of resources (Rosen 2007: 147). However, this does not mean that these coping mechanisms were randomly adopted by groups affected by environmental stress. On the contrary, most researchers agree that the adoption of specific adaptive mechanisms is a conscious choice and that the cost of the adaptation should match the severity of the environmental change. In other words, extremely severe or persistent periods of environmental stress will lead to the adoption of increasingly more costly or disruptive adaptation mechanisms in terms of fitness, energy expenditure, effects on social status, or a combination of these variables (Bates and Lees 1977: 838–9; Braun and Plog 1982: 508; Halstead and O'Shea 1989a: 5; Minnis 1996: 67).

This approach assumes that it is impossible to continuously maintain buffering mechanisms that are able to cope with every possible type of environmental stress. Maintaining such a general mechanism would not only be extremely expensive, but would also require that environmental stress can always be anticipated and/or predicted, which is unlikely to have been the case in any past society. For this reason, it is much more cost-effective to organize buffering mechanisms and adaptation strategies according to a hierarchy, whereby higher-level strategies are only adopted if the strategies below them prove insufficient to deal with the resource shortage being experienced (Tainter 1988: 110; Forbes 1989: 90).

Several researchers have proposed a correlation between the magnitude of resource stress and the social cost of the adopted response (Braun and Plog 1982; Minnis 1996). They suggested that increasingly severe periods of resource stress require the involvement of increasingly larger groups of people, leading to stronger social obligations. In an ethnographic study on responses to subsistence crises in a mid to late twentieth century AD Greek farming community, Forbes suggested that there was a distinction between low-level and high-level responses to environmental stress, which he called Hazard-Response Mechanisms (HRMs). Low-level responses dealt with regularly recurring short-term variations and were characterized by (*a*) continuous or frequent operation, (*b*) high cost in terms of energy expenditure, (*c*) high integration in socio-cultural practice, and (*d*) low visibility to those who are involved and to the observer. High-level responses, on the other hand, dealt with less frequent but more severe resource shortages and were characterized by (*a*) infrequent operation, (*b*) low energy expenditure, (*c*) high social expense because normal social rules are breached, and (*d*) high visibility. Examples of low-level responses observed in the Greek community were storage, diversification of crops, and scattering of farming plots across the landscape, whereas high-level responses included begging and marrying off daughters to partners that would be considered unsuitable under normal conditions (Forbes 1989: 95).

According to this model, low-level adaptation mechanisms create a safety-net against the possibility that environmental stress will result in resource shortages. These safety-net mechanisms are maintained even though the cost in terms of energy expenditure is much higher than required for immediate subsistence needs. By investing in this safety-net in times of abundance, it is hoped that socio-culturally less-acceptable responses can be avoided when a period of environmental stress actually occurs. For example, Forbes observed that there was a tendency to sow higher amounts of grain and to store more foodstuffs than immediately needed, even if this meant that their market price would go down as a result of rotting. However, this behaviour would ensure the presence of foodstuffs in the case of an unanticipated shortfall in resource availability, such as a bad harvest (Forbes 1989: 93). Similarly, Winterhalder et al. were able to show that Peruvian farming households scattered their agricultural plots to a degree that effectively allowed them to eliminate the risk of shortfalls, even though this scattering led to a considerable reduction in the net production that could potentially have been achieved by these households if they had consolidated their fields (Winterhalder et al. 1999: 333–4). Thus, low-level responses lower the physical/nutritional as well as social effects of environmental stress through continuous or regular investment in buffering mechanisms against resource shortages.

High-level adaptation mechanisms, on the other hand, are only irregularly called upon; mainly if low-level responses prove inadequate. It is important, however, that these responses do not necessarily result in permanent social change. According to Forbes, the high-level responses that he observed in the Greek farming community 'can be recognised not simply as social breakdown, but as culturally patterned hard choices in the face of life-threatening hazard, for which lower-order HRMs are inadequate' (1989: 95). These high-level responses are, then, clearly within the recognized spectrum of available responses. Extreme resource shortage therefore does not necessarily lead to the adoption of entirely new responses, but to the acceptance of already known responses that would otherwise be neglected in favour of less costly alternatives. In other words, an increase in the severity of a resource shortage will not necessarily lead to an increase in the range of *available* alternative responses, but to an increase in the range of *accepted* alternative responses.

3.4 Maintaining alternative social responses to environmental change

If social responses to environmental change are hierarchically ordered, with high-level responses only being adopted if low-level responses fail, it is likely that these high-level mechanisms will be only infrequently called upon. Because such high-level responses are only irregularly used, it is necessary to embed knowledge about these adaptation mechanisms in social practice to ensure their continued existence beyond immediate need. McIntosh et al. refer to the concept of social memory to describe mechanisms that aim to preserve vital knowledge of climates and environments of the past in societies that do not have recourse to written records (2000a: 24). Similarly, Salzman (1978a) referred to the 'cultural maintenance of organisational alternatives' that could be activated during periods when normal social structures failed. Through these mechanisms, behavioural alternatives can be maintained that can be activated when low-level responses to environmental stress prove insufficient. The operation and effectiveness of these mechanisms will be discussed below.

A clear example of maintaining knowledge of behavioural alternatives was provided by Minnis (1996: 62) in his discussion of adaptation strategies to resource shortage in the North American Southwest, a region that experienced severe climate changes in the past. He drew attention to the use of famine foods under conditions of environmental stress. Famine foods are foodstuffs that are edible, but that are usually avoided because of their low dietary value or taste, unless there is no alternative. Due to the fact that entire generations can come and go without ever experiencing a crisis requiring the consumption of famine foods, the knowledge of these foodstuffs must be passed on in the absence of actual experience. Minnis observed that in one case this knowledge was maintained through the listing of edible plants in prayers, and through the annual ritual obligation to collect these plants. In this way, it was assured that knowledge of both the plants and their growing locations would be maintained through time and passed on to later generations. The recent decline in this knowledge can be attributed to changes in the subsistence base, and to the inclusion of the region into the wider state and world economy.

A slightly different system was observed among Bedouin living in the Egyptian Eastern Desert. There, during prolonged droughts, the Bedouin had to rely on drought-enduring plant species such as the acacia tree (Hobbs 1990: 39). In fact, through various social mechanisms, the Bedouin even ensured that these trees were protected from over-exploitation during periods without drought. Thus, the Bedouin continuously maintained a religiously motivated prohibition on the cutting of trees, they declared certain areas 'lineage reserves', falling under the protection of an important leader, and they protected individual trees by marking them with a clan's signature, thereby creating a sense of ownership (Hobbs 1990: 105–6). Similar regulations were also maintained for some other shrubs, as well as certain animal species.

The concept of maintaining behavioural or organizational alternatives can not only be applied to single adaptive responses such as famine foods, but also to more broad social structures that strongly affect the way in which social units function and interact with others. Salzman has pointed to the existence of 'social structures in reserve' that can be activated in periods of crisis or severe environmental change (1978a, b). He suggests that such social structures in reserve can especially be expected in 'environments in which there is an alternation of conditions through time, in which changes of circumstance are repetitive and to a degree of the same character', but at the same time points out that maintaining such alternatives provides 'flexibility in adapting to new conditions' (Salzman 1978a: 619). Salzman suggests that several systems exist through which social structures in reserve can be maintained (1978a: 619–21): (*a*) the deviant minority, (*b*) operational generalization, and (*c*) the asserted ideology. Under the first mechanism, there exists a small group adhering to (social) practices that are considered to be inappropriate under normal circumstances, but who are nevertheless tolerated by the society at large. In this way, the minority maintains knowledge of alternative behaviours, to which the society at large can revert in times of crisis. Under a system of operational generalization, a social unit adheres to any of a number of organizational alternatives, whereby the others remain within the 'conceptual repertoire' of the social unit. Salzman suggests that this system is especially relevant for the maintenance of alternatives to short-term activities. Finally, the asserted ideology represents a system under which social units acknowledge the existence and validity of a given social structure, but to which the social units do not actually adhere. However, when faced with circumstances for which the asserted ideology is considered appropriate, social units will put that social structure into place and act accordingly.

In a comparative study on drought responses among pastoralist groups in Kenya and the Negev, Bruins et al. (2003b: 29) documented the historical awareness or social memory of these groups. They found that both the Maasai in Kenya and the Bedouin in the Negev remembered years with severe droughts, and even assigned names to them. The Maasai used a system of naming that provided information on the severity of the drought. Thus, the word *olameyu* was used to indicate drought in general, without any reference to its effects. The words *emboot* and *emperi*, on the other hand, described droughts with particular emphasis on the adverse effects of droughts. An *emboot* was a severe drought resulting in loss of livestock due to lack of pasture. An *emperi* drought was even more severe and affected not only livestock but also endangered the lives of the Maasai themselves. Such a drought could last over a year, resulting in mass migrations. Years with particularly severe droughts were remembered as far back as the 1920s. However, the study also indicated that the word *emperi*, signifying the most severe droughts, was less well known among younger Maasai because modern technology and international food programmes had reduced the impacts of droughts. The Negev Bedouin employed a similar system of naming particularly dry years. This system retained memories of dry years dating back to 1914.[20]

The Maasai and Bedouin cases are not strictly concerned with maintaining knowledge about alternative response mechanisms. However, they do illustrate another important point, namely that knowledge about past climatic fluctuations adds to the successfulness with which new situations can be adapted to. Awareness of the potential variation of climatic fluctuations, even though they cannot be predicted precisely, can help social units to maintain behavioural or organizational alternatives suited to these situations (Binford 2001: 42).

3.5 Modelling social responses to environmental change

Many responses to environmental change can be carried out by individual social units. Thus, an individual household can diversify its subsistence strategies, intensify its production, or store resources for anticipated periods of shortage. In this sense, exchange and the formation of social networks stand apart from other coping strategies because they require the involvement of multiple social units, even though cooperation between social units is regularly observed in the employment of other coping strategies. Similarly, mobility can also be considered a social strategy, especially when the destination area of a relocating social unit is already occupied. Because the present study is particularly interested in the emergence and development of such social responses to environmental change, the interdependence of social structures and environmental change will be discussed in more detail.

Social structures can serve to regulate a social unit's access to scarce resources. This function can be the result of the fact that a resource does not occur in the region inhabited by a social unit, or because a social unit does not produce or extract that resource. Thus, social structures

---

20 It was observed in all regions that the knowledge about past resource shortages had declined significantly in the recent past. In the North American Southwest, this decline resulted from changes in the subsistence base and from the inclusion of the region into the wider state and world economy (Minnis 1996: 63). In the case of the Maasai and the Bedouin, the negative effects of droughts are today more easily countered through better long-distance transport options and the introduction of a monetary economy, allowing additional foodstuffs to be bought on the open market (Bruins et al. 2003b: 34). The waning importance of the term *emperi* in the Maasai case illustrates how this memory evolves under changing social conditions, again emphasizing the social aspect of environmental stress.

can reduce the effects of environmental variability through their ability to forge relations between social units with differentiated access to resources (Rautman 1996: 199). Differentiated access to resources is usually linked to the geographical space that a social unit occupies or uses, and to its (in)ability to exclude others from using that area or resource. Under these conditions, various social structures can emerge that regulate access to scarce resources. Depending on socio-cultural conditions and the nature of the resources that are involved, these structures can include exchange networks, hierarchical relations between groups, territoriality, and conflict (Dyson-Hudson and Smith 1978; Boone 1992; Kelly 1995: 161; Kohler and Van West 1996).

Which social structure regulating access to resources will emerge primarily depends on the resources that are utilized by each group within the network, on the nature of resource availability for each group, and on differences in resource availability between groups. These variables can be expressed along two axes as inter-group correlation of resource availability and intra-group variation in resource availability, resulting in four different configurations of related social structures regulating access to resources (Table 3.1). Inter-group correlation indicates the degree to which groups experience resource shortages and abundances at the same time, with a high correlation indicating strong temporal synchronization of resource fluctuations, and a low correlation indicating that groups usually do not experience resource shortages at the same time. Intra-group variation indicates the degree to which a group is likely to experience resource shortage, with high variation indicating that there is a high probability of resource shortage, but also the potential for producing a surplus. Low variation indicates that the overall production of a group remains stable through time. In other words, production is relatively predictable (Kelly 1995: 198).[21]

It is clear from Table 3.1 that the social structures are largely determined by inter-group correlation, with intra-group variation in resource availability mainly affecting the intensity of the structures. Thus, spatial boundary defence, territoriality, and conflict are more strongly associated with high correlation of resource availability between groups. This situation is represented in configurations (A) and (C). Social boundary defence and exchange, on the other hand, are the main social structures in configurations (B) and (D). Physical storage of resources, on the other hand, seems more strongly associated with intra-group variation and is expected in configurations (A) and (B). These associations point to a fundamental difference in how groups interact with others when faced with resource shortage. This difference can be summarized as the decision by individual social units whether or not it is possible or advantageous to share/exchange resources with other groups.

If neighbouring groups experience synchronous shifts in resource availability, that is inter-group resource availability is highly correlated, resource shortfalls and periods of abundance will occur at the same time. During a period of abundance, every group has sufficient resources and therefore does not need to procure them through exchange, whereas during a period of shortage, exchange is likely to fail because no group has resources to share or exchange. Thus, if social units

	high inter-group correlation of resource availability	low inter-group correlation of resource availability
high intra-group variation in resource availability	(A) -household storage -restricted social access between groups -spatial boundary defence -strong territoriality -conflicts common	(B) -social boundary defence -some storage -intensive and rigid reciprocity between groups
low intra-group variation in resource availability	(C) -passive territoriality -long-distance migration -conflicts rare	(D) -relaxed social boundary defence -differentiated exchange between groups

Table 3.1: Expected social structures regulating access to resources under different conditions of resource availability (based on Kelly 1995: fig. 5-6).

---

21 Although the configurations in Table 3.1 were originally proposed for hunter-gatherer societies, much research on interaction between more complex agricultural and/or pastoral communities has actually been based on the same underlying principles, as is for example the case with the work by Cashdan (1987) and Rautman (1996), both discussed in Section 3.5.2. See also Wendrich and Barnard 2008 on commonalities between research on hunter/gatherers and pastoralist and even settled agriculturalist societies.

with similar subsistence strategies experience similar conditions of environmental stress, they will have less incentive to engage in exchange networks with each other because it is less likely that goods can be reciprocated. Conversely, when groups rely on different resources, or when periods of resource availability are asynchronous, that is the correlation of inter-group resource availability is low, there is much more incentive for exchange, simply because resources are available. Below, these two response configurations will be explored in more detail.

3.5.1 Territoriality, social hierarchy, and conflict

Territoriality is the restriction of access to resources by one social unit and can be achieved through social regulation or through spatial exclusion, known as social boundary defence and spatial boundary defence, respectively (Casimir 1992). In the previous section, it was suggested that territoriality, conflict, and spatial boundary defence are expected adaptive strategies against resource strategies among groups experiencing synchronous fluctuations in resource availability. This synchronicity will often be the result of a reliance on similar resources, or because they are engaged in similar subsistence strategies. Even under these conditions, however, territorial behaviour will not always emerge. The concept of economic defensibility has been used to explain the occurrence of territorial behaviour: 'territorial behavior is expected when the costs of exclusive use and defence of an area are outweighed by the benefits gained from this pattern of resource utilization' (Dyson-Hudson and Smith 1978: 23). Economic defensibility is a function of the predictability and density at which the competed resource occurs, and the time and energy that is needed to defend it. Thus, in terms of the availability of resources, territorial behaviour is most likely to occur when resources are dense and predictable. Furthermore, territoriality is likely to increase together with investments that are made in a competed resource (Stone 1994). In any other constellation, other behaviour may be more appropriate to regulate access to resources (Dyson-Hudson and Smith 1978: 25).[22]

The choice for conflict is based on an assessment of the risks involved, and of those inherent in the alternatives. Given the high risks associated with conflict, it is likely that social units will first explore alternative methods of gaining access to the competed resource. Instead of open conflict, social units seeking access to a resource can accept a lower social status with respect to the social unit controlling the resource, or a social unit can move away. A social hierarchy is only likely to emerge if the social unit controlling the contested resource benefits from the acceptance of subordinate social units, and if social units seeking access to the competed resource are willing to accept a lower position (Boone 1992: 312). Movement is only possible if the resource can be obtained elsewhere (Carneiro 1970: 735). If these requirements are not met, conflict becomes an economically advantageous and sometimes unavoidable strategy to gain access to a resource. In this sense, violence and warfare are not inherently different from alternative methods of resource procurement because the decision to engage in conflict to gain access to resources derives from the same assessment of risks and opportunities as would have alternative responses.

The correlation between conflict and resource availability was investigated by Ember and Ember (1992). They conducted a cross-cultural ethnographic study on the frequency and causes of warfare. Their aim was to investigate whether there was a correlation between warfare and deteriorating environmental conditions. The underlying idea was that resource scarcity, or unequal access to resources, would lead to competition. Ember and Ember used the Human Resource Area Files (HRAF) ethnographic sample to test this hypothesis. They analysed a sample of 186 mainly non-state societies and looked for a correlation between warfare and a number of socio-cultural variables including threat of natural disasters, threat of famine, chronic resource scarcity, and several social conditions that were often assumed to increase the potential for warfare. The threat of natural disasters and famine was argued to be present when the ethnographic literature indicated that societies feared a disaster or famine, even though these events had not taken place during their recent past. This analysis indicated that fear of nature, that is, the perceived possibility that some

---

22   For example, Dyson-Hudson and Smith (1978: 25) used the term superabundance to describe situations in which a resource is so abundant that territoriality is economically and energetically unviable, so that allowing others to use the resource freely is the best solution (cf. Barnard 1992: fig. 4.1).

natural disaster or famine would cause loss of vital resources, and chronic resource scarcity were good predictors of warfare. Further analysis revealed that fear of a natural disaster was probably also a cause of warfare. In those societies where the ethnographic studies revealed a fear of natural disasters, conflict and warfare were often endemic and resulted in the capturing of resources, even when there was no shortage at that time. It seems, then, that in many societies warfare may indeed have been actively used as a buffering strategy against unpredictable resource scarcity (Ember and Ember 1992: 257).

The processes that are involved in the emergence of territoriality and conflict are likely to be intensified by demographic pressure on resources. If the number of people relying on a given resource increases, the per-capita return decreases and eventually a group may decide to actively resist growth and exchange, resulting in an increased potential for conflict (Zhang et al. 2007: 19218). The complex interaction between resource use and the incentive for exchange between social units with similar subsistence strategies has been modelled by Kohler and Van West (1996). According to them, social units base their decision on whether to cooperate with others and share/exchange a resource, or to act individually, on the risk associated with obtaining that resource. If the average production of a group is below the subsistence needs of each individual group member, but a member can potentially produce sufficient resources for himself due to unpredictable variability, that member is better off acting independently. If a member acts independently, there is no risk that he has to share or exchange his potentially above-average production with under-producing group-members. Conversely, if average group production is above the subsistence needs of individual members, group members will likely remain within the group because any unpredictable shortfalls can be compensated by the production from other group members. Thus, if social units with similar subsistence strategies experience similar conditions of environmental stress, they will have less incentive to engage in exchange networks with each other because it is less likely that goods can be reciprocated. Even though sanctions on free-riding will likely exist, that is profiting from a system or network without investing in it, such sanctions will become increasingly strained and may eventually break down, resulting in competition, territoriality, and conflict.

3.5.2 Exchange networks and social structure

Exchange as a buffering mechanism against resource shortage is especially expected among groups experiencing asynchronous fluctuations in resource availability. Exchange networks function as buffering mechanisms against resource shortage because they possess a spatial as well as temporal component (Minc and Smith 1989: 65; Minnis 1996: 9). Through their spatial component, exchange networks allow the distribution of resources from areas where they are widely available to areas of scarcity. The temporal component of exchange networks allows participants to acquire resources without directly reciprocating. In other words, there can be a time-lag between the acquirement of resources and their payment. Especially the temporal component of exchange networks is likely to be influenced by the social relations existing between the participating social units because delayed reciprocity requires social units to trust each other that an exchange will be reciprocated at a later point in time. Manipulation of social relations between social units can therefore affect the nature of exchange taking place between these units.

Whereas Braun and Plog argued that 'sustained increases in the intensity of integration will occur as concomitants of sustained increases in region-wide uncertainty or risks arising from environmental change' (1982: 508), Rautman (1996) proposed a more specific correlation between environmental change and the emergence or reorganization of exchange networks. She suggested that exchange networks will preferentially emerge between locations that are not equally vulnerable to environmental change (Rautman 1996: 201; Whallon 2006: fig. 2). This point of view coincides with configurations (B) and (D) of Table 3.1, characterized by high inter-group variance in resource access. Thus, if regions experienced differences in environmental variability, it might be expected that exchange networks will develop between these regions, or that existing networks are reinforced or strengthened. Similarly, Read and LeBlanc have shown through theoretical modelling how two groups experiencing different levels of resource availability through time and space can attain a higher average production over time through cooperation and therefore have a competitive advantage over groups that experience one of these levels of resource availability exclusively (2003: 72).

However, Rautman assumed that a similar subsistence economy existed in each region, based on the same resources. Under such a scenario, shortages in one region could only be alleviated through imports from elsewhere. If in one region multiple subsistence economies relying on different resources existed, it is likely that social networks will emerge between social units participating in different economical systems. This means that if social units experience environmental stress, they will preferably establish or invest in social relations either with social units that live in regions where there is less environmental stress, or with social units that pursue different subsistence strategies (cf. Kelly 1995: 200–1). Johnson and Earle point to the similarity between the concepts of economy and niche as expressions of subsistence specializations within human societies (2000: 22), thereby reinforcing the distinct nature of each strategy and the potential for exchange between their practitioners. Cashdan (1987) has shown that economic specialization within a limited region indeed leads to stronger exchange relations between specialized groups than between groups practising more generalized strategies.

It has long been recognized that exchange networks are often built on, and couched in the terminology of social structures such as kin, gender, age, class, ethnicity, and ancestry. Depending on the determining structure and the scale of the network, real, perceived, or constructed shared identities between the engaged social units can exist. For that reason, exchange networks and social structures as described above are related concepts that can be difficult to disentangle. This interdependence between exchange networks and other social networks can be described through the concept of reciprocity. Reciprocity is the exchange of goods and/or services, whereby giving creates an obligation to return goods of equal value by the receiving party. Reciprocity can be ranked from generalized through balanced to negative reciprocity, whereby the need to reciprocate goods with equal value and directly at the time of the actual exchange increases from generalized to balanced reciprocity. Generalized reciprocity is characteristic of households or small communities, with reciprocity becoming increasingly balanced as social distance between social units increases. Negative reciprocity includes theft and raiding (Sahlins 1972: fig. 5.1; Johnson and Earle 2000: 47–50).[23]

Several researchers have investigated the correlation between environmental change and the development of exchange networks and social structures (e.g. Braun and Plog 1982; Minc and Smith 1989; Rautman 1996; Rowe 2002). For example, Nettle (1996; 1998) found a correlation between language diversity and environmental risk. An analysis of the distribution of languages across the globe suggested that the number of languages was negatively correlated with the risk of crop failure. In areas with a low risk of crop failure, the number of languages was high, whereas in areas with a high risk of crop failure, the number of languages was low. Because languages play a fundamental role in communication, their distribution can serve as a proxy for social networks existing between spatially separated communities. The negative correlation between language diversity and environmental risk therefore suggests that social networks are indeed partially shaped by environmental conditions.[24]

Cashdan observed an even stronger correlation between environmental variability and the emergence of social structures aimed at the creation of a social environment favourable to exchange (2001a). Using a global cross-cultural sample, Cashdan found that climatic variation and predictability were among the strongest predictors of ethnic diversity. Ethnicity is here defined as an identity that is recognized by the ethnic group itself, as well as by outsiders. Ethnicity is usually based on a common ancestry and often includes perceived or constructed elements of kinship, a common language, and a common past (see Section 7.3.1). In areas with an unpredictable environment, and more particularly unpredictable rainfall, ethnic diversity was low. These conditions also

---

23 Note that even generalized reciprocal exchanges are conducted to achieve personal gain. In fact, personal gain, however defined, is inherent in every reciprocal exchange; otherwise giving or sharing social units would run the risk of being exploited by free-riders attempting to get goods without reciprocating (Boone 1992: 271).

24 Nettle, as well as Cashdan in the following example, compensated for the possibility that lower language diversity, as well as ethnic diversity in Cashdan's case-study, could have resulted from lower population densities.

led to higher intra-ethnic loyalty, that is, the degree to which group-members favour others within the same ethnic group over outsiders (Cashdan 2001b: 761–2). Cashdan argued that this correlation can be explained by the role of ethnicity in 'providing social insurance against unpredictable risks', resulting in the following generalization: 'where environmental unpredictability requires reliance on reciprocal or redistributive mechanisms among a spatially dispersed group, ethnic diversity should be lower' (2001a: 979). Thus, ethnicity functioned as a shared identity aimed at creating a social environment in which reciprocity tends to be more balanced or even generalized (cf. Horowitz 1985: 81; Bentley 1987: 42). This observation, then, strengthens the correlation between language diversity and environmental variability.

3.5.3 Hypothesizing social responses to environmental change

Summarizing, the present discussion has shown that social structures play an important role in reducing the effects of environmental change. More specifically, it is argued that social networks aimed at allowing and regulating exchange will emerge or will be intensified between social units experiencing different levels of environmental risk, either because they live in different environments, or because they employ different subsistence strategies. It is furthermore suggested that the emergence of social networks rests on the willingness of social units to engage in exchange networks. Resources are unlikely to be exchanged if environmental stress among social units within the exchange network happens simultaneously for each social unit. Furthermore, exchange is less likely to take place if individual social units risk having to share with under-producing social units. Under such circumstances, processes that are aimed at regulating access to resources will likely emerge, to ensure that the controlling social unit can meet its subsistence needs when experiencing environmental stress. Alternatively, conflict may arise or intensify.

Based on these considerations, it is now possible to formulate hypotheses regarding the formation, maintenance, and disintegration of social networks at various scales and under various environmental conditions. These hypotheses are the following: (*a*) under resource stress and within similar environments, that is environments with a similar risk regime, competition over scarce resources between social units will increase; and (*b*) under increasing resource stress, cooperation will emerge between social units that experience different regimes of environmental risk, for example due to different environmental conditions, or as a result of different subsistence production strategies. Within the context of the present research focussing on social responses to late third-early second millennium BC climate change in northern Mesopotamia, these hypotheses can be rephrased as follows: (*a*) it is to be expected that, as a result of environmental stress, social relations between sedentary agricultural communities will develop along competitive lines in order to improve each communities' chance of survival and access to scarce and therefore contested resources, and (*b*) sedentary and (semi-)nomadic groups will develop cooperative mechanisms ensuring mutual survival and maximization of their respective subsistence strategies. The archaeological correlates of each of these hypotheses will be discussed below.

3.6 Archaeological correlates of competition and cooperation

In the present study, two main proxy records will be employed to study the degrees of competition and cooperation between social units: the settlement record of northern Mesopotamia, and the historical record that provides details about the constellation of northern Mesopotamian society during the third-early second millennium BC. The settlement record will serve as a proxy of competition and cooperation between sedentary communities, whereas the historical record will document the formation and reorganization of social relations between sedentary communities and (semi-)nomadic communities. The choice for these records will be discussed below, as will several additional proxies that can potentially support the argument made here.

Human dispersal across the landscape is not random; it reflects social structure as much as it reflects the environment itself. Settlement systems are organized according to a number of environmental and social principles (Roberts 1996: fig. 1.5). Depending on which principle is given priority, a number of settlement systems can emerge, ranging from fully nucleated to fully dispersed, and including everything in between (Roberts 1996: 19). Economic practicality often dictates that households focusing on subsistence agriculture live on or next to their land in order to reduce transport and travel costs, resulting in a dispersed settlement pattern. However, social relations, obligations with other households, or safety reasons may lead households to live in nucleated communities or at least close together (Roberts 1996: 35; Peterson and Drennan 2005: 6). Furthermore, environmental variables make certain places in the landscape more suitable for habitation than others, depending on for instance the availability of water, good soils for agriculture or grazing, presence of game animals, or defensibility. In many cases, the choice for a particular site will be the result of a compromise between these social and environmental variables.

It can therefore be assumed that settlement patterns are the result of conscious human behaviour, and that settlement analysis can be undertaken to gain insight into that behaviour. Archaeologists have routinely employed various settlement pattern analyses to investigate the issues addressed here. Among the more well-known analyses are nearest neighbour analysis and rank-size analysis (e.g. Johnson 1980; Meijer 1990; Ristvet 2005). In nearest neighbour analysis, the distance between each settlement and the surrounding settlements is calculated, and these figures can be used to calculate average, minimum, and maximum distances between settlements. Nearest neighbour analysis can be used to detect settlement clusters, which in turn can be interpreted in terms of social relations existing between nearby settlements. Rank-size analysis ranks sites according to their size, and the resulting graph can be used to posit interdependence between sites, and to explore issues of population nucleation and dispersal.

Although nearest neighbour and rank-size analyses are clearly interrelated methods, they require different data in order to be effective. Nearest neighbour analysis assumes that distance between settlements is a meaningful socio-cultural category and in some way correlated to social relations and economic interdependence. Conventional nearest neighbour analysis calculates distances between sites as if they were located on a flat, featureless plain. It is much more likely that the actual travelling distances were a more socially meaningful category, but these economic distances are difficult to calculate in the absence of detailed information on the past landscapes. The results of nearest neighbour analysis should therefore be particularly suspect in unstable environments such as river valleys, where shifting river beds may continuously eradicate and create opportunities for movement.

Rank-size analysis assumes that the size of a settlement is a relative indication of the importance and position of that settlement in a larger settlement network or wider region. The necessary data may be entirely drawn from archaeological sites, and may therefore be more useful in areas with relatively unstable environments, although the processes that modify the landscape in general may also affect individual sites. As will be argued in Section 4.2, the degree to which Early and Middle Bronze Age sites were affected by post-depositional and environmental processes seems modest when compared to other archaeological periods, suggesting that rank-size analysis indeed has some value in evaluating settlement systems dating to the northern Mesopotamian Early and Middle Bronze Ages. Chapter 4 will discuss in more detail the process of generating and interpreting rank-size plots, and Chapter 5 will present the data on which the analyses will be carried out.

Exchange networks serve to disperse goods and/or information across the landscape. As with human settlement, this dispersal is not random and reflects conscious choices, needs, and opportunities of the participants. Such networks can be studied directly through sourcing a given material and identifying its distribution, or indirectly through the study of the distribution of containers within which perishable goods were transported and exchanged. The spatial distribution of iconographic styles or motifs can be tentatively used as a proxy for social contacts. Similarly, the distribution of luxury items is commonly argued to reflect social interactions (e.g. Whallon 2006). Although these distributions highlight the existence of some dispersal mechanism, they do not identify the processes by which materials moved across the landscape. These approaches, then, are problematic for the social network that is studied here, particularly because the hypothesized exchange goods, that is, wool, foodstuffs, and possibly information, leave almost no traces in the archaeological record. A further problem is the fact that historical sources for the early second millennium BC in northern Mesopotamia are generally not concerned with economy, including exchanges between sedentary and (semi-)nomadic communities, although references to such interactions do occur in certain periods.

However, texts for the late third-early second millennium BC provide information on a different aspect of exchange networks, namely the social context in which such networks operate. This context consists of the social identities of the participants, that is, those aspects of social identity that are shared by each participant and that form the basis for reciprocity. In Section 3.6, it was suggested that these identities can function as a proxy for the existence of exchange networks. It is not argued that the existence of a shared identity is equal to the existence of an exchange network, but that such networks will more easily emerge between social units sharing common identity traits, and that these networks will be more effective in alleviating environmental stress. According to this view, shared identities will be emphasized by social units participating in the same exchange networks. Chapter 7 will discuss the formation and spatial distribution of shared identities in northern Mesopotamia, and evaluate which social units were involved.

3.7 Conclusions

It can be argued that, despite the large variety of responses to environmental stress, these responses can be grouped according to one of two different principles: the associated social and/or energetic cost or disruptiveness of a given response, and the nature of resource availability. The first principle dictates that responses with a low cost will be adopted before responses that have a higher cost or that are more disruptive. Low-level responses deal with regularly occurring small-scale shortages, whereas high level-responses deal with increasingly severe fluctuations, often at an increasingly greater social cost or involving larger social groups. Whereas low-level responses are strongly embedded in social practice, the infrequent operation of high-level responses requires the existence of effective mechanisms for the transmission of knowledge about these responses through periods when they are not required. Most importantly, high-level coping mechanisms should not be considered as unique responses, but as socio-culturally structured behavioural alternatives for periods when the regular accepted range of strategies turns out to be inadequate.

The second ordering principle dictates that under specific conditions of resource availability, certain social responses are more likely to emerge than others. Social responses to environmental change, that is those responses involving multiple social units, are fundamentally concerned with the decision whether or not it is advantageous to exchange resources with others. It is expected that social units will exchange in resource networks if there are differences in resource availability. If resource availability is synchronous across multiple social units, acting independently is a better strategy to buffer against resource shortage, and territoriality and conflict are much more likely to emerge.

4 Methods for the reconstruction of regional settlement trends

4.1 Introduction

This chapter is concerned with methods for the reconstruction of regional demographic trends through the study of the material record yielded by archaeological excavations and regional archaeological surveys. In archaeology, surveying is a type of research whereby an area is subjected to surface inspection with the purpose of identifying archaeological remains. Through their regional coverage, surveys complement the picture emerging from single, excavated sites. Surveys have thus proven an invaluable tool for archaeologists, allowing them to address a whole range of research questions that are beyond the scope of single site excavations. Surveys have been especially successful in highlighting demographic changes, settlement patterns, and landscape use in general (Wilkinson 2000a; Banning 2002; Wilkinson 2003). Like excavated data, survey data can only be interpreted through a large body of middle range theory. There is a lively debate on the limitations of survey data, and how gaps in the survey record can be recognized and compensated for. The question as to which problems can be addressed by survey data, and what their relation is with other data sources, is less clear. Many of the inherent flaws of surveys can be compensated for through the application of statistical methods, and by the use of survey data with as many other sources as possible (Bintliff 2000: 214). Therefore, many researchers hold a positive view of survey data as a reliable source for the reconstruction of the past (e.g. Bintliff et al. 2000; Wilkinson 2000a; Banning 2002).

Since the early to middle twentieth century, archaeological surveys have been a vital part of the toolkit of archaeologists working in greater Mesopotamia. Starting with the pioneering work by archaeologists like Braidwood, Jacobsen, and Adams (Braidwood 1937; Jacobsen and Adams 1958; Adams 1965), a large number of systematic archaeological surveys has been carried out with various underlying strategies and aims (see for a recent overview Wilkinson 2000a, 2003). Surveys are no longer just a means to select a site suitable for excavation but have become a research field in their own right, with a separate methodology and sometimes driven by very specific research questions (Banning 2002).

The early survey work in the Jezirah is characterized by, but not limited to, the fieldwork of Poidebard, Mallowan, and van Liere and Lauffray. Regional investigation of archaeological remains in the Jezirah started with the aerial reconnaissance by Poidebard (1934) in the 1920s and early 1930s. He photographed numerous tells and landscape elements and created an aerial photographic record devoid of the modern overburden of agricultural and construction activities, that is useful up to this day. In the 1930s Mallowan surveyed and excavated numerous sites in the Upper Khabur and the Balikh Valley, including Brak and Chagar Bazar (Mallowan 1936, 1946). Using French aerial photographs from the 1920–30s, van Liere and Lauffray recognized and mapped the radial lines extending from many tells in the Upper Khabur (van Liere and Lauffray 1954).

Lower Mesopotamia has always been a major focus for fieldwork, but the political events of the last decades have brought major changes, resulting in the switching of much research to Syria. Thus, the 1960s and 1970s saw many surveys being conducted in southern Mesopotamia, whereas in southeast Turkey and the Jezirah survey research with strict methodologies only really started in the following decades, except for the Euphrates Valley, where dam construction necessitated intensive archaeological research from the 1970s onward. The survey and excavation bias toward easily accessible river valleys where the largest sites can be found, resulted in the one-sided view that ancient Mesopotamian civilization, or North Syrian for that matter, principally focused on these valleys, the *Kranzhügel* phenomenon being a notable exception. Recent surveys aimed to correct this bias.[25]

This chapter aims to discuss the strengths and weaknesses of survey evidence, suggesting that statistical analysis can and indeed must fill in some of the gaps inherent in archaeological survey data. It starts with an outline of the limitations and potential of survey data, focusing on Near Eastern survey practices. This discussion focuses on three key issues: (*a*) the visibility of archaeological periods in surface artefact collections (Section 4.2), (*b*) the problems associated with chronology construction (Section 4.3), and (*c*) the contemporaneity problem (Section 4.4). There follows a discussion of the theory that is necessary to translate survey data into historical population levels. Three steps are considered here: (*a*) simulation of site contemporaneity, (*b*) estimation of total occupied area, and (*c*) reconstruction of on-site population densities. It is argued here that this procedure is suited to document trajectories of growth and decline, rather than individual settlement histories. Thus, it has a high comparative value, and it will be used for that purpose here. Complementarily to this demographic reconstruction, rank-size analysis will be carried out on the studied settlement systems in order to study their organizational aspects. Section 4.6 will discuss the methodological aspects of rank-size analysis, and consider the various explanations that have been offered for different rank-size distributions. In Chapter 5, the demographic and rank-size analyses will be performed on datasets from three surveys in northern Mesopotamia, namely the Birecik-Euphrates Dam Survey (B-EDS), the Balikh Survey (BS), and the North Jazira Survey (NJS).

4.2 Interpreting survey evidence: limitations and possibilities

Straightforward interpretation of the results of archaeological surveys is impossible. Many studies have highlighted the difficulties of interpreting archaeological surface collections in terms of site numbers, sizes, and occupations of sites (e.g. Ammerman 1981; Redman 1982; Wilkinson 2000a; Banning 2002). These difficulties stem partly from the acknowledgement that the composition of particular surface collections is shaped by a large variety of socio-cultural processes and that surface collections as perceived by archaeologists are the result of very complex taphonomical processes. Any attempt to interpret archaeological surface collections must therefore take into account the possibility that (*a*) the sites and landscape features that are recognized do not represent the whole spectrum of sites and features that existed in the past; (*b*) the surface collection does not reflect the true diversity of materials and periods that is present at a surveyed location; and (*c*) the surface collection does not necessarily provide insight into the socio-cultural processes leading up to the deposition.

Early surveys focused on tell-sites that are easily identified from the ground, often from the seat of a car, and on aerial/satellite imagery and stand out as major foci of human activity. The most important consequence of this strategy is that smaller, non-mounded sites, such as pastoral campsites, will invariably be underrepresented. Several survey projects have included an off-site transect-walking component to assess the degree to which the landscape between tells is filled with archaeological remains (e.g. Wilkinson and Tucker 1995; Ur 2004; Wilkinson 2004). Apart from the recognition of many non-habitation features, and in fact a continuous archaeological landscape, such intensive surveys have also resulted in the recognition of low settlement sites that would not have been found by less intensive surveys (Redman 1982: 377). A further bias toward large sites can be introduced by the geomorphology of a region. In river valleys, alluvial sedimentation can for example easily cover entire flat sites and even smaller tells. Heavy erosion may result in the complete removal of entire occupation layers from the flanks of settlement mounds (Postgate 1994: 50). The inclusion of geomorphological research in survey methodology has allowed the

---

25   Examples are the Wadi Ağığ survey (Bernbeck 1993), the surveys in the area between the Euphrates and Balikh Valleys (Einwag 1993; Danti 1997), the Western Khabur survey (Hole 1997a), the survey of the Jebel Bishri to the south of the Euphrates (Lönnqvist 2006), and the Wadi Hamar Survey around Tell Chuera and Kharab Sayyar (http://web.uni-frankfurt.de/fb09/vorderasarch/survey.htm (Accessed 16 November, 2007)).

reconstruction of anthropogenic influence on the landscape through time, and the evaluation of problems of site preservation and visibility. A further step is the recognition of landscapes of survival and landscapes of destruction, resulting in an apparently uneven distribution and survival of sites and periods across space (Wilkinson 2000a: 229).

It was recently noted that there is a large discrepancy between the results of field surveys conducted in Greece and Italy, and in the Near East (Wilkinson et al. 2004). A review of multiple recent surveys in both areas showed that the number of recorded sites per square kilometre for Near Eastern surveys is far below that for modern surveys in the Mediterranean, mainly Greece. At first sight, this discrepancy appeared to result from the relatively low intensity of Near Eastern surveys, as compared to surveys elsewhere. However, comparison of the results and strategies of several northern Mesopotamian surveys suggested that other factors than post-depositional processes or survey biases may have been responsible for this discrepancy. Firstly, the northern Mesopotamian surveys that included a fieldwalking component found relatively few sites that would not otherwise have been found with less intensive strategies. Secondly, investigation of sections resulting from modern digging activities suggested that alluviation was not or barely obscuring sites outside river valleys. Finally, it was suggested that it is more efficient to make new mudbricks than to reuse old ones, resulting in relatively minor destruction of old sites in the search for building materials. These points suggest that a survey that combines intensive field-walking with satellite/aerial imagery analysis can assemble a relatively good record of archaeological features in a region (Wilkinson et al. 2004: 196).

On the level of the individual site, tell formation processes will likely cover early periods with the overburden of later occupations. As a result, pottery from the earlier periods will be less represented in the surface assemblage, and any interpretation based on the number of sherds and their occurrence on the surface may be biased. However, there is some reason to believe that this overburdening may be less problematic for Early/Middle Bronze Age sites than for earlier periods. Wilkinson (2000a) has plotted the ratio of occupations on tells versus occupations at low sites of less than 5 m high through time for the Tell Beydar Survey area, and found that particularly the third millennium BC represents a period during which settlement nucleated on high tells. After that period, the significance of tell-based settlement decreased, in favour of lower and smaller sites. This development suggests that third millennium BC occupations in particular will be highly visibly in the Jezirah. Evidence from rescue excavations in the Birecik-Euphrates Dam area suggests that there, too, third millennium occupation is particularly well represented on tells and therefore highly visible (see Section 5.2.1). The discrepancy of tell-based versus low site occupation is less pronounced during the Middle Bronze Age, suggesting lower visibility for this period, but with the caveat that overburden from later periods is also relatively limited.[26]

There is also discussion about the degree to which the archaeological surface assemblage reflects the chronological components of a site. Surveys usually assume that there is a positive correlation between the surface assemblage and the archaeological contents of a site. Yet it cannot be ascertained that all periods present in a tell will be equally represented on the surface. For example, surface assemblages may be subjected to stronger erosion than buried artefacts, meaning that large or high-fired sherds will dominate the record (Postgate 1994: 50). The preference of individual surveyors to pick up highly visible sherds may leave certain unremarkable types underrepresented. Furthermore, re-use of mudbricks, or construction of mudbricks from earth taken from tells, may result in the displacement of archaeological material. However, research at sites that are surveyed as well as excavated seems to suggest that there is some agreement between later ceramic periods inferred from survey, and those actually excavated. Pre-pottery periods, on the other hand, often go unnoticed and their presence can only be ascertained through excavation. Similarly, the on-site distribution of sherds on the surface is correlated to past settlement and activity areas of the site.

---

26 A similar transition from nucleated, tell-based occupation to dispersed occupation on non-mounded sites has been noted for the Amuq Valley, but there the transition occurred between the Iron Age and the Seleucid periods (Casana 2007: 203–4).

In both cases, the correlation is not very strong, but sufficient to use it as an indication of site size (Lyonnet 2000: 7, cf. Redman and Watson 1970 and Flannery 1976, but see Kohlmeyer 1981 and Schiffer 1987: 126, 353).

The results from archaeological surveys are thus nothing more than a sample. Just how representative this sample is, and how much of the total assemblage is actually recovered, are the main issues that researchers have to address to correctly interpret their data. The comments made above suggest that the site recovery rate for Near Eastern surveys can be relatively good, depending on the methodology that has been adopted. There is furthermore reason to believe that the presence of material on the surface dating to a specific period reflects use of the site during that period. Additionally, the surface assemblage seems to reflect, if anything, the minimal occupation of a site. Therefore, when the surface assemblage is combined with information on site morphology, site sizes may be reasonably estimated. Based on these considerations, it can be suggested that regional surveys that include off-site field-walking and/or analysis of aerial/satellite imagery have the highest probability of recording archaeological features. Furthermore, in order to achieve reliable site-size estimates per period, the surface of the site has to have been divided into sub-areas during surveying to allow recognition of shifting patterns of occupation through time. As regards the problem of interpreting artefact densities and distributions, it will here be assumed, as is actually done by all surveys analysed here, that the recorded sites are settlements rather than special-purpose sites, unless there are strong indications to the contrary.[27] This assumption may, and in fact likely will, introduce a major bias in the present reconstruction, but unless more fieldwork is undertaken there is no way of correcting it. The present analysis is therefore undertaken in the full awareness that it can only reconstruct very crude trends that are subject to many biases and that will likely be corrected as new evidence becomes available.

However, it must also be realized that for some regions, archaeological surveys remain the single most important source of information and provide a much-needed contextualization of the few excavations carried out in these regions. Despite its evident limitations, survey evidence is too valuable to be completely excluded from any regional analysis. Concluding, then, it seems that combining evidence from surveys and excavations has the greatest potential of highlighting regional trends through time.

4.3 Problems of phase-based chronology

Studies based on survey data rely on material that has been collected from the surface, rather than obtained from excavations. For late Neolithic and later sites, pottery fragments are generally the most abundant material, and therefore the most suitable chronological markers. It is assumed that material present on the surface of a site in some way reflects the use and occupation of that site. Based on this assumption, surveyed archaeological sites are dated to phases of which artefacts are present on the surface of the site.[28] Without stratigraphic data obtained from excavation, site occupation phases cannot be more precisely established than the time-depth of the prevailing archaeological chronology allows. If a given pottery type can for example be dated to within a century, the century becomes the smallest unit of time within which occupation signified by that type of pottery can be dated if only survey evidence is available.

Pottery, as the most abundant chronological marker in survey collections, is generally dated through stylistic variation, and this will likely lead to the imposition of phases upon culture change, or rather its material remains. The delineation of phases initially helps to bring order into the vari-

---

27 It is for example possible that smaller sites represented temporary camps or agricultural stations that were not occupied year-round, but as Horne has shown (1993: 47), it may be extremely difficult to distinguish between such stations and permanent dwellings.

28 In the following discussion, and irrespective of how a chronological unit is called elsewhere, the term phase is used to designate the smallest chronological unit that can be identified in archaeological material culture development. A phase is usually distinguished by delineating a particular set of features, attributes, or artefacts appearing together for a certain amount of time. The term phase is preferred above similar notions such as stratum, or level as these are stratigraphic units that are strictly speaking unrelated to cultural development and certainly cannot be applied to material yielded by surface surveys.

ation observed in the archaeological record, and allows archaeologists to develop a vocabulary to discuss this variation. At a later stage, however, the drawbacks of phase-based chronology become apparent (Plog and Hantman 1990). The first drawback is that a phase-based chronology implies that culture change is step-wise, rather than continuous. By imposing phases on culture change, it is assumed that phases were stable periods without variation, interspersed by short periods of change. Secondly, phases can mask variety, especially when their duration is very long. If a long period of time is divided in few phases, it becomes difficult to study the correlation between culture change and other diachronic processes, for example climate change (Plog and Hantman 1990: 440–1). Thirdly, pottery types that are in use for an exceptionally long period are often neglected in the process of phase delineation in favour of types with shorter life-spans that better indicate diachronic changes. This may be true from a perspective of chronology construction, but what these long-living types signify is exactly the opposite, namely that alongside change, there is considerable continuity between phases. It is only the archaeologist's predilection for with change that neglects this continuity (cf. Porter 2007: 72). Finally, there is the site contemporaneity problem, which will be discussed in the following section. Plog and Hantman are, however, aware that: 'Phase based chronologies are thus often a compromise between what is needed and what is attainable, between a desire to understand as much of the observed variation as possible and the necessity of conveying information to one's colleagues' (1990: 440).

4.4 The contemporaneity problem

Many studies of survey data explicitly or implicitly assume that sites dated to the same chronological phase were contemporaneous and were occupied throughout the entire phase. The assumption of site contemporaneity may be false when sites were only occupied for a shorter period of time within a phase, and not during the entire phase. As excavations often show interspersed periods of occupation and abandonment, it is reasonable to assume that sites were periodically deserted, regardless of the cause. Especially if a phase lasted for multiple centuries, it is likely that sites were established or abandoned within that phase, but material culture alone would not be sufficient to detect this shorter lifespan. If all sites from a particular phase are considered contemporaneous, there is a risk of counting more sites than actually existed at any point in time during that period, because some sites existed sequentially rather than synchronically. Consequently, population estimates based on site numbers will be too high, and site catchment and territory analyses will be flawed. Already as early as 1965, Adams recognized in his study on settlement in the Diyala Basin the problem of overpopulating the landscape with sites in the case of sequential instead of contemporaneous occupations (1965: 124). This observation is variously labelled the problem of overestimated maps (Ammerman 1981: 77), or the contemporaneity problem (Schacht 1984).

The obvious solution to the mentioned problems is to increase chronological discrimination up to the point that sites can be assumed to be occupied during an entire phase. Chronological discrimination of phases, however, is an ongoing process that only solves the contemporaneity problem in the long term. In the absence of better chronologies, other methods have been employed to compensate for the contemporaneity problem.

Plog suggested life-histories for each site (Plog 1974 quoted in Weiss 1977). He divided phases into smaller time units, and assigned total site size to the middle unit of each phase, and smaller sizes to the initial and final units of each phase. Total site size for the entire phase was then calculated by adding the site sizes of each time unit together. Thus, sites grew and declined within a single phase. Weiss (1977) suggested dividing the total amount of occupied hectares in a phase by the length of the phase in years, as a measure of settlement intensity. However, both approaches still assume that sites existed during an entire phase, although the approach outlined by Weiss is likely to deflate actual occupied cumulative site size per phase. Sumner (1990) identified contemporaneous settlement by assuming that settlements occupied during two consecutive phases were inhabited during and around the transition between these phases, and by assuming that sites occupied during three consecutive periods were occupied during the mid-point of the middle phase. Chapman (1999: 72) proposed that clusters of sites represent sequential occupation of an area by the same small group, favouring residential mobility of a small group over a large sedentary population as an explanation for the great number of observed sites during specific phases. This solution assumes, however, that there was a high degree of mobility, and does not compensate for the duration of individual phases.

4.4.1 Simulating site contemporaneity

Building on earlier approaches, Dewar (1991) suggested a simulation method for establishing contemporaneous site occupations. This method uses assumptions similar to Sumner's propositions outlined above. Following the recognition that the transitions between phases represent key points in the occupational history of sites, Dewar developed a site typology based on the presence or absence of archaeological material from consecutive phases. Without stratigraphic evidence, the relation between archaeological finds from consecutive phases remains unknown. Also, the date of the find within the phase to which it is dated remains unknown. Thus, a find from a hypothetical phase X could belong to the beginning of phase X, whereas a find from subsequent phase Y could belong to the end of that phase. The assumption that these finds signify occupation for the entire lengths of phases X and Y runs counter to the presence of significant occupational gaps at the site (Fig. 4.1).

If a phase lasted for multiple centuries, it cannot be assumed that occupation lasted for the entire duration of the phase, and occupation for only part of the phase must be reckoned with. As the phase already represents the smallest unit of time that can be distinguished in the archaeological record, it is impossible to know for how long a site existed within that phase. As a result, it is also impossible to know which sites existed simultaneously.

The key to solving this problem is to be found in the transition points between archaeological phases. Two assumptions are made to solve the contemporaneity problem: (*a*) if two consecutive phases are present at a site, it is assumed that the site was occupied at least at the transition point between these phases (cf. Sumner 1990); and (*b*) if a site was occupied at the beginning and end of a phase, it is assumed that the site was also occupied during the rest of that phase. Assumption (*a*) is based on the idea that sites with material from two consecutive phases have a higher *probability* of being occupied during the transition between these phases, than sites with material from only one of these phases. This concept of probability is then extended in assumption (*b*), resulting in a 'dynamic probability model' of settlement abandonment and establishment (Wilkinson 2000a: 249).

These assumptions can be represented graphically by including the starting and end points of any given phase as discrete points in time, t_1 and t_2 (Fig. 4.2). Sites are then classified according to whether or not they were occupied at any one of these points. In this way, four different occupation types *a–d* can be distinguished for hypothetical phase Y. A type *a* site is occupied at t_1 but not at t_2. This means that the site already existed in phase X, but was abandoned somewhere during phase Y. A type *b* site is occupied at t_1 and t_2, and by assumption (*b*) it follows that the site was occupied during the entire length of phase Y. A type *c* site was occupied at point t_2 but not at t_1, so occupation must have started somewhere in phase Y and continued into the next phase. Occupation of a type *d* site is not attested at either point t_1 or t_2, and establishment and abandonment of type *d* sites must have taken place entirely within phase Y. Changes in the ratios between these occupation types can potentially indicate long-term settlement pattern changes. For example, a high ratio of type *d* sites may indicate a high settlement flux, whereby settlements are regularly relocated. Conversely, a large number of type *b* sites could indicate long-term settlement stability (Dewar 1991).

	phase X	phase Y	phase Z
real occupation	—— gap —————— gap ——		
reconstructed occupation	- -		

Fig. 4.1: Discrepancy between real periods of occupation at a site for hypothetical phases X, Y, and Z, and the period of occupation as it would be reconstructed from survey finds from that site.

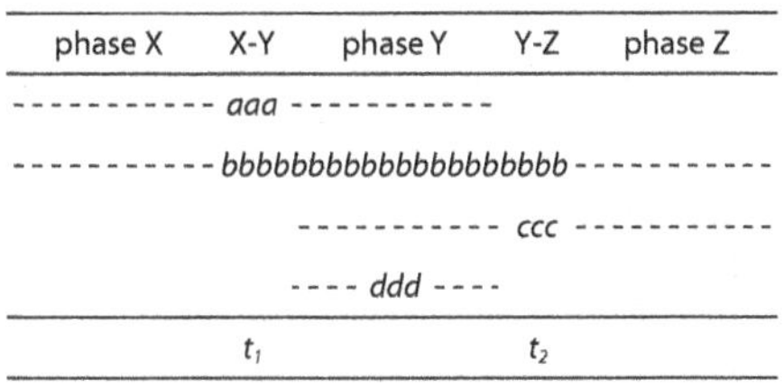

phase X	X-Y	phase Y	Y-Z	phase Z
---------- *aaa* ----------				
---------- *bbbbbbbbbbbbbbbbbbbbbb* ----------				
		---------- *ccc* ----------		
	---- *ddd* ----			
t_1			t_2	

Fig. 4.2: Occupation types *a–d* for hypothetical phases X, Y, and Z, and transition points t_1 and t_2. Letters *a–c* indicate the period of time during which these site types were certainly occupied. Letter *d* indicates a floating period of time in phase Y during which the site was occupied. Dashed lines indicate a period of time during which the site may have been occupied (after Dewar 1991: fig. 3).

This procedure of classifying sites can be repeated for consecutive periods and for multiple sites, whereby t_2 of any given phase equals t_1 of the following phase. The resulting classification of sites according to occupation type can be used to generate equations of site abandonment and establishment rates. It is assumed that settlement abandonment and establishment are relatively continuous processes and that every site has an equal probability of being abandoned or established. The site abandonment rate A_{occ} of a given phase is then calculated as the number of sites that is abandoned during a phase, that is, all type *a* and *d* sites, divided by *p*, the duration of the phase in years:

(1) $A_{occ} = (a + d) / p$

The site establishment rate E_{occ} is similarly calculated as the number of sites established during a phase, that is, all type *c* and type *d* sites, divided by *p*:

(2) $E_{occ} = (c + d) / p$

It is now possible to simulate the average number of simultaneously occupied sites during a phase. This simulation starts with the number of sites existing at t_1, $a+b$, and simulates for every year until t_2 the probability that a site is abandoned and/or established. The site abandonment probability equals A_{occ}, whereas the site establishment probability is equal to E_{occ}. Because of the method used to calculate A_{occ} and E_{occ}, the number of sites at t_2 will invariably equal or approach $b+c$. For every year, the number of simultaneously existing sites is noted and these numbers are used to calculate the average number of simultaneously existing sites during that phase (Dewar 1991). In further calculations, this average number is taken to be representative for point $t_{1.5}$, that is, the middle of a phase. The number of simultaneously existing sites is now known for points t_1 and t_2, respectively the starting and end points of a phase, and point $t_{1.5}$, the middle of that phase.[29]

4.4.2 Evaluating the site contemporaneity model and its results

The site contemporaneity model has several weaknesses resulting from its mathematical and theoretical assumptions. First, the model cannot deal with multiple sequential occupations during a single phase, as can be illustrated by the following example. When the data on real occupation from Fig. 4.1 are entered into the model's format of type *a–d* occupations, it would result in the site being type *c* during phase X, type *b* during phase Y, and type *a* during phase Z if there were no other occupation periods. This example shows that Dewar's model is unable to detect the occupation gap in phase X. More generally, this means that the model fails to detect multiple occupations within a

---

29 The simulations were computed with the SetDyn2 computer program that is a slightly modified version of the program provided in Dewar 1991. The main difference is that the SetDyn2 version uses a Poisson distribution instead of a simple probability for the calculation of the chance that a site is abandoned or established. The program allows the simulations to be repeated 1000 times per phase to check the stability of the estimates.

single phase (Dewar 1994: 150). Thus, this reconstruction is still only an approximation of the real occupation, but it better reflects the uncertainties about occupation length than the assumption that the site existed for the entire length of every phase for which material is evidenced.

Furthermore, failure to identify an occupation hiatus from survey data changes the occupation type of a site. Take a site for which survey evidence indicates a continuous sequence for two successive phases that, when excavated, shows a destruction layer between these phases. This would mean that a site that was formerly considered a type c site in the first phase, and a type a site in the second, is now a type d occupation in both phases. Input of these different occupation types in the contemporaneity simulation will result in slightly lower estimates of contemporary sites in the case of two type d sites as compared to a type c and type a site. Also, the abandonment rate in the first phase will be slightly lower, and the site establishment rate in the second phase will be slightly higher (Dewar 1991: 610–11). Considering the relatively large number of sites, the effects on the estimates will be minor.

A third problem involves the assumption that sites with material of two consecutive phases are occupied during the transition between these phases. This assumption contradicts the intuitive idea that this transition is a period of change, possibly including, or resulting from, abandonment or the replacement of one cultural group by another. This idea is, however, not necessarily correct and derives from a conceptual misunderstanding of what a phase represents. An archaeological phase represents a particular set of features, attributes, or objects appearing together for a certain period in time. This combination is subject to change, and once significant change has occurred, it can be designated a new phase. A transition between two phases is, however, only rarely accompanied by a change of the total associated inventory or all of its attributes. Only if such strong variation is observed, might the transition between two phases may actually coincide with a possible settlement abandonment followed by reoccupation by another cultural group. More commonly, however, one or more types will carry over the first into the second phase. The distinction between phases is therefore primarily a product of the current archaeological discourse, its history, and its conception of change, and only rarely a true reflection of settlement abandonment and reoccupation of a site. In other words, there is no *a priori* reason why a site may not be occupied at the transition between two phases.

Fourthly, the site contemporaneity model assumes constant rates of site abandonment (A_{occ}) and site establishment (E_{occ}) for each phase. This may not necessarily reflect reality as there may have been moments of regional destruction, for example resulting from earthquakes or conflict. In these cases, the reconstructed number of simultaneously occupied settlements would be flawed. Such regional destructions are often reconstructed using a combination of stratigraphical, historical, cultural, and radiocarbon chronologies. The combination of radiocarbon dates with other types of evidence may, however, lead to 'suck in and smear effects'. An abrupt event is smeared out over time through the inherent imprecision of radiocarbon dates. Precisely dated historical events, on the other hand, tend to 'suck in' more poorly dated evidence, especially if that evidence is radiocarbon dated. Through this process, false event horizons may be created (Baillie 1998). An example of such an event horizon is the attribution of destructions at several sites to the Akkadian kings Sargon and Naram-Sin, recorded in texts, evidenced by excavations and supposedly supported by radiocarbon dates. The use of historical texts only to pinpoint destructions is hampered by the fact that rulers often claimed more deeds than they actually carried out. Finally, cultural chronologies are rarely sufficiently detailed to date destructions, even at neighbouring sites, more precisely than 100 years.

Fifthly, Kintigh (1994) proposed to revise Dewar's model in order to remove several other inherent flaws. According to Kintigh, Dewar's simulation of the number of contemporary sites does not take into account information from type d sites, and the number of contemporary sites is calculated independently of phase length. He also showed that the standard deviation resulting from the simulation may be unrealistically low due to the program's built-in limit of only one site abandonment and/or establishment per year. Kintigh's revisions included a fundamental difference between type b sites on the one hand, and type a, c, and d sites on the other hand. His model was built around the assumption that type b sites are fundamentally different than the other types because their occupation extends beyond the duration of a phase. Occupation length of the other sites is essentially the same, only the point in time during which they were occupied varies. Kintigh therefore offered new calculations for occupation length of these two types. His estimations of the number of contemporary sites are generally higher than those derived from Dewar's simulation.

In a response, Dewar (1994) argues that Kintigh's revisions to the original model derive from a different understanding of the concepts underlying the model. Instead of arguing that type b sites on the one hand and type a, c, and d sites on the other are fundamentally different as regards occupation length, Dewar conceives these sites as representing different occupation spans on a continuous distribution. In such a distribution, type b sites represent sites with the longest occupation spans. Also, Kintigh's revised formulas for calculating site occupation spans sometimes fail to meet an important criterion for type b sites, namely that their occupation length must be longer than the duration of a phase. Finally, whereas Dewar's simulation provides a standard deviation of the estimated number of contemporary sites, Kintigh's revised formulas do not. The standard deviation partly reflects the number of type d sites during a phase and signifies changes in site abandonment and establishment rates. Summarizing, then, there is no need to apply Kintigh's revised formulas, although they do represent a viable alternative to Dewar's original simulation. Both authors agree, however, that there is a great need for models of site contemporaneity in the study of survey data (cf. Wilkinson 2000a: 248–9).[30]

A final objection to Dewar's simulation may be that it introduces a number of unverifiable variables, leading to what could be called a floating reconstruction of contemporary sites. These variables are the assumption of a steady rate of A_{occ} and E_{occ}, as well as the derived probability that a site is established or abandoned. However, the use of these variables does not lead to a reconstruction that is entirely unsupported by the data. At the transition points between phases, that is, points t_1 and t_2, the number of contemporaneous sites is entirely based on the presence or absence of archaeological material. The number of reconstructed sites at points t_1 and t_2 is therefore not probabilistic or floating. For point $t_{1.5}$, the number of contemporaneous sites is partly based on the presence of archaeological material, that is, the type b sites, and partly on a simulation using A_{occ} and E_{occ}. A distinction can therefore be made between the non-probabilistic nature of points t_1 and t_2 on the one hand, and the partly probabilistic nature of point $t_{1.5}$ on the other hand. Thus, the degree to which the reconstruction is floating is limited, and there remain points in the reconstruction that directly reflect the underlying archaeological data.

The results from the simulation represent a probable average number of contemporaneously occupied sites per archaeological phase. In other words, there is always a chance that the actual number of occupied sites was either higher or lower. In order to allow a better interpretation of the simulation results, the potential maximum and minimum number of sites per phase will also be indicated. The maximum number of occupied sites per phase is considered to be equal to the maximum number of sites that a survey detected for that phase. The minimum number of occupied sites per phase is considered equal to the number of type b sites per phase, as these sites have the greatest probability of being continuously occupied. The differences between the simulation results and the potential range will be presented separately for each survey region (Section 5.3).

4.4.3 Estimating total occupied area

Settlements go through cycles of growth and decline, and different states of abandonment, depending on a number of variables, including environment, demography, and social and cultural factors. It is therefore to be expected that settlement size will vary through time, and that not all parts of the settlement will be occupied at the same time. Houses, or parts thereof, can be both permanently and temporarily abandoned (cf. Otto 2006: 258). Pfälzner (2001: 34) notes that household cycles emerging from, for instance, birth, mortality, and marriages are a strong motivation for the modification and/or rebuilding of houses, in turn influencing settlement organization as a whole. These cycles, however, appear to be relatively short in duration, perhaps lasting no more than one

---

30 Previous applications of Dewar's model to survey data from various parts of the Near East can be found in Neely and Wright 1994, Kouchoukos 1998, Pollock 1999, Lyon 2000, and Ristvet 2005.

generation. Environmental cycles may of course last longer (Horne 1994: 79). Horne (1994: 79) suggests that especially small settlements are likely to suffer from such cycles, due to the relatively severe impact of demographic changes, as compared to settlements with larger populations.

Settlement location and expansion can be limited by the availability of land for building houses. If a settlement is located on a tell, the slopes of the tell can limit expansion of the settlement. City walls can have a similar effect. A study of the village of Qdeir, Syria, indicated that expansion of this village was impeded by the presence of a cemetery, good quality arable land, and the necessity to reserve open land for the pasturage of animal herds (Aurenche and Desfarges 1983). This suggests that, when good soils are scarce, and therefore in high demand, they will not always be used for settlement. Rather, such areas may remain primarily reserved for agricultural or pastoral production. Settlement will then either be moved to less fertile soils, or to non-arable lands. Also, when arable land is in high demand, land prices or their equivalents may be so high as to restrict house sizes and settlement sizes in general (Kramer 1982: 164). Such an environment would result in highly compacted settlements. Within such an environment, occupants of a site may consider reusing old parts of a settlement as a viable option when the need arises, for example as a result of population growth, instead of expanding the settlement onto useful arable land.

The architectural composition of a settlement is very fluid, as is clearly illustrated by Aurenche and Desfarges' case-study of Qdeir (1983). Houses are constructed, demolished, rebuilt, or modified according to the whims and needs of the occupants. These activities change the layout of the settlement through time, and may also impact on the overall population density within the settlement. An interesting observation is the difference in on-site population densities based on the duration of occupation of that site. Post-medieval, pre-modern Arabic architecture is essentially agglutinative in nature, leading to cumulative filling in of open spaces in a settlement, especially because many post-medieval Arabic settlements lacked a central authority regulating the use of space. This leads to the old town centres being the most densely built-up parts of settlements, with the highest population densities. Aurenche and Desfarges observe that expansion into previously unoccupied areas only occurs when all land within the village is built up. This expansion may, however, at the same time be accompanied by abandonment of older parts of the village, presumably because the newer parts prove more pleasant to live in (Aurenche and Desfarges 1983: 183–5).

When a site's total occupied area is estimated from archaeological data, the nature of the site and the post-depositional processes acting upon the material from which the estimate is derived must be taken into account. Most sites in the surveys analysed here are tells, meaning that older periods may be obscured by later depositions. This means that parts of or the entire period may be absent from a surface collection, which in turn influences the estimate made for the occupied area during that period. There is, however, reason to believe that this problem is less applicable to the north Mesopotamian Bronze Age than to other periods. Bronze Age occupations at many tells seem to have been by far the most extensive of all periods, reducing the chance that they were completely obscured by later occupations. Estimations of Bronze Age occupation sizes seem, then, more reliable than for many other periods (see Section 4.2).

The estimation of individual site sizes is problematic when Dewar's method for establishing site contemporaneity is used. This method establishes the number of contemporaneously occupied sites for three points in every phase, t_1, $t_{1.5}$, and t_2. However, point t_2 of any phase is equivalent with point t_1 of the following phase. Because site size estimates vary per phase, it is possible that there are two different estimates for every transition point between phases. There are no objective criteria to select any of these estimates. Furthermore, the number of simultaneous sites at point $t_{1.5}$ is an average, and it is not known which individual sites were occupied at this point, except the type b sites. Thus, it is not possible to calculate the total area that was occupied at this point in time. In a study on late fourth to early third millennium settlement in Mesopotamia, Kouchoukos (1998) proposed several additions to Dewar's model in order to solve these problems. Kouchoukos' model will be discussed below.

The calculation of total occupied area of all sites within a survey region at point $t_{1.5}$ is hampered by the fact that not all sites that were occupied at that point in time are known. This particular problem of the simulation model can be addressed by dividing the settlement record in size classes and assigning every site per phase to one of these classes. The site contemporaneity simulation is then performed independently for each size class. The result is an estimate of the number of contemporary sites at points t_1, $t_{1.5}$, and t_2 per size class per phase. By assigning an average size to each size class, it is then possible to calculate the amount of hectares that was simultaneously occupied at points t_1, $t_{1.5}$, and t_2 of every phase and for every class. Finally, by adding up the size estimates for every class, one arrives at the total simultaneously occupied area for the entire survey region. In order to come to meaningful size-classes, Kouchoukos analysed long-term datasets from the Nippur Warka Survey, the Susiana-Deh Luran Survey, and the North Jazira Survey to see if there were regularities in site sizes. He suggested that sites in these three surveys tended to cluster in ranges of 1–3 ha, 3–8 ha, 8–15 ha, 15–40 ha, and 40+ ha. Given the fact that $c.$ 80 % of the sites in the present analysis fall within Kouchoukos' 1–3 ha size class, new, more fine-grained size classes will be determined from the surveys used here (see Section 5.3).

The second problem involves differences in site size estimates for any two subsequent phases. Because point t_2 of the first phase is equivalent to point t_1 of the following phase, two different estimates exist for the same point in time. This problem can be avoided by calculating the average of these two estimates at this point, and using this average in the calculation of total occupied area per size class. It is argued that this average represents a gradual growth or decline of the site, and does not necessarily lead to a loss of information.

4.5 Reconstructing population levels

The estimation of population sizes on the basis of Near Eastern archaeological data is notoriously difficult (van de Mieroop 1997: 95–7). There are several methods to reconstruct population sizes, based on a variety of archaeological, historical, and/or ethnographical data, including artefact densities through space and time, regional carrying capacities, burial ground sizes, population censuses or tax records, settlement sizes, and house counts. Preferably, multiple sources are used to corroborate the respective results of each individual source (Hassan 1981; Bintliff 1997; Chamberlain 2006).

It has become increasingly clear that the number of sites per period alone is not an accurate proxy for the total population during that period, as a few large sites may have a considerably larger aggregate population than numerous small hamlets (Osborne 2004: 170). In the present analysis, data on aggregate settlement size will be used to reconstruct population levels. Settlement size as determined by the occurrence of diagnostic sherds on the surface of the site will be the determining variable of total population. It will be assumed that there were no significant changes in settlement sizes within a period. As suggested earlier, it will furthermore be assumed that all archaeological material in the survey reflects human occupation, rather than specialized activities involving a temporary stay at a certain location. Where there is doubt about as to the function or permanent occupation of a site, it is removed from the investigated sample, as is for example the case with a cultic centre in the Birecik-Euphrates Dam Survey region, and several supposed Early Bronze Age burial grounds recorded by the Balikh Survey (see Sections 5.2.1, 5.2.2).

Settlement size can only be used as a proxy for population size if both variables are positively correlated in some way. Various researchers have explored this correlation. Sumner analysed a sample of 110 Iranian villages and found a good correlation between settlement size and number of occupants (Sumner 1979: 165). De Roche (1983) studied villages of Mexican subsistence farmers and found the assumption that settlement size correlates positively to population size to be true as well. Kramer's analysis of Iranian villages and cities also found a particular strong correlation (Kramer 1982: 158, 163), and an analysis of another dataset provided by Aurenche (1981) confirms this result. All these calculations suggest that population size and settlement size are to some degree positively correlated, although there must be additional variables that determine settlement size. Both Sumner (1979: 165) and De Roche (1983: 190) suggested on the basis of their calculations that the correlation between settlement size and population size is not strong enough to be applied to individual settlements, but that it could safely be used on a regional level. There seems therefore to be a basis to use regional aggregate site sizes as a crude proxy for the settled population living in that area.

Whereas Sumner did not list the individual settlements that he used for his analysis, Kramer (Kramer 1982: tables 5.3, 5.6) and Aurenche (1981: table 3) did provide information on size and population of individual settlements, allowing the exploration of the relationship between settlement size and population density. The correlation between these two variables is significant at the 1 % level, but nevertheless weak ($r_s=0.29$), as has also been observed by Kramer (1982: 178). In fact, variability in population densities is slightly larger in towns and cities than in villages.[31] Thus, although there is a statistically significant correlation between settlement size and population density, population is obviously not the only variable influencing settlement size. Other variables may include the need for defence, or the need for space to carry out various activities within the settlement.

The next step involves the calculation of a population density that can be applied to the aggregate settlement areas for every survey region during each phase. Usually, researchers apply a range of 100–200 persons/ha to the total settlement area, based on ethnohistorical parallels (Kramer 1982). For third and second millennium BC northern Mesopotamia, several excavations allow a reconstruction of on-site settlement densities using data from these excavations, rather than ethnographically derived estimates. Such a reconstruction must take into account the possibility that parts of a settlement were abandoned and that there were differences in population density due to social or economic distinctions. Below, several previous studies addressing demography and population density in Mesopotamia will be discussed, thereby highlighting the problems associated with demographic reconstructions, and the great diversity in their results.

Postgate (1994) used a three-step approach to arrive at a range of possible population densities for the late Early Dynastic (2600–2400 BC) layers of Abu Salabikh (Iraq): (*a*) estimate total settlement area, (*b*) estimate the total residential area of the settlement and the number of houses therein, and (*c*) estimate the number of occupants per house. He argued that the 10 ha main mound was primarily shaped by the Early Dynastic city wall, so that the currently visible mound is a good indicator of the past walled area, even though the earliest Early Dynastic layers themselves were 2 m below the current surface of the plain. The area taken up by the city wall was then subtracted from the total area of the main mound, resulting in 9.5 ha of urban space. This urban space can be divided in public space and residential space. Public space consists of streets, alleys, and squares, whereas residential space consists of private houses, which in turn can be divided in house walls, courtyards, and roofed space. The subdivision into these categories was made on the basis of a 2500 m² area excavated in the Main Mound of Abu Salabikh. This subdivision suggested that 10 % of the total urban space was public space (0.95 ha), and the remaining 90 % was residential space (8.55 ha).[32] For the South Mound, a residential space of 6.75 ha was calculated. A similar analysis of an Old Babylonian residential area in Ur revealed comparable percentages. Postgate then examined six almost completely excavated houses at Abu Salabikh to determine the dwelling space within each house, that is, kitchens and living rooms. By assuming an average range of dwelling space per person (4–7 m²/person), the range of persons per house can be determined. By estimating the number of houses in the residential space of Abu Salabikh, it is possible to calculate a range of persons per hectare of residential space (248–1205 persons/ha). By applying this number to the total urban space, Postgate arrived at a possible range of *c.* 3800–18400 persons for the ED III occupation of both the Main Mound and the South Mound (1994: table 5).

Stone attempted a correlation between Old Babylonian contracts from Nippur, and architectural modifications of the houses in which the documents were found (1981). She found that the term used for house in these contracts, e_2-du_3-a, had the more specific meaning of roofed floor space, suggesting that the actual houses including courtyards were considerably larger. The picture emerging from the texts and architectural modifications as revealed by excavation is that house ownership was very fluid and that houses were modified according to the needs of the owners.

---

31 For the group of villages, SD=43.8 and V=0.51, whereas for the towns, SD=86.9 and V=0.74. All settlements listed by Aurenche (1981: table 3) and those in Kramer 1982: table 5.3 are considered villages, whereas those in Kramer 1982: table 5.6 are considered large settlements.

32 Note, however, that the figures used by Postgate are somewhat problematic. According to Postgate 1994, table 1, 87.5 m² of the 2500 m² area is taken up by the city wall, but fig. 7 in the same article, illustrating the excavated area on which the figures in table 1 are based, shows no evidence for a city wall.

Stone also tried to link house types to different family units. She suggested that square houses with rooms on four sides of a courtyard were used by extended families, whereas linear houses with rooms on one or two sides of a courtyard were used by nuclear families. Analysis of the number of living rooms per house suggested that each nuclear family required on average one living room, two other rooms, and courtyard space. When these data are translated into actual numbers of occupants per house using ethnographic parallels to determine the size of a nuclear family, they correspond remarkably well with the number of occupants per house that could be expected on the basis of the textual evidence alone. The number of occupants per house as reconstructed from the archaeological/ethnographical data translated into 5.31 m² of roofed space per person. Although not part of Stone's analysis, this figure can be extrapolated to a population density of *c.* 680 persons/ha using Postgate's data on the relative proportions of roofed versus other residential space.[33]

Pfälzner used a similar method for establishing the population of archaeological sites (2001). For several sites, he determined the total site size and the area taken up by public buildings such as temples. He then used Postgate's rule of thumb to determine the actual area taken up by houses. He studied numerous houses at several sites in northern Mesopotamia and calculated an average house size, from which the total number of houses per settlement could be determined. Finally, this number was multiplied by the average number of persons in a household, which he found to be 5–6 persons, based on ethnographic parallels. In this way, he calculated the number of occupants of the third millennium BC sites of Selenkahiye, Halāwa A, Chuera, and Taya. The resulting population densities ranged between 300–500 persons/ha.[34]

Battini-Villard (1999) analysed the Old Babylonian domestic quarter of area AH at Ur. In an attempt to calculate the number of residents of this area, she reconstructed the average family size in order to apply this value to the number of houses in area AH. Two texts, from the Old Babylonian and Middle Assyrian periods respectively, provide information on the number of family members. The first text comes from Mari during the reign of Zimri-Lim and deals with the deportation of Hurrian families from several villages that were probably located in the Upper Khabur or Tur-Abdin region. A total of 469 people making up 96 families were deported from five villages, resulting in an average family size of 4.9 persons. The Middle Assyrian list also documents the deportation of Hurrian families. In total one counts 211 individuals, divided across families comprising from as few as two to as many as eighteen members. The average family consists of 7.8 persons. Battini-Villard reanalyzed the houses excavated in area AH and reconstructed a total of 55 houses, leading to an estimate of 270–430 persons for this area. Area AH measured 6750 m², resulting in a population density of 400–635 persons/ha. The area enclosed by the city wall, but excluding the temple area, is 56 ha in size. Assuming that this area was fully built up, these figures would lead to a population estimate for Ur of *c.* 22300–35600 persons (Battini-Villard 1999: 388).

Similarly wide-ranging estimates of the number of persons living in a house were made by Waetzoldt (1996). Using Ur III lists of house inventories and inheritance documents, he found that houses ranged between 30 and 200 m², although most families lived in houses smaller than 100 m². Even larger houses were also documented, but these were invariably associated with workshops or storerooms. As far as can be reconstructed, and taking into account that information on the number of children of a married couple or head of family is often lacking, between two and ten persons lived in a house, sometimes including slaves. In some cases, as many as 40 people may have lived in a single house.

The reconstructions discussed above all concern relatively large, urban sites. As the survey record analysed here contains numerous small sites as well, it is necessary to reconstruct densities that may have pertained to such sites. One of the few small settlements to have been exposed over a large area is Raqa'i level 3 (2600 BC). This level has been exposed over *c.* 1400 m², representing

---

33 The plans of Nippur published in Stone 1987 seem to support Postgate's dense urban fabric where roofed space is 40 % of the total house area, and houses make up 90 % of the total urban area.

34 In a study on urbanism at Tell Mozan, Pfälzner and Wissing (2004: 81) assumed a density of 400 persons per hectare, based on the results of Pfälzner 2001. For Selenkahiye, Meijer (2001: 112) suggested a total population of *c.* 860–1700 persons, based on a size estimate of 10 ha, the observation that *c.* 1/7th of the excavated area was public space, resulting in a built-up area of 8.6 ha, and an ethnographically derived population density of 100–200 persons/ha.

over 45 % of the hypothesized 0.3 ha rural settlement. This undoubtedly atypical settlement was centred on the large Rounded Building in the south, from which streets and buildings radiated outwards to the northwest, north, and northeast. As the southern part of the Round Building was eroded, the expansion of the settlement in that direction could not be ascertained. Several districts could be distinguished: a temple area, several storage/industrial areas, and a living quarter (Schwartz and Curvers 1992). Pfälzner (2001: 305, pl. 30) identified fifteen separate houses at Raqa'i, predominantly with one or two rooms. Under the assumption that each house was inhabited by a nuclear family of 5–7 persons, 75–105 people would have lived in the excavated area, translating into a population density range of 535–750 people per hectare. If this density is extrapolated to the entire site, then 160–225 people would have lived at Raqa'i. There is, however, considerable room for debate regarding this figure. Pfälzner included the supposed temple as a house, as well as several areas that can be clearly identified as industrial areas. Furthermore, Pfälzner's house 2, consisting of rooms 20 and 83, has a total floor area of less than 10 m^2. It is difficult to see how such a small structure would have housed a nuclear family or similar household unit. Nevertheless, this house is among the clearest examples of distinct architectural house units. However, even when these houses are excluded, the reconstructed population density for Raqa'i remains relatively high and well above densities recorded in ethnohistorical sources.

All approaches to on-site population densities have to determine the amount of non-residential space, including open space, city walls, and public buildings at a settlement. Here, Postgate's proportions as calculated from Abu Salabikh are often cited, but there is reason to assume that these figures are not entirely representative. The ratio of residential to public space of 9:1 is calculated on the basis of a relatively small area of 2500 m^2 at Abu Salabikh that is almost entirely filled with residential structures. If this ratio is extrapolated to the entire site, this leaves little room for public buildings such as temples or palaces, as Pfälzner also realized. A further possible distorting factor is that Abu Salabikh is located outside the study area, and that urban layout may have been different in the Jezirah.

The Late Bronze Age *Weststadt* of Tall Bazi on the Syrian Upper Euphrates may serve as an example. The *Weststadt* at Tall Bazi has been excavated over a larger area than Abu Salabikh, and the excavation was supplemented by magnetometric inspection. The *Weststadt* represents a planned extension of the older town on the main mound. The extension was built on a natural terrace and bordered on the north side by terrace walls, clearly delimiting the built-up area. The quarter existed between 1250–1180 BC before it was destroyed by fire. This short life-span of the settlement meant that large parts of the settlement were used contemporaneously and were relatively undisturbed by later activities. The settlement was structured by several streets and a central square. The quarters in between were densely built up, but there is evidence for empty parcels, as well as for houses that were temporarily or permanently abandoned. Some fifty houses were wholly or partially excavated, and magnetometric investigations indicated that non-excavated parts of the *Weststadt* had a similar architectural layout. The entire *Weststadt* may have had eighty houses on a 3.5 ha area (Otto 2006). Assuming an average household size of 5–7 persons, and one household per house, this would have resulted in a population for the *Weststadt* of 400–560 persons, and a population density of 115–60 persons/ha. This range is much more in line with ethnohistorical population densities than the Early and Middle Bronze Age estimates. The very crude estimates for site size (3.5 ha) and number of houses (80), and the average plot size of 132 m^2 (Otto 2006: fig. 169) allow the calculation of the ratio of residential space to public space, which amounts to 3:7. This ratio is supported by the plan of the excavated area, which shows wide streets and a large central square (Otto 2006: fig. 9). The figure of *c.* 30 % of the total area of the *Weststadt* being devoted to residential built-up space is by no means very precise, but it is a considerably lower figure than Postgate's estimate of 90 %.

Table 4.1 and Figs. 4.3 and 4.4 summarize the archaeological reconstructions discussed above. The ethnographic data from Kramer (1982) and Aurenche (1981) have been included as well. The following observations can be made. First, population densities for Early and Middle Bronze Age settlements in northern Mesopotamia are structurally higher than those observed in (pre)modern villages and cities. The method used in these reconstructions, that is calculating site size, counting the number of houses, and assuming an average household size, is equally capable of producing

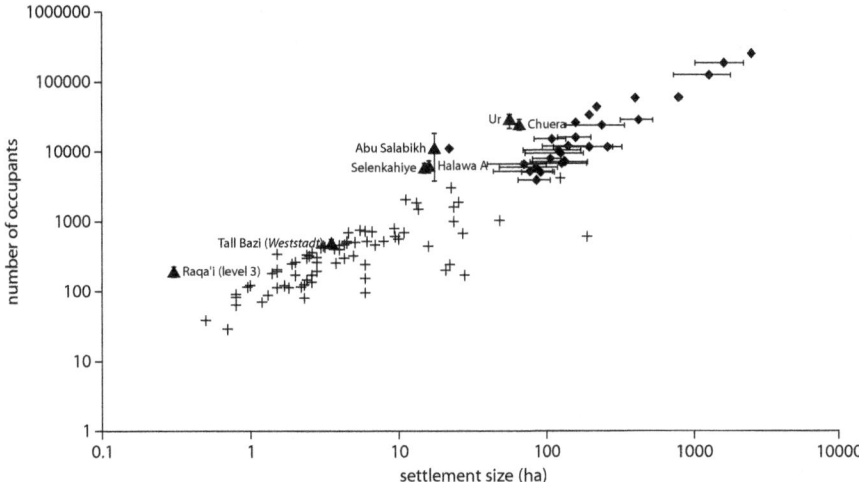

Fig. 4.3: Scattergram indicating the correlation between settlement size and number of occupants. Crosses indicate the villages drawn from Aurenche 1981, table 3, and Kramer 1982, table 5.3. Diamonds indicate the towns drawn from Kramer 1982, table 5.6. Triangles represent archaeological sites. Where necessary, the range of different site size and population estimates is indicated. Number of occupants and settlement size are given in logarithmic scale.

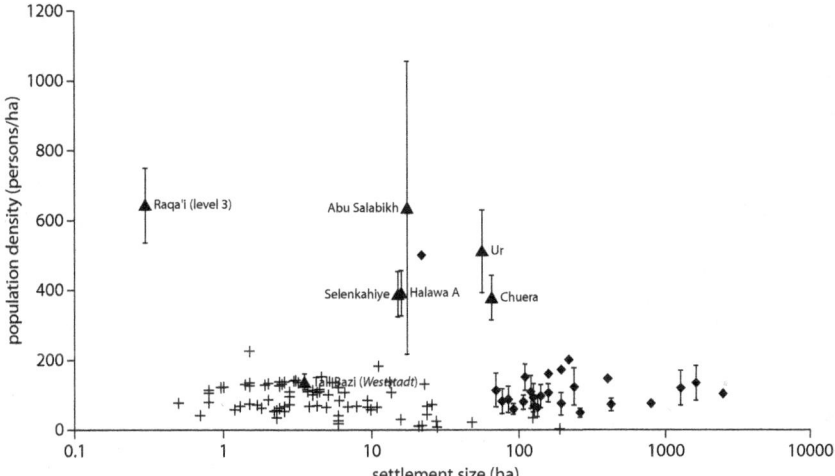

Fig. 4.4: Scattergram indicating the correlation between settlement size and population density. Crosses indicate the villages drawn from Aurenche 1981, table 3, and Kramer 1982, table 5.3. Diamonds indicate the towns drawn from Kramer 1982, table 5.6. Triangles represent archaeological sites. Where necessary, the range of possible population densities resulting from different site size and population estimates is indicated. Settlement size is given in logarithmic scale.

name	period	settlement size (ha)	number of occupants	persons/ha
Abu Salabikh	ED III	17.5	3800–18400	215–1050
Chuera	EJ II–III	65	c. 24600	380
Halawa A	EB IV–MB I	16	c. 6260	390
Nippur	OB	no data	no data	680
Raqa'i (level 3)	EJ II	0.3	160–225	535–750
Selenkahiye	EB IV	15	c. 5820	390
Tall Bazi (*Weststadt*)	LB II	3.5	400–560	115–60
Ur	OB	56	22300–35600	400–635

Table 4.1: Settlement size (ha), number of occupants per settlement, and population density for selected excavated sites in greater Mesopotamia dating to the third and second millennium BC.

results that do fall within the range of ethnographically recorded densities, as is clear from the reconstruction of the *Weststadt* at Tall Bazi. Second, there is no clear difference in population density between sites of different sizes, although this may be partly attributable to the small sample that was analysed here. Third, it is clear that the results of the population density reconstructions fluctuate widely, and that they do not easily allow the extrapolation of an acceptable average range. However, the results also indicate that it is quite possible that ethnohistorical parallels do not apply to third and second millennium BC settlements in northern Mesopotamia. Even if only half of the area of a settlement would have been occupied at a given time, resulting population densities would still be in the upper range of the ethnographically recorded densities.

4.6 Rank-size analysis

The reconstruction of ancient populations on the basis of survey data allows the reconstruction of regional population levels. The data on settlement sizes also allow the exploration of the organization of the settlement system. For this purpose, the reconstructed settlement systems will be subjected to rank-size analysis. Traditionally, rank-size analysis compares an observed settlement pattern with an expected pattern. This expected pattern is subject to the rank-size rule, which holds that when sites are ranked according to size, the size of any site with rank N equals 1/Nth of the size of the largest site.[35] Thus, in a settlement system whose largest site is 30 ha, the next-largest site would be 15 ha, the third-largest 10 ha, and so forth. When this pattern is plotted on a log-log rank-size plot, it appears as a straight line from the upper left to the lower right corner of the plot. This pattern is called the log-normal distribution (Fig. 4.5A). If the curve falls off more sharply than the log-normal distribution, there are fewer large or more small sites than expected, resulting in a concave or primate distribution (Fig. 4.5B). If there are more large or fewer small sites than expected, the resulting distribution is called convex (Fig. 4.5C). Various scholars have proposed combinations of these distributions (e.g. Falconer and Savage 1995), for example the primo-convex distribution in which the upper part of the curve is primate, but the lower part conforms to a convex distribution (Fig. 4.5D).

4.6.1 Inspecting rank-size distributions

Rank-size analysis is most profitably carried out when the spatial extension of the settlement system is fully known, and when all sites are known within the region covered by the system. This is rarely the case for archaeologically observed or reconstructed systems. The failure to recognize the boundaries of a settlement system or the absence of sites from the system can significantly distort the shape of the size distribution. Archaeologists must therefore be particularly careful when

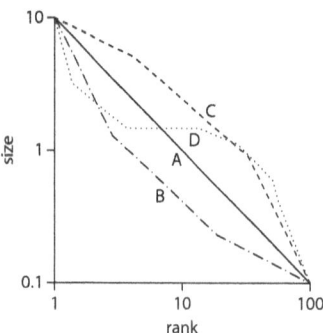

Fig. 4.5: Rank-size plot showing (A) log-normal, (B) primate, (C) convex, and (D) primo-convex distributions.

---

35   The actual rank-size rule, formulated by Zipf, is expressed as $r \cdot P^q = K$, where r is a site's rank, P is the site's size, K is the size of the largest site, and q is the ratio between the force of unification and force of diversification. Archaeologists routinely assume that q=1, which actually represents a simplification of the original rule (Savage 1997: 233).

applying rank-size analyses to a dataset. The degree to which all sites are known within a given region is difficult to assess. It appears, however, that this is a relatively minor problem for the settlement systems analysed here. Because the size distribution is primarily shaped by the largest sites, the overall shape will probably remain intact as long as these large sites are included in the system that is analysed (Johnson 1981: 176). Although it is certain that the archaeological surveys that are analysed in the following chapter failed to record some sites, it is likely that the missed sites will be relatively small. Their absence from the size distribution will therefore have only a minor effect on the overall shape of the distribution.

Defining the boundaries of a settlement system is more difficult. It is clear that many surveys in northern Mesopotamia were not or only partially bounded by landscape features such as rivers or mountain ranges, which would have likely confined ancient settlement systems as well. It is therefore highly likely that the settlement patterns observed in survey regions represent only a sample of a larger system and that there are no clear criteria to assess this sample in terms of its place in the overall size distribution of the system. This issue is intricately linked with the actual surveys and their environmental settings that are analysed here. The issue of defining boundaries of the settlement system will therefore be dealt with in Section 5.3.2, where the individual rank-size analyses are discussed.

The problems of site size estimates and site contemporaneity apply to rank-size analyses as well. The issue of reliably estimating site sizes has been extensively dealt with above and will not be repeated here. Just as overfilled landscapes will inflate population estimates, it is likely that such landscapes will result in the pooling of settlement systems that existed one after another rather than at the same point in time. As Johnson has shown (1977: 498), pooling of multiple settlement systems will likely result in a convex system even though the constituent systems originally had other distributions. It is therefore necessary to include a chronological analysis of the settlement system before it is subjected to rank-size analysis. For that reason, the rank-size analysis will be applied to the datasets that are corrected for site contemporaneity, rather than to the original datasets.

The statistical significance of observed deviations from log-normality in rank-size plots is usually tested with the Kolomogorov-Smirnov or K⁻ test. There are, however, several problems with this test when it is applied directly to site size distributions (Savage 1997: 236): (*a*) the K⁻ test is normally applied to continuous distributions, whereas site size distributions are discrete, (*b*) the K⁻ test assumes that the expected distribution is defined independently from the investigated dataset, whereas the expected distribution in rank-size analyses is derived from the dataset itself, and (*c*) the value at which the K⁻ test finds a deviation to be statistically significant can be incorrect if some parameters of the sample are unknown and must be estimated.

In order to address these issues, Falconer and Savage wrote the RankSize simulation program to investigate the chance that a random sample drawn from a theoretical site universe deviates more strongly from log-normality than the observed archaeological dataset. The number of sites in the theoretical site universe is determined from the archaeological dataset by first defining the site recovery rate, or more generally an estimation of the percentage of sites that is known from the actual settlement system, and by inferring the actual number of sites that must have existed in the settlement system. Under the assumption that the largest site is known, the site universe is then created with a log-normal distribution. It contains the inferred actual number of sites in the settlement system. From this site universe, the program uses Monte Carlo sampling to create random samples containing as many sites as are present in the archaeological dataset. This process is repeated a number of times, resulting in a percentage that indicates how many times the random samples deviate more strongly from log-normality than the archaeological dataset. If this percentage is high, it is likely that any deviation from log-normality in the archaeological dataset is not statistically significant (Falconer and Savage 1995; Savage 1997).

Drennan and Peterson (2004) recently argued that Monte Carlo sampling is not suited to investigating statistical significance of rank-size distributions. In order to address this issue, they constructed a theoretical site universe consisting of a spatially bounded region containing a large number of sites whose size distribution conforms to a Poisson distribution. They then randomly placed spatial sampling blocks of increasing size in this site universe and analysed the size distributions in each of these sampling blocks. Drennan and Peterson found that the size distributions of the sampling blocks, including a 66 % error range, never coincided with the size distribution of the site universe as a whole, even though this would be expected under normal statistical assumptions. This discrepancy remained when site samples were randomly selected from the site universe, instead of generated from a spatial sampling block. Even when some measure of organization was instilled in the site universe by introducing a hexagonal pattern of first- and second-order large sites, Monte Carlo sampling was unable to approximate the size distribution of the site universe as a whole. Drennan and Peterson therefore concluded on the use of Monte Carlo sampling blocks of different sizes that 'These are simply different analyses of spatial patterning at different scales; it makes no sense to try to think of the smaller blocks as useful for characterizing the larger population' (2004: 539). In other words, the chance that a small sample from a settlement system will adequately represent the entire system is very low.

Instead of the K⁻ test in combination with Monte Carlo sampling, Drennan and Peterson presented the A value with error range as an alternative method for testing the statistical significance of deviations from log-normality in observed site size distributions. The A value is created by first scaling the rank-size plot so that the plot has a square shape and the log-normal line a 45° angle. In this way, the areas above and below the log-normal line are of equal size and they can be assigned a value of 1.0 and -1.0, respectively. The A value represents the proportion of this area that is taken up by the area between the log-normal line and the line of the observed size distribution (Fig. 4.6). In this way, convex size distributions will have a positive A value (Fig. 4.6A) whereas primate distributions will result in a negative A value (Fig. 4.6B). If the observed site size distribution approaches the log-normal line, A will be close to 0. The maximum value of A is 1.0, when all sites are of equal size. The minimum value of A can be lower than -1.0 if observed sites are smaller than the expected smallest settlement. The A value does not, however, give information on the shape of the size distribution because very different size distributions can produce similar A values. Furthermore, a size distribution with both convex and primate aspects can produce an A value close to 0, even though the distribution itself deviates markedly from the log-normal line (Fig. 4.6C). Overall, then, the A value can give a good indication of how much a size distribution departs from log-normality, although visual inspection of the size distribution remains necessary. Even more importantly, because the A value is a coefficient of the area between the log-normal line and the observed site size distribution, it can be used to compare site size distributions from different periods and with different numbers and sizes of settlements (Drennan and Peterson 2004: 534–5).

In order to test the statistical significance of the A value, Drennan and Peterson suggested the use of the bootstrap resampling approach (Drennan and Peterson 2004: 539–41). The great advantage of the bootstrap method over the K⁻ test is that the bootstrap does not require knowledge or assumptions about any expected distribution in order to test the significance of the observed distribution. The bootstrap method takes numerous samples from the original dataset. Each sample has as many observations as the original dataset, but duplicates the results of some observations whereas others are omitted. The distribution of the resulting A values are plotted and readjusted so that they can be used to express a confidence range within which the A value of the original

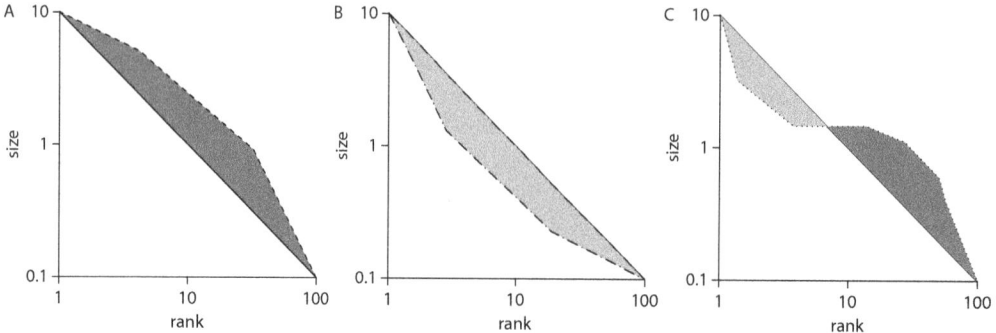

Fig. 4.6: A values for (A) convex distribution, (B) primate distribution, and (C) primo-convex distribution. Positive and negative components of A value have dark and light shading, respectively (based on Drennan and Peterson 2004: fig. 2).

dataset will probably have fallen. If this confidence zone is wide, it must be recognized that the original dataset may only approach reality to a limited degree. If the confidence zone is narrow, the original dataset is probably a good reflection of reality. Furthermore, by calculating different error ranges for A values of site size distributions that have to be compared, an error range can be calculated for which the A values do not overlap anymore, and this range indicates the confidence with which any observation on change in site size distributions can be made. The comparison of A values with error ranges of site size distributions of different periods thus allows observations to be made on the possibility that change actually took place.

4.6.2 Interpreting rank-size distributions

After the statistical significance of the observed distributions has been assessed, the distributions can be interpreted. Various explanations have been offered for the different types of rank-size distributions (Savage 1997: table 1). Johnson proposed that different rank-size distributions result from the degree and mode of system integration (Johnson 1980). System integration is defined as 'the statistical interdependence of change in the population sizes of the settlements of a system, such that the probability of change in a given settlement is conditionally related to the probabilities of change in other settlements of the system' (Johnson 1980: 245). This interdependence can be expressed both vertically, that is, between sites at different levels within a hierarchical system, and horizontally between sites that are on an equal functional level. In this sense, convex distributions indicate low overall integration or strong horizontal integration, whereas primate distributions indicate strong vertical integration. In log-normal distributions, both horizontal and vertical integration are strong. More generally, Johnson argued that low integration implied the existence of relatively autonomous sites within the system, whereas high integration would indicate strong interdependence and centralization (1980: 245).

A number of explanations focused on the fact that a settlement system may not have been identified correctly, resulting in a rank-size analysis of only a small part, which in turn leads to flawed results similar to those generated by Monte Carlo sampling (see Section 4.6.1). Johnson pointed to the possibility that a primate distribution resulted from a partially identified settlement system, whereas a convex distribution could be the result of the pooling of multiple systems (1977: 498). Thus, system boundaries were not correctly identified. This problem is particularly relevant for archaeologists who often work with data from arbitrarily defined survey regions. Other explanations considered the urban history of a given region as critical factor, with primate distributions emerging in areas with a short history of urbanization. Still others linked rank-size distributions with core-periphery models, arguing that different distributions will develop in cores as opposed to peripheries. Falconer and Savage pointed to the possibility that different systems operated at different organizational levels, for example resulting in a primo-convex distribution where a centralized system is superimposed on a lower central place system (1995). Because several of these explanations contradict each other, it is necessary to combine rank-size analyses with as many other data as possible, in order to be able to argue why a given interpretation is preferred over another possibly contradicting explanation.

In the present study, site size is expressed in hectares or absolute population figures, so that a rank-size distribution is actually a measure for the organization of a population in a given landscape. At a very basic level, a convex size distribution indicates that there are many large settlements compared to the number of small settlements. Convexity also indicates that these larger sites are of roughly similar size; otherwise the distribution would tend towards log-normality. Convexity, then, is an expression of population dispersal across the landscape in sites that are of roughly similar size. A primate distribution indicates that there are one or very few large centres with a large population, and a great number of smaller settlements. The central issue that has to be addressed here is why households choose to live in a certain community, and not in another. If a primate distribution is observed, the advantages of living in a large community clearly outweigh those connected with living in smaller, dispersed communities. Conversely, in a settlement system with a convex distribution, there are clear advantages to living dispersed over the landscape.

The theoretical understanding of site-size distributions as reflections of system integration, and the actual observation that rank-size distributions reflect population organization can be combined within the framework of competition and cooperation presented in Chapter 3. If there is competition between communities, integration between these communities would be low, and a convex site size distribution may be expected. This understanding of competing communities is consistent with the observation that convexity often results from the pooling of more than one settlement systems. In other words, convexity indicates the existence of several settlements or settlement systems that function more or less independently. Conversely, if there are cooperative mechanisms in operation in a given region, the site size distribution would eventually tend towards log-normality. Primate distributions are more difficult to interpret. Such distributions indicate strong centralization, which can arise from cooperation or can be a reflection of the dominant position of one site over many others achieved by force. Concluding, then, rank-size analysis can point to certain processes operating within a given region, but the analysis must be combined with other data in order to fully understand the data at hand.

4.7 Conclusions

The present chapter started by highlighting the problems associated with the interpretation of surface survey results. In the present context this discussion cannot be more than a brief overview. The discussion has nevertheless shown that some confidence may be placed in survey results, especially when they are combined with data recovered from stratigraphic excavations. The chapter continued by pointing out the need for a model that explores the degree to which sites within a chronological phase were contemporaneous, and has detailed and discussed one such model that fits the data at hand. This model focuses on the identification of contemporary settlements through the calculation of average settlement life-spans per chronological period. This model will be applied to the data presented in Section 5.2.

The next part of this chapter dealt with the reconstruction of on-site population levels and densities. Based on this discussion, the following conclusions can be drawn: (*a*) there is considerable variation in modern on-site population densities, (*b*) there is no reason to assume that such variation did not exist in the past as well, (*c*) the extrapolation of ethnohistorical population densities to late third-early second millennium BC northern Mesopotamian settlements must be carefully considered, and (*d*) the available evidence suggests that third to early second millennium on-site population densities in northern Mesopotamia were different to, and more specifically on average higher than, ethnohistorically recorded densities. Nevertheless, a reasonable guess as to actual absolute population numbers would be 150–250 persons/ha, including all built-up and open space. The evidence discussed here suggests that this density may be considered as representing the lower end of possible ranges for the late third-early second millennium population densities. The previous discussion has, however, also highlighted the possibility that on-site densities fluctuated through time, as is made clear by the Tall Bazi *Weststadt* example. The present discussion furthermore highlighted the fact that settlement sizes, especially aggregated settlement sizes for entire regions, seem to be a good indicator of changes in the overall population of a region. This observation further justifies the use of regional settlement data provided by surveys in the reconstruction of demographic patterns.

Finally, Section 4.6 dealt with rank-size analysis as a method to characterize settlement systems and their organization. A method for determining the statistical significance of observed rank-size distributions was presented, and a framework was given within which observed distributions can be interpreted. This framework does not focus on whether or not an observed rank-size distribution coincides with an expected pattern, but aims to identify diachronic changes in the rank-size distribution. Additionally, the framework aims to explain the direction of these changes, that is whether the settlement system develops into a more convex or primate direction.

5 Regional settlement trends

5.1 Introduction

The method for the reconstruction of simultaneously occupied sites imposes certain requirements on the regional data set that is used. These requirements are: (*a*) a survey strategy that is comparable to the strategies employed by all other surveys that are used, (*b*) a sound chronology with fixed, absolute dates for every recognized pottery phase, and (*c*) a clear indication of site sizes per phase. After the selection of surveys according to these requirements, each survey is described, and additions or changes to each dataset are presented (Section 5.2). The results of the site contemporaneity simulation and the demographic and rank-size analyses are presented in Section 5.3. Finally, Section 5.4 will discuss other surveys in northern Mesopotamia, as well as evidence from historical records, in order to contextualize the evidence from the selected surveys, and to find whether or not the reconstructed patterns represent unique, local developments, or may be indicative of larger, regional trends. The relation between the reconstructed settlement patterns, climate change, and the expected social responses will be discussed in Chapter 8.

5.2 Selecting the surveys

The procedures for simulating site contemporaneity, reconstructing population levels, and performing rank-size analyses all require detailed information that is not available for every survey in the Jezirah. For most surveys, at least preliminary reports exist, but these only occasionally contain data on individual site sizes. A selection has to be made based on the availability and quality of the data. Furthermore, survey strategies have to be similar to some extent to allow data comparison between surveys. There are two qualitatively assessable variables that allow selection of suitable surveys, namely comparability of the data, and suitability for the anticipated reconstruction. Inspection of multiple survey reports suggests that three surveys in and around the Jezirah provide relatively reliable details on site sizes per period for the third and early second millennium BC; the Birecik-Euphrates Dam Survey (B-EDS), the Balikh Survey (BS), and the North Jazira Survey (NJS) (Fig. 5.1, Table 5.1). The BS and NJS used comparable strategies, although details may vary (Table 5.1). Table 5.1 suggests that the survey strategies of the BS and NJS were most similar, with the B-EDS differing on several points. The B-EDS, which was part of an archaeological salvage project, was much coarser in resolution. However, the B-EDS has the advantage that numerous sites were subsequently excavated, thereby providing a measure of control for the sizes originally assigned to the sites based on the survey results.

The most important characteristic in this context is that all these surveys applied a controlled sherd collection programme to each site, whereby sites were divided into distinct sampling sectors from which material was collected separately. This suggests that site size estimates for each survey region were made on a similar material base. Where possible, the information from the three original surveys will be complemented with data available from subsequent research.

The relatively precise chronology of the BS was largely based on the excavated stratified sequence from Tell Hammam et-Turkman. Although difficulties with this sequence remain, and high-precision radiocarbon dates that would allow the relative chronology to be fixed on an absolute time-scale are currently lacking, this chronology may be considered sufficiently sound to be employed without major adjustments. The other two surveys used much coarser chronologies (Table 2.4). In the case of the B-EDS, later excavations have led to a considerable refinement, as well as the filling-in of an apparent chronological gap in the settlement record as reconstructed by the survey. The results of these excavations have therefore been incorporated in the settlement

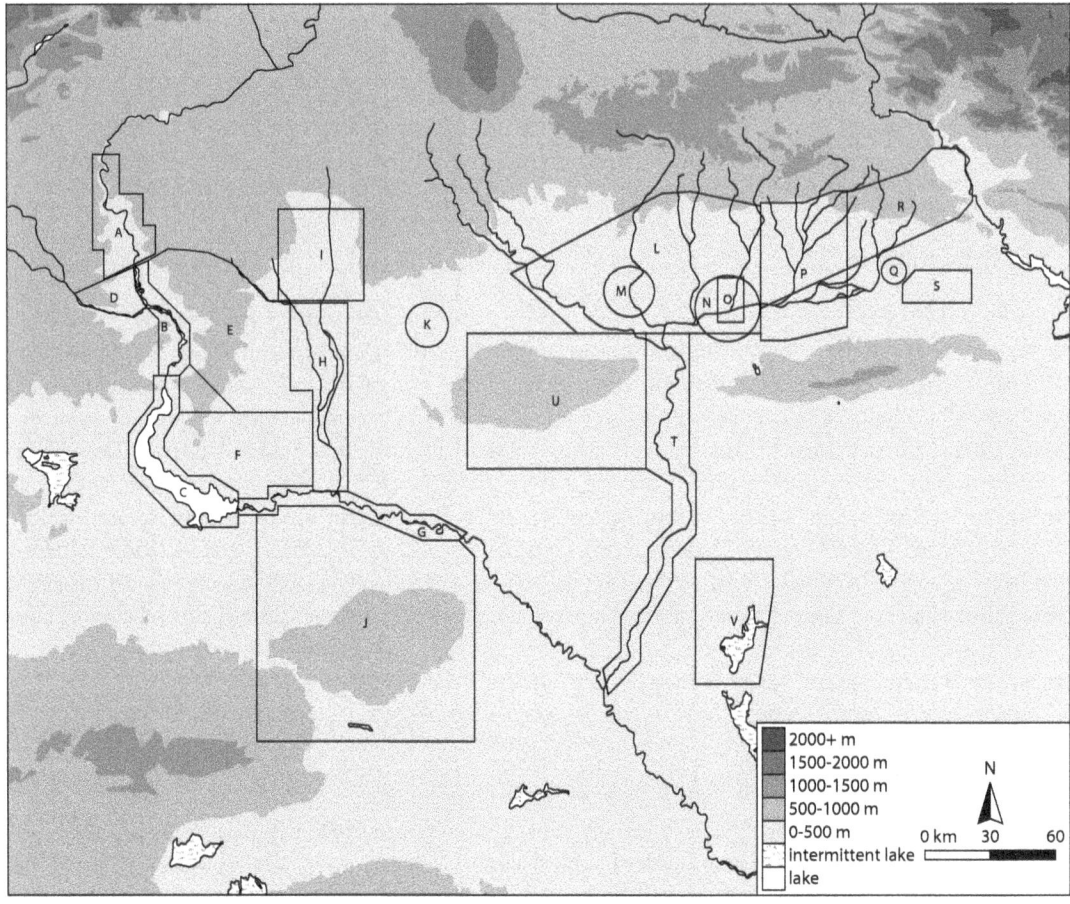

Fig. 5.1: Map of northern Mesopotamia indicating spatial coverage achieved by recent surveys. (A) Birecik-Euphrates Dam Survey (B-EDS). (B) Tishrin Dam surveys. (C) Tabqa Dam surveys. (D) Land of Carchemish survey. (E) Westgazira survey. (F) Balikh-Euphrates uplands survey. (G) Middle Euphrates survey. (H) Balikh Survey (BS) and Wadi Qaramogh survey. (I) Harran survey. (J) Jebel Bishri survey. (K) Wadi Hammar survey. (L) Upper Khabur Survey. (M) Tell Beydar Survey (TBS). (N) Brak sustaining area survey. (O) 1988 Tell Brak Survey. (P) Leilan Regional Survey (LRS). (Q) Tell Hamoukar Survey (THS). (R) Northeast Syria survey. (S) North Jazira Survey (NJS). (T) Lower and middle Khabur survey. (U) Khabur Basin Project. (V) Wadi Aǧiǧ survey.

map	survey	years	number of seasons	number of sites	area (km²)	remote sensing imagery	main sources
A	B-EDS	1989	1	85 (28)	186	no	Algaze et al. 1991, 1994
H	BS	1983	1	210 (86)	450	yes	Akkermans 1984, 1993; Curvers 1991; Bartl 1994
S	NJS	1986–1990	4	186 (68)	475	yes	Wilkinson 1990; Wilkinson and Tucker 1995

Table 5.1: Survey strategy details for the BS, B-EDS, and NJS. Number of sites refers to the total number of surveyed sites, with number of sites with evidence for the period discussed here in parentheses. Letters in the left column refer to Fig. 5.1.

record that is used in the present study. The chronology of the NJS was modified from that used at Tell al-Hawa, which in turn derived from material excavated in the area of the Saddam Dam salvage project. This pottery chronology was supplemented through surface associations from demonstrably single-period sites (Wilkinson and Tucker 1995: 89).

5.2.1 The Birecik-Euphrates Dam Survey

The Birecik-Euphrates Dam region differs from the other survey regions analysed here in its environmental setting. The region is not part of the north Syrian steppe plateau but extends into the Taurus Mountains via the Euphrates Valley (Fig. 5.1A). Concurrent with the Birecik-Euphrates Dam Survey, a geomorphological assessment of a small area representative of the whole survey region was carried out (Algaze et al. 1994, 6–8). This assessment illustrated the complex geomorphological history of the area, characterized by several alluviation and erosion phases. Five different zones can be recognized within the valley. The upland plateaus bordering the valley are 600–750 m above sea-level. Although they contain good quality arable soil, settlement today is marginal due to the very deep water table. The pediments lack perennial water sources, and only one place was observed where a high water table allowed the use of wells. This area contained a number of sites. The high terraces at 395–405 m above sea-level resulted from several Pleistocene alluvial and colluvial phases. The low terraces are 6–18 m above the low water level of the Euphrates. These terraces are incised by small tributaries of the Euphrates. The soil is generally suited to cultivation, with good access to water. Typical current land use includes a combination of pistachio trees and dry and irrigated crops (Algaze et al. 1994: 7). Geomorphological observations around Şaraga Höyük (B-EDS 73), combined with investigations of archaeological deposits, suggested that the current Euphrates floodplain was shaped in the third millennium, paralleling observations at Jerablus Tahtani (Sertok and Kulakoğlu 2001: 476; Sertok et al. 2007).

As part of the Güneydoğu Anadolu Projesi (GAP), several dams have been constructed in the Euphrates, including the Birecik and Carchemish dams. Archaeological surveys were carried out to assess the archaeological potential of the areas to be flooded. The Birecik and Carchemish basins were surveyed in 1989. The survey was limited to a 60 km stretch of the Euphrates Valley, from Carchemish in the south to the modern town of Halfeti in the north. The surveyed area included the Euphrates Valley and extended into the uplands up to a maximum elevation of 400 m above sea level, *c.* 10 m above the projected level of the artificial lake. A total area of *c.* 186 km² was investigated. The survey started by locating the principle settlement mounds. Around these sites, pedestrian survey was carried out along transects to detect dependent settlements.[36] Small sites were located with the help of local villagers, and by transect-walking areas with environmental conditions similar to areas where sites were already known to have existed. Areas along wadi courses and around permanent water sources were also transect-walked. This strategy resulted in the recognition of 85 sites, 28 of which yielded third or early second millennium BC material. Most sites were visited once to collect diagnostics, that is, rims and decorated body sherds. Depending on the size, complexity, and morphology of the site, diagnostics were collected in multiple sectors to detect possible settlement shifts through time. Neither aerial nor satellite imagery seem to have been used during this survey (Algaze et al. 1991, 1994). Periodization suffered from a lack of good single-period contexts. The surveyors recognized thirteen periods, ranging from the Pre-Pottery Neolithic to the thirteenth century AD. Palaeolithic remains were also recovered, but not studied as part of this survey. A later survey in the same region focused on the Palaeolithic remains (Taşkıran 2002).

Occupied area estimates were derived from maximum site size in the case of single period sites, and whenever possible from the results of the individual sampling sectors in the case of multi-period sites. When this proved impossible, an arbitrary size of 0.5 ha was assumed for these sites. It was suggested that this estimate probably resulted in a significant underestimation of total occupied area, but that all periods would be equally affected, leaving the resulting trends through time in place. Although this is not necessarily true given the fact that the proportion of single-period versus multi-period sites may be different for every chronological period, there seems to be no method to correct this bias (Algaze et al. 1991: n. 6).

The absence of evidence for material dating to the second part of the Middle Bronze Age suggested to the surveyors that habitation of this region was limited to a few strategically located sites outside the survey area (Algaze et al. 1994: 17). This settlement gap was excluded from the chronology, which jumped from period 8, Late EBA/Early MBA, to period 9, LBA. For the purpose of the present analysis, which requires a continuous chronological framework, this gap will here be called Late MB. Subsequent excavations have shown that the B-EDS area was indeed occupied during this period, and the results of these excavations will be incorporated in the present analysis.

---

36 The spacing between transects is not mentioned. See Redman (1982: 377) on the impact of transect spacing on site detection.

The settlement record provided by Algaze et al. (1994) serves as the basis for the data set that will be used here. Rescue excavations at several sites have, however, necessitated some modifications of the picture as gleaned from the survey. These excavations have especially highlighted the apparent continuity between Late Uruk and Early EBA at several sites, and the presence of Late MBA assemblages. As far as can be judged from the excavation reports, the survey seems to have somewhat underestimated the occupations at individual sites. The individual modifications will be discussed below; the resultant data set of occupied sites per phase is presented in Table 5.3 and Figs. 5.2 and 5.3.

The recent re-dating of some of the diagnostic ceramics used by Algaze has led to a re-evaluation of the apparent Late MBA settlement gap (Peltenburg 2007a: 17, 2007b: 258). Especially the use of the 'fruit stand' stem as a type-fossil for the Late EBA/Early MBA has led to an over-estimation of the number of sites for that period, because 'fruit stands' seem to have gone out of use before the final quarter of the third millennium (Sertok 2007: 242–3). A similar argument can be made for corrugated Hama goblets. Their use continues into the Late EBA/Early MBA period, but these goblets were especially dominant before that period. This suggests that sites where they were found could not only date to the Late EBA/Early MBA, but also to earlier periods (Peltenburg 2007b: 258–9). The re-dating of these diagnostic types in the B-EDS would affect a significant number of sites, as shown in Table 5.2.

Due to the uncertainties in the dating of these diachronic types, it is difficult to assign sites with 'fruit stands' to either the Early or Middle EBA period, whereas those with Hama goblets could still be dated to the Late EBA/Early MBA. In order to overcome this issue, Peltenburg suggested re-dating sites with 'fruit stand' to the mid-third millennium BC (2007b: 259). This suggestion will be followed here for those sites where no excavations have taken place, and where no Hama goblets were present, as these could indicate Late EBA/Early MBA occupation as well. In other words, the occupations at Near Zeugma 2 (B-EDS 18), Near Hacınebi 1 (B-EDS 27), Kefri Höyük (B-EDS 38), Şara Mezar Harabe (B-EDS 60), and Near Akarçay 2 (B-EDS 68) will be treated as Middle EBA, rather than Late EBA/Early MBA. For the excavated sites, the results from the excavations will also be taken into account, and these are discussed below.

site	name	'fruit stand' stem	corrugated Hama goblet
1	Kalemaydanı Höyük		x
2	Karaçalı Höyük		x
10	Horum Höyük	x	x
16	Tılmusa Höyük	x	x
18	Near Zeugma 2	x	
27	Near Hacınebi 1	x	
34	Tılvez Höyük		x
38	Kefri Höyük	x	
44	Zeytinli Bahçe Höyük		x
49	Mağara Höyük		x
51	Mezraa Höyük		x
60	Şara Mezar Harabe	x	
65	Şavi Höyük		x
68	Near Akarçay 2	x	
69	Gre Virike	x	x
72	Akarçay Höyük		x
82	Tıladir tepe	x	x
85	Carchemish	x	

Table 5.2: Occurrence of 'fruit stand' stems and corrugated Hama goblets on sites in the B-EDS region (Peltenburg 2007b: table 1).

Horum Höyük (B-EDS 10) has been excavated as part of the international cultural heritage rescue operation preceding the flooding of the Birecik and Carchemish lakes. The site consists of a 6.4 ha main mound and a 21 ha lower town. A significant part of the main tell has been eroded by the Euphrates, probably already at the end of the second millennium BC, and Marro gives an estimate of 10 ha for the third millennium site (2007b: 384). The lower town seems to have been occupied only in the Roman period. The excavations were carried out on the southeastern slope and the top of the main mound, and confirmed the periods detected by the American survey, as well as adding the Early and Middle EBA and Late MBA periods. Architectural features were found for the EB I and EB IV/MB I periods, whereas the EB II–III periods were attested by substantial digging activities. The excavators note that the early EB deposit at Horum Höyük is *c.* 5 m thick, similar to that uncovered at Zeytinli Bahçe Höyük (B-EDS 44) (Marro et al. 1997, 1998, 2000; Marro 2007b). Based on these considerations, the size of Horum Höyük will be conservatively estimated at 8 ha during the Early EBA, Middle EBA, and Late EBA/Early MBA periods.[37]

Excavations at 12 m high 1.2 ha Tilbes Höyük (B-EDS 14) confirmed occupation for the EB I and EB III–IV periods, and suggested a possible EB II and MB I occupation. There seems to be some scant evidence for MB II activities, but the nature of these activities remains unclear. EB I buildings were found on the southern slope. There is also evidence for outdoor activity areas, where metal smelting may have been carried out, and some stone-lined tombshave been found. Radiocarbon dates confirm a late fourth–early third millennium date. EB II occupation seems to have been much more restricted than the preceding and following phases, as it was not found in all trenches. An EB III burned building is interpreted as a building with cultic connotations. This building had several predecessors on the same spot, but with different internal layouts. Extensive EB IV construction activities are evidenced in all trenches (Fuensanta et al. 1998, 2001, 2002a). Based on these data, an Early EBA and Late EBA/Early MBA occupation of 1.0 ha will be assumed, as well as a 0.5 ha Middle EBA and 0.1 ha Late MBA occupation.

The large site of Surtepe Höyük (B-EDS 33) was occupied during the LC 5 period, after which there is evidence for a conflagration. The site was probably immediately re-occupied during the EB I period. This is also suggested by the similarities in pottery technology for both periods. There is evidence for at least two large mudbrick platforms at the site that may have functioned in both the LC 5 and EB I periods. The excavators estimate the LC 5 occupation at 6 ha and suggest that the site may have extended to 20–50 ha during some periods, judging from surface sherd densities (Fuensanta et al. 2002a, 2006). The periods that are represented by these sherd scatters are, however, not mentioned. In the absence of more detailed information on these estimates, it seems quite possible that these densities represent sherd scatters resulting from off-site activities, for example manuring, in which case the dimensions of the site itself would have been much more restricted during these periods. Size estimates for the EB I period range from 3 ha (Rothman and Fuensanta 2003: table 3) to 12 ha (Fuensanta 2007: 143). Based on the LC 5 estimation, the apparent occupational continuity into the EB I period, and the presence of large mudbrick platforms, the EB I occupation at Surtepe Höyük will be estimated at 4 ha. For EB IV, an estimate of 12 ha is given (Fuensanta 2007: 143).

Tilvez Höyük (B-EDS 34) is a 3.5 ha site that is *c.* 9 m high and measures 1.2 ha at the top. The earliest material found by the survey dates to the late EB period, but excavations brought to light, among other things, champagne cups, cyma recta bowls and reserved slip ware, suggesting a transitional Late Chalcolithic/EB I occupation as well. EB III is also present in the mound, but there seems to have been a hiatus during EB II. EB IV architectural remains consist of stone walls, and there is some evidence for MB I occupation (Fuensanta et al. 2001, 2002a, 2005). Based on these results, Fuensanta (2007: 143) estimates the size of this site at 3 ha during EB I, 1 ha during EB III, and 3 ha during EB IV. These estimates will be retained here.

---

37 Marro (2007b: 396) stressed the strong continuity observed in funerary and ceramic traditions at Horum Höyük, and suggested that the EB/MB cultural scheme imposes upon the Birecik region a transition that is in reality difficult to discern in the archaeological record.

As part of the archaeological rescue operations, Zeytinli Bahçe Höyük (B-EDS 44) was resurveyed twice, including an intensive survey using 10x10 m collection units. Subsequent excavation revealed an extremely rich Early Bronze Age I sequence. The Late Uruk–EBA transition seems to have been very gradual. There are a number of architectural changes indicating different types of occupations, but these nonetheless seem to have succeeded each other rapidly, as there is no apparent break in pottery. The EB I sequence is characterized by extreme continuity in architecture and pottery. Architecture consisted of small, simple dwellings. During the later part of the EB I period, the architecture is of a larger and planned scale, with large rectangular buildings that are rebuilt several times following the same layout. After this occupation episode, the site is abandoned until EB IV. During the final part of the Early Bronze Age, as well as during the Middle Bronze Age, settlement was restricted to the main mound. The EB IV settlement is characterized by irregularly laid out mudbrick structures whose functions change through time. According to the excavators, this area may have been on the outskirts of the settlement, again emphasizing the limited extension of the settlement in that period. Alluvial deposits were occasionally encountered in the stratigraphy of the site, suggesting that high Euphrates floods may have been responsible for the size reduction from EB I to EB IV. On top of this, a late Middle Bronze Age wall was unearthed, interpreted as a fortification wall. Slightly to the east, at the base of the upper mound, a domestic activity area was uncovered (Deveci and Mergen 1999; Frangipane and Bucak 2001; Frangipane et al. 2002, 2004; Balossi et al. 2007). Zeytinli Bahçe Höyük will be considered as a 0.8 ha settlement during the Early EBA period, and as a 0.2 ha settlement during the Middle EBA, Late EBA/Early MBA, and Late MBA periods.[38]

At Yarım Höyük/Yarım Tepe (B-EDS 50), excavations brought to light Late Uruk/LC 5, EB I/Kurban V, and Hellenistic levels. The Uruk and Kurban V levels were highly disturbed by later digging activities. The shallow Uruk/Kurban V deposits indicated a relatively short occupation span, perhaps 200–300 years. For both periods, parts of several stone houses were excavated, but the fact that the Euphrates cut away part of the tell suggested that the settlement may have been larger than that which still stands today. Botanical and zoological material from the site indicated a subsistence economy including agriculture and animal husbandry geared toward local consumption (Rothman et al. 1998; Kozbe and Rothman 2005). The rather disturbed deposits at Yarım Höyük, and the fact that an unknown part of the site has eroded away, make any size estimate problematic. Based on the site map and the interpretation of the site by the excavators as a small self-sufficient farming community, Yarım Höyük will be considered as a 0.3 ha Early EBA community.

At Mezraa Höyük (B-EDS 51), a second survey found evidence for the Late Chalcolithic, Early EBA, and MB II periods, which the survey by Algaze et al. did not detect. Cleaning of a large section in the east slope of the mound resulted in the recognition of eight occupation levels, of which VI to VIII span the Early to Middle Bronze Age. Level VI is dated to the MB II period, but no architectural features could be assigned to it. The fact that MB II pottery types occurred on most parts of the site surface, suggests that there was some activity during this period. Both the survey results and the evidence from the cleaned section suggest that the EB III/IV period was the most extensive, whereas the LC and EB I/II periods were more restricted (Ökse and Tekinalp 1999; Ökse et al. 2001; Yalçıklı and Tekinalp 2002). The occupation of Mezraa Höyük will be estimated at 0.25 ha during the Early and Middle EBA, at 0.5 ha during the Late EBA/Early MBA, and at 0.1 ha during the Late MBA.

Şavi Höyük (B-EDS 65) consists of a seven separate mounds along the Euphrates, numbered I–VII from south to north. A narrow valley separates Şavi Höyük I from the other mounds. Resurvey of Şavi Höyük yielded Halaf sherds, no Uruk sherds, few Early EBA, more Middle and

---

38 The presence of MB I material at Zeytinli Bahçe Höyük is uncertain at the moment. The clear architectural break between EB IV and MB II suggests either complete abandonment at the site, or the destruction of MB I architectural levels by the MB II fortifications. Alternatively, the fortification of the site could point to a change in the function of the site, rather than abandonment followed by reoccupation (Balossi et al. 2007: 376). Radiocarbon dates from *in situ* wooden beams from the MB II layers provided an age range of 2028–1980 cal BC. These beams may have been recycled from older structures that are as yet unidentified at the site (Balossi et al. 2007: 375). Further, the regional MB pottery assemblage remains problematic. For example, Marro (2007b: n. 25) notes that the pottery assemblage supporting the MB II date for Zeytinli Bahçe Höyük closely resembles the ceramic repertoire defined as MB I at Horum Höyük.

Late EBA, and EB/MB transitional sherds. There were possible LBA sherds as well. Settlement remains dating to the third and early second millennium BC have only been encountered at Şavi Höyük I and II, whereas the other mounds seem to have functioned as cemeteries during this period. Excavation at Şavi Höyük I revealed twenty occupation phases, ten of which represent an almost continuous sequence from late EB to late MB. The earliest phases 20–19 (EB III/IV) consisted of stone foundations, followed by simple domestic structures during phases 18–13 (EB III/IV–MB I). No architecture was recovered from phase 12, but the large amount of pottery suggests at least continuous use of the site. Phase 11 (MB II) consisted of simple architecture, whereas from phase 10 onwards there was again evidence for stone foundations. At Şavi Höyük II, a Late Uruk building was identified, on top of which Early EBA walls were found that probably belonged to several buildings. (Dittmann et al. 2001, 2002; Bucak et al. 2004). Based on these considerations, a 0.4 ha size will be assumed for the Middle EBA, Late EBA/Early MBA, and Late MBA occupations at Şavi Höyük I, and a 0.1 ha size for the Early EBA occupation at Şavi Höyük II.

Excavations at Gre Virike (B-EDS 69) confirmed Algaze's suggestion (1994: 54) that this was a special-purpose site. The excavations also confirmed the EB date of the site, but failed to detect any signs of permanent occupation. Instead, the site seems to have had a purely cultic function during much of the third millennium BC. The site consisted of a large mudbrick platform with several smaller structures. During period I (EB I–II), there were several plastered and stone-lined pits, a stone-lined channel, and a stone stairway leading away from the terrace and into a subterranean stone-lined chamber.[39] The stone-lined pits contained grains, ashes, unbaked clay figurines, and miniature stone axes. During period IIA (EB III–IV), several chamber tombs were constructed on the platform. Grave goods included numerous vessels, and stone and metal objects. During period IIB (EB IV), the platform was used for child burials. Associated finds included bronze pins and numerous vessels. There is also some evidence for outdoor activities on the mudbrick platform. Gre Virike is interpreted by the excavator as a cultic centre serving the small EB settlements in the Euphrates basin. There is a change of focus between periods I and II, but the continuity between IIA and IIB is clear, and strengthened by the re-use of some IIA tombs in the following IIB period (Ökse 2005, 2006, 2007). This evidence suggests that Gre Virike never saw permanent occupation, and it will not be used in the present analysis of population dynamics in the Euphrates basin.[40]

The surface of Şaraga Höyük (B-EDS 73) has been re-investigated with the aim of excavating the site. Virtually all the pottery that was recovered from the surface of the tell dated to (post-) Roman periods, except in exposed sections of the site. These sections showed the presence of Bronze Age buildings. Early Bronze Age material dated to the mid to late third millennium BC.[41] The transition to the Middle Bronze Age seems to have been gradual at the site.[42] During the Middle Bronze Age, a large building existed with rooms that are interpreted as storage rooms, based on their inventory of large storage jars. The building was renovated several times, demonstrating a long period of use. Adjacent structures were also well-built, and included a stone stairway. The pottery assemblage yielded a large range of MB types, including types of the later MBA. Excavation in the western part of the site revealed what the excavators believe to be a fortification

---

39 The stone stairway is interpreted as leading to a now defunct underground spring (Ökse 2006). Note also the parallels with the later underground structure unearthed at Tell Mozan and identified as a gate to the netherworld (Kelly-Buccellati 2002).

40 Gre Virike is not the only non-habitation site in the B-EDS area. Early Bronze Age cemeteries were found near Birecik (Sertok and Ergeç 1999a), and possibly at the site of Harabebezikan Höyük (B-EDS 75) (Bilgen 2001). Boztarla Tarlası #2 (B-EDS 24) may also represent an EBA tomb (Algaze et al. 1994: 37). These cemeteries, as well as Gre Virike, testify to the lively regional social and religious traditions that must have existed in the Birecik area during the Early Bronze Age.

41 Based on the presence of a nearby EBA cemetery, the excavators also inferred a notable Early EBA occupation at Şaraga Höyük (B-EDS 73) (Sertok et al. 2007: 343).

42 In two areas of Şaraga Höyük, the excavators noted that the Early and Middle Bronze Age layers were separated by thick Euphrates flood deposits. Historical records of Euphrates floods document that, once the snow melted, the water level could rise to as much as 3.5 m above the normal water level in the Birecik area. Thus, before the construction of dams controlling the flow regime of the Euphrates, significant floods may have been the norm rather than the exception. In fact, EBA/MBA flood deposits were not encountered in the eastern part of the mound, and the event responsible for them apparently did not completely disrupt life at Şaraga Höyük (B-EDS 73) (Sertok et al. 2007: 348).

site	name	Early EBA	Middle EBA	Late EBA/ Early MBA	Late MB
1	Kalemaydanı Höyük			0.5	
2	Karaçalı Höyük			0.25	
10	Horum Höyük	8	8	8	
14	Tilbes Höyük	1	0.5	1	0.1
16	Tılmusa Höyük			0.5	
18	Near Zeugma 2	0.5	0.5		
21	Tepecik			0.5	
27	Near Hacınebi 1		0.5		
31	Tılöbür Höyük			1.44	
33	Surtepe Höyük	4		12	
34	Tılvez Höyük	3	1	3	
38	Kefri Höyük	0.5	0.5		
40	Büyük Tılmiyan Höyük	0.5		0.5	
44	Zeytinli Bahçe Höyük	0.8		0.2	0.2
49	Mağara Höyük	0.32		0.32	
50	Yarım Höyük/Tepe	0.3			
51	Mezraa Höyük	0.25	0.25	0.5	0.1
55	Danaoğlu H öyük			0.5	
56	Elifoğlu Höyük		0.5	0.5	
60	Şara Mezar Harabe		0.5		
63	Cısırın Höyük			0.5	
65	Şavi Höyük I		0.4	0.4	0.4
(65a)	Şavi Höyük II	0.1			
68	Near Akarçay 2		0.4		
71	Şaraga Höyük		0.5	0.5	
72	Akarçay Höyük			0.5	
82	Tıladir Tepe			12.2	
85	Carchemish	4	30	30	30

Table 5.3: Size estimates (ha) for B-EDS sites with an EB–MB component (based on Algaze et al. 1994, with modifications). Note that no site number was assigned to Şavi Höyük II by the B-EDS.

wall, probably dating to the early part of the MBA (Sertok and Ergeç 1999b; Sertok and Kulakoğlu 2001, 2002; Sertok et al. 2004). The size of Şaraga Höyük will be estimated at 0.5 ha during the Middle EBA and Late EBA/Early MBA periods.

The site of Carchemish (B-EDS 85) consists of three parts: the Acropolis mound (3 ha), the Inner Town (44.5 ha), and the Outer Town (102 ha) (Algaze et al. 1994: 61). Algaze et al. included Carchemish at an estimated minimal size of 0.5 ha, but the site was not surveyed due to its location on the modern Turkish–Syrian border (1994: 61). This estimate was at the lower end of the range of estimates made by various researchers, summarized by Cooper as falling between 0.5 and 44 ha (2006: 56). Material from the excavations by Woolley has recently been re-analyzed, resulting in a new scenario for the growth of Carchemish from a small village to a large urban centre with an upper and a lower town (Falsone and Sconzo 2007). The evidence from the Acropolis clearly suggests that this part of the site was occupied during the early third millennium BC. If the Acropolis was also occupied during the second half of that millennium, or the early second millennium, any remains have been destroyed by later levelling and construction activities. On the other hand, the

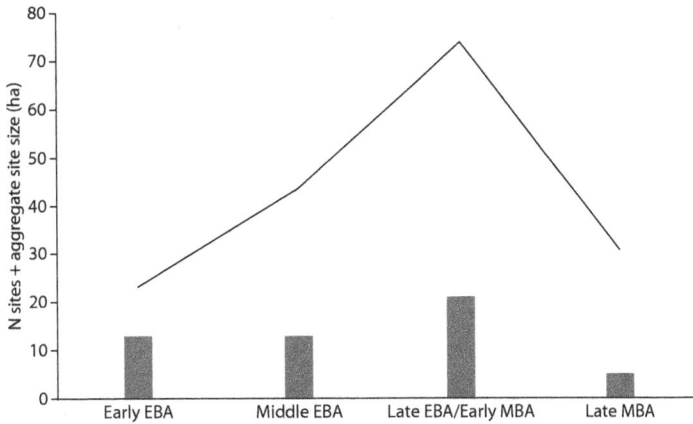

Fig. 5.2: Total number of sites (bars) and aggregate site size in hectares (line) for every B-EDS period without correction for site contemporaneity.

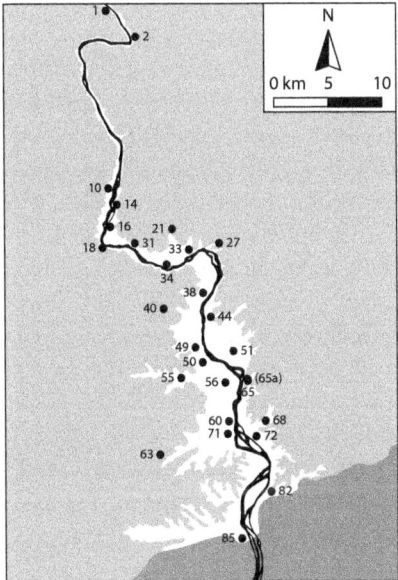

Fig. 5.3: Map of the B-EDS region indicating all sites with an Early EBA, Middle EBA, Late EBA/Early MBA, or Late MBA occupation (based on Algaze et al. 1994: fig. 2).

Inner Town now provides clear evidence for EB III–IV occupation, especially in the area of the Great Staircase (Falsone and Sconzo 2007: 88–9). The construction of the wall around the Inner Town cannot be dated with certainty, but Falsone and Sconzo suggest that this may have occurred around the same time that the Inner Town became occupied, based on the occurrence of EB III–IV wares in the fill of the rampart (2007: 90). Peltenburg in turn used this argument to suggest that the greater part of the Inner Town must have been occupied during the final quarter of the third millennium, since it seemed unlikely that the earth for the fill would have come from a large distance (Peltenburg forthcoming). The evidence for Late MBA occupation is equally scanty, with no material of this period attested on the Acropolis, and only some minor occurrences in the Inner Town (Falsone and Sconzo 2007: table 5.1). During the Mari period, Carchemish seems to have formed an alliance with several nearby cities, but the city eventually appears to have lost its independence to Halab (Hawkins 1980). Based on this evidence, the size of Carchemish will be estimated at 4 ha during the Early EBA, that is, the Acropolis mound only. The Middle EBA, Late EBA/Early MBA, and Late MBA occupations will be conservatively, and very provisionally, estimated at 30 ha. This estimate acknowledges both the attested occupation in the area of the Great Staircase and the uncertainty of its expansion toward the Inner Town rampart, as well as the continued importance of Carchemish as a political centre.

5.2.2 The Balikh survey

The Balikh Valley took shape during the Upper Pleistocene, but the river acquired its present course only at the end of this period due to tectonic activity (Mulders 1969: 41). Tectonic movement has resulted in a gradual rise of the northern part of the Balikh basin, whereas the south experienced a gradual sinking (Mulders 1969: 36). On average, the valley is 2–3 km wide, with a maximum of *c.* 6 km. Within the Balikh basin, two terraces can be distinguished; a lower Holocene terrace (BT1) and a remnant higher Upper Pleistocene terrace (BT2). The Holocene terrace consists of brown loam, over 3 m thick with some local marl outcrops (Mulders 1969: 49). Remnants of the higher Pleistocene terrace are located on the western side of the Balikh Valley and consist of limestone overlain by gypsum and gravel layers. Limestone outcrops can be found on higher hilltops (Mulders 1969: 50). The Balikh cuts a floodplain 150 m wide and 4–5 m deep. Both depth and width of the Balikh floodplain decrease after Tell es-Seman due to tapping water for irrigation purposes. After the confluence of Balikh and Wadi al-Kheder, the floodplain cuts 3 m deep into the lowest terrace (Mulders 1969: 8–9). Alluvial sedimentation since the fifth millennium BC has resulted in a considerable rise of the flood plain and may have obscured low-relief archaeological features (Wilkinson 1998: 65).[43]

The Balikh cuts its way through a landscape of undulating plains with predominantly calcareous and gypsiferous soils. To the west of the valley, and in the northern part of the east bank area, marl deposits predominate, covered by a gravel layer of variable thickness. The southern part of the east bank area consists of gypsiferous soils, generally covered by a thin layer of gravelly loam (Boerma 1988: pl. 2). In these areas, low soil aggradation has resulted in high visibility of archaeological features. Visibility decreases towards the Balikh Valley (Wilkinson n.d.). Highly gypsiferous soils are unsuitable for plant growth, but wadi soils generally have a lower percentage of gypsum, resulting in local, slightly more favourable circumstances. Calcareous soils are more favourable for plant growth.

The Balikh is a relatively small perennial stream compared to the Euphrates and Khabur (see Section 2.2). The Balikh receives most of its water from the spring at 'Ain al-Arûs, just south of the present-day Turkish–Syrian border. Additionally, the Balikh is fed by a number of wadis, including the Wadi Qaramogh and Wadi al-Kheder. Most wadis east of the Balikh Valley drain towards the Balikh and run east–west. In the south, wadis drain towards the Euphrates. These wadis can transport considerable amounts of water, as evidenced, for example, by large limestone blocks found in the lower course of the Wadi al-Burj. The Wadi Qaramogh drains a vast area to the west of the Balikh Valley. The Wadi al-Kheder is fed by the Wadi al-Hemar, coming from the northern limestone plateau, and the Wadi al-Burj. The depth of the groundwater table varies from 5–10 m for the lower Balikh Valley to 30 m for the surrounding limestone and gypsum regions (Mulders 1969: 48–9).[44]

Archaeological interest in the Balikh Valley is a recent phenomenon. Sachau, one of the few early travellers to have actually visited the Balikh Valley, mentions that there are no ancient monuments in the area (1883: 241). Le Strange mentions some Islamic monuments (1905: 104). Sykes, travelling in the early twentieth century, mentions ancient irrigation canals between Tell as-Saman and Raqqa and observes of the Balikh that 'The numerous ruins along its banks show that a very large population must at one time have subsisted on both sides of the river' (1915: 314). Interest in the archaeology of the Balikh Valley started with Mallowan's soundings at Aswad, Hammam Ibn-Shehab, Jidle, Mefesh, and Sahlan (Mallowan 1946). In Mallowan's words, finds from these sites 'merely amplify and confirm the evidence already obtained from the Habur' (1946: 111), and further led him to conclude that 'a country such as the Balih is likely to have kept within its confines only the more primitive and least enterprising forms of society' (1946: 115). Compared to the scale of settlement in the Khabur area, the Bronze Age sites of the Balikh Valley are indeed modest. As

---

43 The effect of this alluviation on Early and Middle Bronze Age settlements is as yet unknown. Given the preferential location of especially mid–late third millennium BC settlements on tells, a case could be made that the high visibility of this settlement location at least partially offset the effects of alluviation.

44 Recent canalization of the Balikh and the large-scale use of mechanized pumps have drastically altered the hydrological qualities of the wider Balikh drainage.

a result, prehistoric layers at multi-period sites are more exposed, and later surveys of (parts) of the Balikh Valley showed that the area was home to multiple prehistoric societies (Cauvin 1970; Abu Assaf 1978/9; Copeland 1979, 1982). A more ambitious survey, targeting the entire Syrian part of the Balikh Valley, was undertaken in 1983 by a team from the University of Amsterdam as part of the excavation project of Tell Hammam et-Turkman (Akkermans 1984). Following this survey, surveys of (parts of) the Balikh Valley, as well as the Harran plain to the north, were undertaken by Córdoba (1988), Yardimci (1993) and Wilkinson (1998) (see below). Excavated material for the third and early second millennium BC comes from Tell Bi'a (BS 1) (Einwag 1998; Miglus and Strommenger 2002 and references therein), Tell Hammam et-Turkman (BS 175) (van Loon 1988; Meijer 1989, 1996), and from small soundings at Sahlan (BS 247) and Jidle (BS 275) (Mallowan 1946).[45]

The 1983 Balikh Survey (BS) (Fig. 5.1H) has been partially published (Akkermans 1984; Curvers 1991; Akkermans 1993; Bartl 1994). The purpose of the survey was to locate as many archaeological sites as possible, and to reconstruct ancient settlement patterns and their development. The survey also created a regional context for the stratified ceramic sequence of Tell Hammam et-Turkman. Recognition of sites did not take place in the field, but was based on the screening of available maps and aerial photographs taken in the 1960s. This resulted in the recognition of 210 archaeological sites, as well as some mounds that were non-archaeological in nature. These sites were then visited and sherds and other materials were collected in a sampling manner. Large sites were divided into multiple sampling areas based on morphological characteristics of the site. No surveying was undertaken away from the sites that were identified prior to the actual field walking. It is important to note that this technique is biased towards large, recognizable mounds. Small sites, representing only one period or other limited activities, are likely underrepresented (Akkermans 1984; Curvers 1991).

Based on the survey results, Curvers (1991: 180) estimated site sizes by dividing each site into sectors and adding up the sectors that yielded sherds of a specific period. It seems that this division was based on the divisions used during the survey. The survey used a chronology adapted from the Hammam et-Turkman sequence, in order to avoid any chronological or cultural implications attached to the adoption of a chronology from another region (Curvers 1991: 181). The chronology used here is modified after the periodization presented by Curvers (Table 2.4).

In 1986, a survey of the Balikh basin was carried out by a Spanish team from the Universidad Autónoma of Madrid. This survey was part of a project on 'the origins, development and extinction of the Hurrian culture and people', and specifically aimed at finding a suitable site for excavation. The survey was restricted to twenty-six tells that were identified on maps and Landsat satellite images. Many of these sites were previously visited by Akkermans in 1983, but some new sites

BS	Córdoba	name	BS periods	Córdoba periods
2	1	Tell Zaidan	VIab, VIcd	EB IVA
83	7	Tell as-Saman	VIab, VIcd, VIIa, VIIb, VIIc	EB III, EB IVA, EB IVB, MB I
147	11	Tell as-Sawwan	VIab, VIcd, VIIb	EB III, EB IV, MB
152	13	Tell ez-Zkero	VIIa, VIIb, VIIc	EB IV, MB I
175	14	Tell Hammam et-Turkman	VIab, VIcd, VIIa, VIIb, VIIc	EB III, EB IVA, EB IVB, MB I, MB II
247	19	Tell Sahlan	VIab, VIcd, VIIa, VIIb, VIIc	EB III, EB IVA, EB IVB, MB
275	24	Tell Jidle	VIab, VIcd, VIIb, VIIc	EB III, EB IV, MB
276	25	Tell Hammam Ibn-Shehab	VIIb	MB
280	26	Tell Abyad	VIab, VIcd, VIIa, VIIb, VIIc	EB IV, MB
297	23	Tell Breri	VIab, VIcd, VIIb, VIIc	EB IVA, EB IVB, MB

Table 5.4: Third and second millennium BC chronological concordance for sites surveyed by the BS and Córdoba. Site numbers in the second and third columns refer to those published in Curvers 1991 and Córdoba 1988, respectively.

---

45 Surrounding areas that were recently surveyed include the upland area between the Euphrates and the Balikh (Fig. 5.1E, F) (Einwag 1993; Danti 1997, 2006–7), the Wadi Hammar (Fig. 5.1K) (http://web.uni-frankfurt.de/fb09/vorderasarch/survey.htm (Accessed 16 November, 2007)), the Euphrates Valley west and east of Raqqa (Fig. 5.1G) (Kohlmeyer 1984), and the Harran plain (Fig. 5.1I) (Yardimci 1993).

located on the Wadi Qaramogh were surveyed as well. No information on on-site collection strategies is provided (Córdoba 1988). Generally, their findings seem to confirm those of Akkermans for the tells that were visited by both teams, except that the early EB period is conspicuously absent from Córdoba's data (Table 5.4). Lack of familiarity with local EB pottery traditions, or the overburden from later periods, particularly Balikh VIcd, possibly combined with a low-intensity survey strategy may explain this absence. Unfortunately, data on tell morphology and settlement size per phase are lacking, preventing inclusion of this survey in the present analysis.[46]

The Western Jazira Archaeological Landscape Project investigated the Balikh valley from 1992 to 1995, focusing on the northern part of the valley between Tell Sahlan (BS 247) and Tell as-Saman (BS 83) (Wilkinson 1998). The project was aimed not only at finding new sites, but especially off-site features relating to irrigation and land-use. Maps and aerial imagery were studied to identify archaeological and environmental features, focusing especially on those areas and features that had not been studied by earlier surveys. These features were then visited in the field, together with features that were discovered during fieldwork. Some intensive fieldwalking around selected sites was carried out as well. Irrigation features were only sporadically encountered and were generally of Byzantine or later date, if they could be dated at all. Only one feature possibly dating to the Early Bronze Age was found. Among the other features that were found were hollow ways, and field scatters of sherds. The hollow ways seemed to date mainly to the Iron Age and early Islamic period, whereas the field scatters consisted of Hellenistic to early Islamic sherds. Several new, small sites that were recognized were given BS numbers in accordance with the system used by Akkermans. The survey found several Halaf and Ubaid sites and to a lesser extent also Bronze Age sites. Only three new rural EB/MB sites were found, two of which may have been special purpose sites not associated with agricultural activities or permanent occupation. This finding significantly strengthens the picture of third and early second millennium occupation in the Balikh Valley as gained from the 1983 Balikh Survey (Wilkinson 1998: 71).

Lyon visited numerous sites in the Balikh Valley as part of the Western Jazira Archaeological Landscape Project, and presented data on site sizes for periods VII–IX that are sometimes at odds with Curvers' estimates (Lyon 2000: table 2). In some instances it can be clearly seen that Lyon used Curvers' estimates, in which case they are in accordance. At several sites, however, there are considerable differences between Curvers' and Lyon's estimates for period VII occupations (Table 5.5).

site	name	estimate Curvers (ha)	estimate Lyon (ha)
69		no occupation	2.1
73		0.6	1.3
83	Tell as-Saman	9.0	17.0
84		10.0	no occupation
100	Tell Jaruhe	0.2–1.0	2.2
112		0.6	2.5
139	Tell Khadriya	0.5	6.3
182		0.2	1.3
205		1.8	no occupation
219	Tell Qardana	0.6	1.8
247	Tell Sahlan	7.5	16.0
316	Tell Mafraq Sluk	0.3	1.6
334		0.6	1.2

Table 5.5: Comparison of widely varying site size estimates by Curvers and Lyon for Balikh period VII.

46 Work in the Wadi Qaramogh area has recently been resumed with the archaeological investigation of Tell Balabra East and West/Tell Barābira 1 and 2 (http://www.jezireh.org/balabra_excavation.html (Accessed 5 March, 2009)).

Furthermore, Lyon mentioned two Balikh period VII sites that were not found by the 1983 survey. These are BS 393 and JDL-1, estimated at 1.5 and 1.8 ha, respectively. In the absence of information on the assumptions according to which site sizes were estimated by Lyon, it is difficult to accord Curvers' and Lyon's estimates with each other. It seems that Lyon's estimates are consistently higher than those of Curvers, although there are some notable exceptions. A possible explanation may be that Lyon's higher estimates resulted from the lumping together of period VII material, which Curvers subdivided into three phases. Given these uncertainties, it seems most prudent to use Curvers' data, as these data are at least internally consistent.

Before site contemporaneity and population reconstructions could be carried out, a site gazetteer was compiled from Curvers 1991, listing all sites with an Early and/or Middle Bronze Age component, that is, Balikh periods VI and VII. This list was then screened for sites that may not have been in use as habitation sites. These sites were removed from the sample. For some other sites, revised size estimates are presented. The modifications and additions that were made to the original data set are described below; the resultant data set is presented in Table 5.6 and Figs. 5.4 and 5.5.[47]

Tell Bi'a (BS 1), ancient Tuttul, at the confluence of the Euphrates and the Balikh, was not surveyed in 1983, but an earlier pre-excavation survey indicated that the entire tell was littered with Ur III and Old Babylonian sherds (Strommenger 1981: 26). An important size indicator is the city wall, parts of which have been excavated at the northern, western, and southern limits of the tell (Miglus and Strommenger 2002). At least three phases have been excavated. From the earliest phase onward, the wall was rebuilt at the same location, possibly so that the older wall could serve as a foundation for the new structure. The dating of these phases is difficult, as few finds could be linked to the actual construction of the walls. The earliest material from hill E, datable to the ED period, is not found in association with the earliest phases of the wall, suggesting that the first phase dates to the late ED period. Pottery from stratigraphic contexts between the first and second phase of the wall also dates to the ED period, whereas an Akkad-period grave dug into the second phase provides a *terminus ante quem*. The available evidence therefore suggests that the oldest and middle phases were constructed late in the ED period. The third phase seems to date to the late Isin-Larsa/early Old Babylonian period. If it is assumed that the city wall partly shaped the contours of the tell, as is suggested by the excavated parts, then the wall would enclose an area of 30–35 ha. Only a fraction of this area has been excavated, with a focus on public/official

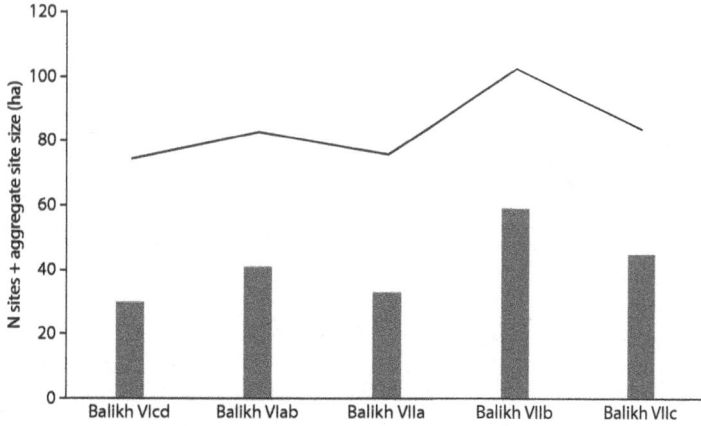

Fig. 5.4: Total number of sites (bars) and aggregate site size in hectares (line) for every BS period without correction for site contemporaneity.

47 No attempt has been made here to systematically transcribe Arabic site names according to one system. For well-known sites, the most widely used transcription is adopted, e.g. Leilan instead of Laylān. Otherwise, the names of sites, as far as known, are directly taken from the respective publications, to allow easy referencing between these publications and this study. The *APUM* project provides a commendable attempt to collect, categorize and systemize site names in northern Mesopotamia (Anastasio et al. 2004: 2–3).

site	name	VIab	VIcd	VIIa	VIIb	VIIc
1	Tell Bi'a	20	20	15	15	15
2	Tell Zaidan	7	0.5			
12					0.1	
21	Tell Mahlas	2	3.2			
27		1	1			
34				0.8	1.1	1.1
35	Merj Abu Sharib	1.3	1.1			0.1
56				0.2	0.8	0.8
66				0.2	1.5	1.5
67				0.5	0.5	0.5
73				0.6	0.6	0.6
81	Khirbet al-Bassal			0.5	1.2	1.2
83	Tell as-Saman	2.7	4.5	9	9	9.7
84		10	10	10	10	
85			0.4		0.4	
86			0.3	0.3	0.3	0.3
95	Tell ar-Rahyar	0.5	0.5			
96		0.1	0.1			
100	Tell Jaruhe	0.8	1	0.2	1	1
101	Helwet'Abid		1.4			
105					1	
106					2.1	1.5
112				0.6	0.6	0.6
113					1.1	
139	Tell Khadriya	1.4	0.4		0.5	0.5
147	Tell as-Sawwan	2	2		0.3	
152	Tell ez-Zkero			0.5	0.5	0.5
171	Khirbet al-Hajaje		0.4			
172	Khirbet el-Ambar		1			
175	Tell Hammam et-Turkman	3	3.7	6.6	6.6	6.6
180		1.5	1.5			
181		1	1			
182				0.2	0.2	
186					0.2	0.2
195				2	2	2
200				0.3	0.3	0.3
203				0.7	0.7	0.7
205				1.8	1.8	1.8
207				0.2	0.2	0.2
211	Tell Jital			4.2	4.2	4.2
212	Tell Eftaim	1.3	3.5		3	1.7
214					0.2	

Table 5.6: Size estimates (ha) for BS sites with a third millennium BC (Balikh VIa–d) or early second millennium BC (Balikh VIIa–c) component (based on Curvers 1991, with modifications).

site	name	VIab	VIcd	VIIa	VIIb	VIIc
219	Tell Qardana				0.6	0.6
235	Tell Gharair/Tell Wazgöl		0.8		0.1	
236			1			
240				0.2		
241			0.5			
242	Tell ed-Dubaghiye				0.3	
244						0.6
246		1.6	3	1.4	3	1
247	Tell Sahlan	2.3	2.3	7.5	7.5	7.5
254					0.2	
264	Tell Shrey'an	1	2.4		2	2
265		0.2			0.2	
267			0.3			
268			0.4			
270	Tell al-Abed	0.7	0.3	0.7	0.7	0.7
275	Tell Jidle	1.5	1.5		1.5	1.5
276	Tell Hammam Ibn-Shehab				0.9	
280	Tell Abyad	3	4.8	4.8	4.8	4.8
281						0.2
282			0.6		0.6	0.6
284	Tell Ksas	1	1			
297	Tell Breri	3.8	3.8		3.8	3.8
304	Maslahat Tell as-Saman		0.3	0.3	0.7	0.7
305	Khirbet al-Halawa	0.2	0.2			
308	Tell Sabi Abyad V			0.2	0.2	0.2
310					0.5	0.5
311					0.1	
316	Tell Mafraq Sluk				0.3	
317					0.2	
322			0.3		0.3	0.3
323				1.1		
324		1	1	1	1	
326			0.3			
327	Tell Abbara			1	1	1
328					0.1	0.1
330					1.4	1.4
331				1.4	1.4	1.4
334		0.6	0.6		0.6	0.6
337				1.6	1.6	1.6
340		0.5				
342		1.3				

Table 5.6: Continued.

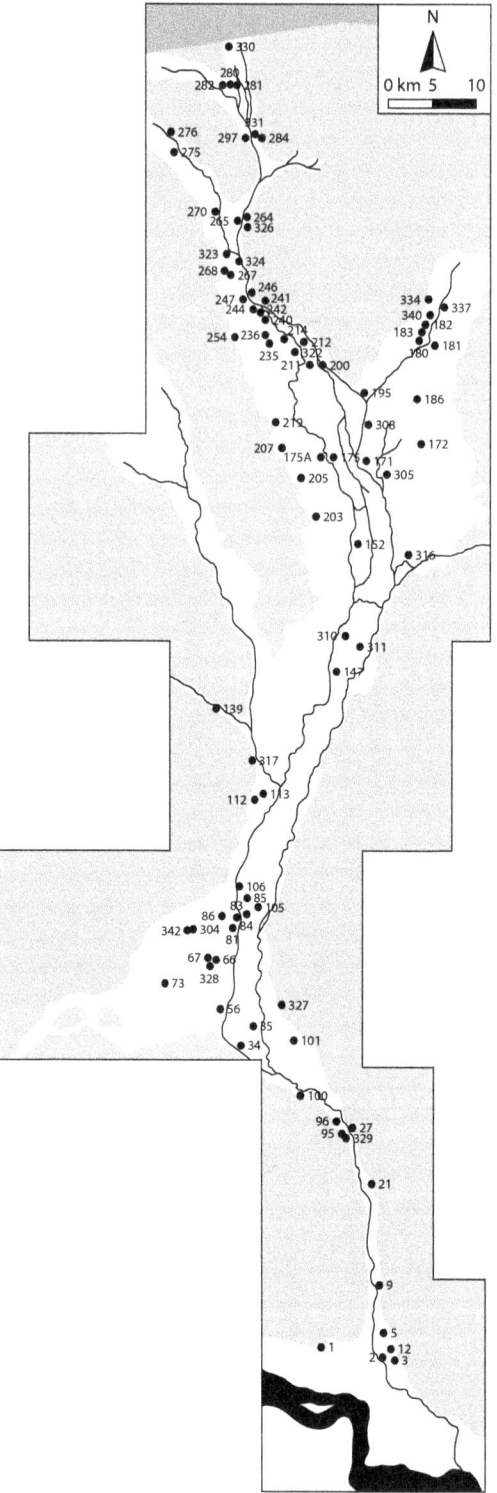

Fig. 5.5: Map of the BS region indicating all sites with a Balikh VIab, VIcd, VIIa, VIIb, or VIIc occupation.

buildings, complicating the estimation of built-up area within the city wall. Based on the dating of the various palaces and temples, it seems reasonable to assume that Tell Bi'a was largest during the mid-third millennium BC, that is, late Balikh VIab. Comparison of Tell Bi'a with urban sites upstream of the Euphrates, notably in the Tabqa/Tishreen Dam areas (Cooper 2006: table 3.1), suggests that the maximum size of Tell Bi'a may have been 15–30 ha. Therefore, a conservative maximum size of 20 ha is estimated for the Balikh VIab and VIcd periods. The Balikh VIIa–c occupation is estimated at 15 ha, reflecting both the repeated destructions of palace A by Yahdun-Lim and Zimri-Lim, and the continuing importance of Tell Bi'a as a regional capital during the early second millennium.

Tell Zaidan (BS 2) was surveyed by Kohlmeyer (1984: 108), the 1983 Balikh Survey (Curvers 1991: 183), and Córdoba (1988: 156). The major occupation of this site probably fell in the Halaf and Ubaid periods. Whereas Kohlmeyer and Córdoba suggested a possible early EB occupation, Curvers assumed full Balikh VIab occupation and a minor Balikh VIcd occupation. A possible early EB burial ground has been identified near the site (Kohlmeyer 1984: 108). Copeland (1979: fig. 15) reported a possible Khabur ware sherd, but as none of the other surveyors found this pottery, it may have been either a stray object or a misidentification. Based on the drawing published by Kohlmeyer (1984: fig. 4), Zaidan can be estimated at *c.* 10 ha, which will here be translated in size estimates of 7 and 0.5 ha for the Balikh VIab and Balikh VIcd periods, respectively.

Strommenger (1981: 34) mentioned an unnamed site northwest of Tell Bi'a, where Old Babylonian sherds were found. However, the site is not indicated on the map of the surroundings of Tell Bi'a published by Schirmer (1987: figs. 1, 3). Curvers mentioned this site as BS xx, but it was apparently not visited by the 1983 Balikh Survey (1991: 183). As no location or any other details are known for this site, it will not be included in this study.

The BS recorded several sites that are interpreted as burial grounds rather than settlements on the basis of the presence of disturbed graves and their location on the edge of the valley. All supposed burial grounds date to the Balikh VI period, that is, the third millennium BC, and include BS 3, BS 5, BS 9, BS 105, BS 106, BS 112, BS 175A, and BS 329. Extramural burial grounds for this period are known from the Upper Syrian Euphrates area (Akkermans and Schwartz 2003: 251–3). Meijer (2007: 319) noted the conspicuous absence of EB burials at Hammam et-Turkman, suggesting that the dead may have been buried outside the settlement. It has been proposed that under certain circumstances extramural burial grounds may be used as a substitute for otherwise invisible settlements (Renfrew 1972: but see Cherry 1979). The vicinity of the Balikh Valley burial grounds to known contemporary settlements suggests, however, that it is unnecessary to add sites to the data set on the basis of the reported burial grounds.[48]

The archaeological nature of some sites identified by the BS was questioned by the Western Jazira Archaeological Landscape Project. BS 221 consisted of pale grayish brown fine sandy silts with very few sherds, suggesting that this site represented the result of digging activities rather than a true habitation site. The same was suggested for BS 243 where virtually no pottery was found (Wilkinson n.d.). Therefore, these two sites were removed from the sample.

Further modifications of the original Bronze Age data set include the removal from the sample of the Balikh VII sites Khirbet esh-Shenef (BS 170) and Tell Sabi Abyad I and III (BS 189 and BS 191, respectively), as excavations suggested that Balikh VII was not present at these sites, although Balikh VIII was (Akkermans 1993).

5.2.3 The North Jazira Survey

The area consists of east–west orientated anticlinal ridges, of which the Jebel Sinjar is the largest. The entire area slopes, resulting in runoff from rainfall and subsequent wadi formation once the soil is saturated. These wadis are deeply incised and severely restrict movement, as does the moderately incised Tigris River to the north and east of the study area. To the north, the area becomes hillier towards the Taurus/Zagros Mountains. Modern climate is a variant of the warm temperate semi-arid climate. The long-term mean annual rainfall for the north Jezirah, based on records from Mosul, is estimated at 370 mm. On average, there is a drought every ten years, resulting in less than 250 mm annual rainfall. The distribution of rainfall across the year, however, allows farmers to

---

48   BS 3, 5 and 9 may be associated with either Tell Zaidan (BS 2) or Tell Bi'a (BS 1), BS 105 and 106 with Tell as-Saman (BS 83), BS 175A with Hammam et-Turkman (BS 175), and BS 329 with Tell Mahlas (BS 21). Only BS 112 seems to have no nearby settlement counterpart although the site could possibly belong to Tell as-Saman (BS 83). Alternatively, the burial grounds may have been used by a (semi-)nomadic population of the region.

site	RJP	name	Ninevite 5	Late third millennium	Khabur
1		Tell al-Hawa	24	66	66
5			0.4		
9		Tell Kuran	4.2		6.5
10			2.2		4.5
11			0.7		
12			1.5		
13			2		0.1
14			3.8		3.8
15		Mowasha	1.3		1.3
18			1.3		1.3
19	21	Tell al-Butha 1	6		6
20			2.8	0.1	2.8
22					1.6
23		Tell Wardan	4.2		4.6
26	25		2.6		
28			0.5		
29	19		7.2	7.2	8.3
30			1.9		3.4
37			0.1		0.2
39		Khirbet Gar Sur	1.5		
40				0.1	
42	28	South Garsur	0.1	0.1	5.9
43		Kharaba Tibn	5.5	17.2	18.3
45	26				2
48			2.1	3	6.1
49			1.9		
50			1.4		
52				0.1	
54				0.1	
55			0.1		
57				0.1	
58		Tell Warada	5	0.4	0.1
60			1.6		
71		Tell Hamide	0.5	1	5.6
72			0.1		
73					1.6
74					2
76				0.5	
79			0.1		

Table 5.7: Size estimates (ha) for NJS sites with a third–early second millennium BC component (based on Wilkinson and Tucker 1995 and Ball et al. 1989 (Tell al-Hawa), with modifications). Second column shows concordance with RJP site numbers as listed by the Jazirah Salvage Project (Altaweel 2007: table 1).

site	RJP	name	Ninevite 5	Late third millennium	Khabur
80			2.8		
83	17				0.1
86		Tell al-Hilwa	1		
87			2.8		
90		Tell Abu Hajira	6.1	6.1	6.1
91		Tell Talab	5	7.5	10
92		Tell Uwaynat	1.5	1.5	3
93		Tell al-Samir	10	19	19
96		Tell Ghubain			1
99					1
108					0.5
115	23	Tell Kiber 2			1
121					1
123		al-Kibar			1.8
126					1
127		Abu Kula	5	10	10
131				0.9	0.9
132					1.1
136					2.1
138					0.5
140		Tell Mana'a		1	8
152				0.5	
154					0.9
155		Tell Hayal		2	2.4
159	2				0.7
160					1.6
169					2.2
175				1	1
177				1	

Table 5.7: Continued

predict at a relatively early stage whether there will be sufficient rain for crops. Should precipitation appear insufficient, they can take precautions such as switching failing agricultural land to pasture, or selling herd animals to buy or trade other foodstuffs (Wilkinson and Tucker 1995).

The North Jazira Survey (NJS) (Fig. 5.1S) was initiated as a side-project of the Tell al-Hawa excavation, and focused on the immediate hinterland of this large site. The shape of the survey area was not dictated by archaeological or landscape considerations, but by conservation issues, because the area was part of a planned large-scale irrigation project. During four seasons, 184 sites were recorded, several of which were excavated. Site detection was performed by car and on foot along transects spaced 500 m apart. This allowed the recognition of sites as small as 100 m in diameter and 1 m high. Pedestrian survey to collect sherds was carried out in selected off-site areas. This strategy was checked against available aerial and satellite images, indicating that only a few

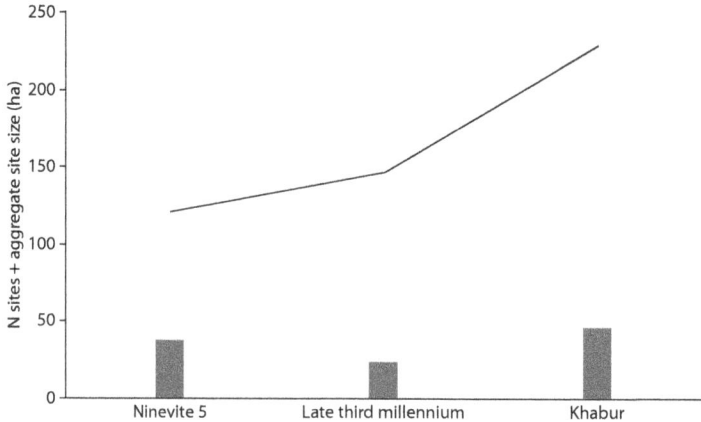

Fig. 5.6: Total number of sites (bars) and aggregate site size in hectares (line) for every NJS period without correction for site contemporaneity.

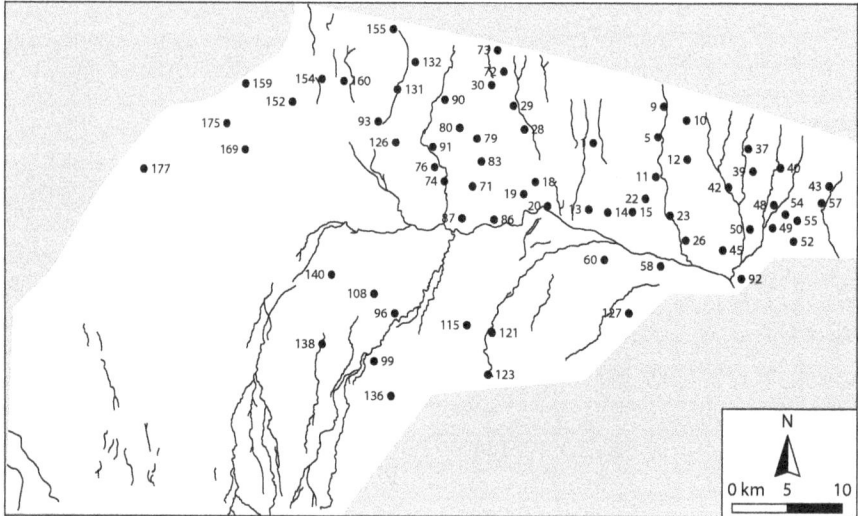

Fig. 5.7: Map of the NJS region indicating all sites with a Ninevite 5, Late third millennium, or Khabur occupation (after Wilkinson and Tucker 1995).

sites were missed during the actual survey. Large sites were divided in sampling sectors in order to detect differences in settlement. These units were either individual mounds, or topographic units (Wilkinson and Tucker 1995).

Site size estimates are taken from Appendix C: Site catalogue in Wilkinson and Tucker 1995. Where site size is listed as 'trace' in the primary catalogue (Wilkinson and Tucker 1995: 125), it is here interpreted as an occupation of 0.1 ha. The largest site in the NJS area, Tell al-Hawa, has been the subject of regular excavations (Ball et al. 1989), and the results are incorporated in the site's size estimate. Some of the other sites surveyed by the NJS have been sounded. The results of these operations, of which only few details are available, seem to have been included in the site size estimates made in Wilkinson and Tucker 1995. The results of Iraqi excavations within the NJS area are now being published (Altaweel 2006, 2007), and relevant results will be discussed below. The resulting dataset is presented in Table 5.7 and Figs. 5.6 and 5.7.

As part of the Jazirah Salvage Project, a number of sites within the NJS area have been sounded by Iraqi archaeologists.[49] Excavations at Tell al-Butha 1 (NJS 19/RJP 21) seem to confirm the results of the NJS for the third and early second millennium BC. The sounding at South Garsur

---

49 In the first publication on these data, the project was called the Ray Jazirah Project, abbreviated as RJP (Altaweel 2006). In the second publication, the name for the project was corrected to Jazirah Salvage Project, but the abbreviation RJP was retained in the interest of maintaining consistency between publications (Altaweel 2007: 117).

(NJS 42/RJP 28) consisted of a single 10x10 m square at the top of the mound. Six levels were identified, of which the lower four were dated to the Khabur period. In each of these levels, walls were found, but no conclusions as to the size of the building, or the size of the site as a whole, could be drawn from these results due to the limited exposure. No changes have been made to the size estimate based on the results of the NJS. At Tell Kiber 2 (NJS 115/RJP 23), Iraqi archaeologists carried out a surface survey, and excavated four 6x6 m squares. The results of the surface survey confirmed those of the NJS. No excavated squares yielded only storage pits, suggesting that the site may have been seasonally occupied during some of the later occupation periods. Whether this was also the case for the Khabur period is unclear, as the excavated squares were not located in the area with the highest density of Khabur ware on the surface. Given the fact that the site as a whole measured 4.5 ha, the original estimate of 1 ha for the Khabur period seems conservative and will be used here. Two long but shallow trenches were excavated in Tell Hujiera 2 (NJS 130/RJP 18). Architectural remains were only encountered in one of these trenches, due to significant disturbance at the site. Some possible Khabur ware sherds were reported (Altaweel 2007: 122), but they did not come from well-defined stratigraphical layers. The NJS found no Khabur ware at this site, and it is therefore questionable whether the site was ever occupied during this period. As a result, it will not be included in the present analysis.

5.3 Reconstructing regional settlement trends

The data presented above can be used to reconstruct the diachronic population development for each survey region by applying the simulation method that was presented in Section 4.4. Step-by-step application of this method to the BS, B-EDS, and NJS data sets will be discussed here. In order to estimate the amount of simultaneously occupied hectares per phase, it is necessary to distinguish between different site size classes (see Section 4.4.3). These size classes are derived from the combined BS, B-EDS, and NJS data sets (Tables 5.3, 5.6, and 5.7). The distribution of site sizes for this combined data set is shown in Fig. 5.8. Visual inspection of the distribution suggests

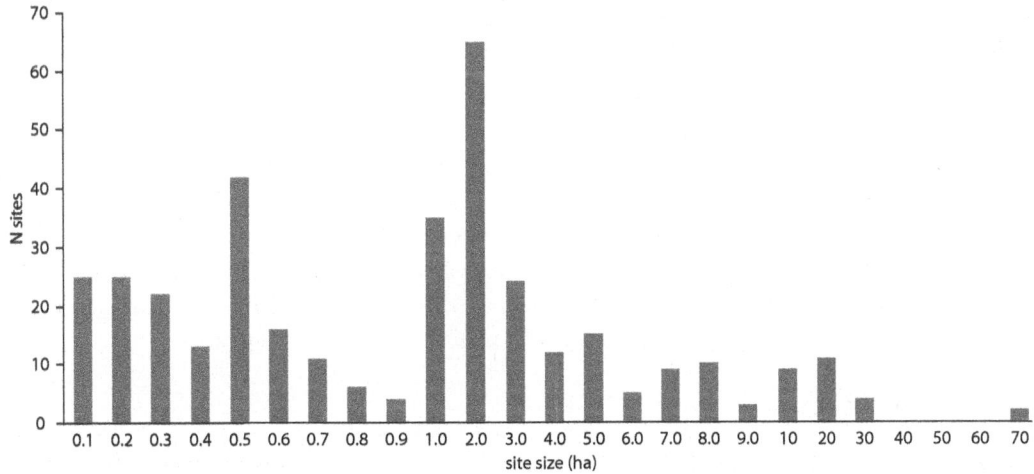

Fig. 5.8: Distribution of site sizes (ha) across all periods for the combined BS, B-EDS, and NJS data sets (*n*=368).

site size class (ha)	*n* sites	mean site size (ha)	standard deviation
0.1–0.3	72	0.2	0.08
0.4–0.9	92	0.6	0.13
1–6	156	2.2	1.3
7–9	22	7.3	0.9
10–30	24	16	6.3
30+	2	individual	0

Table 5.8: Site size classes, number of sites per class, and mean size per class in hectares (*n*=368).

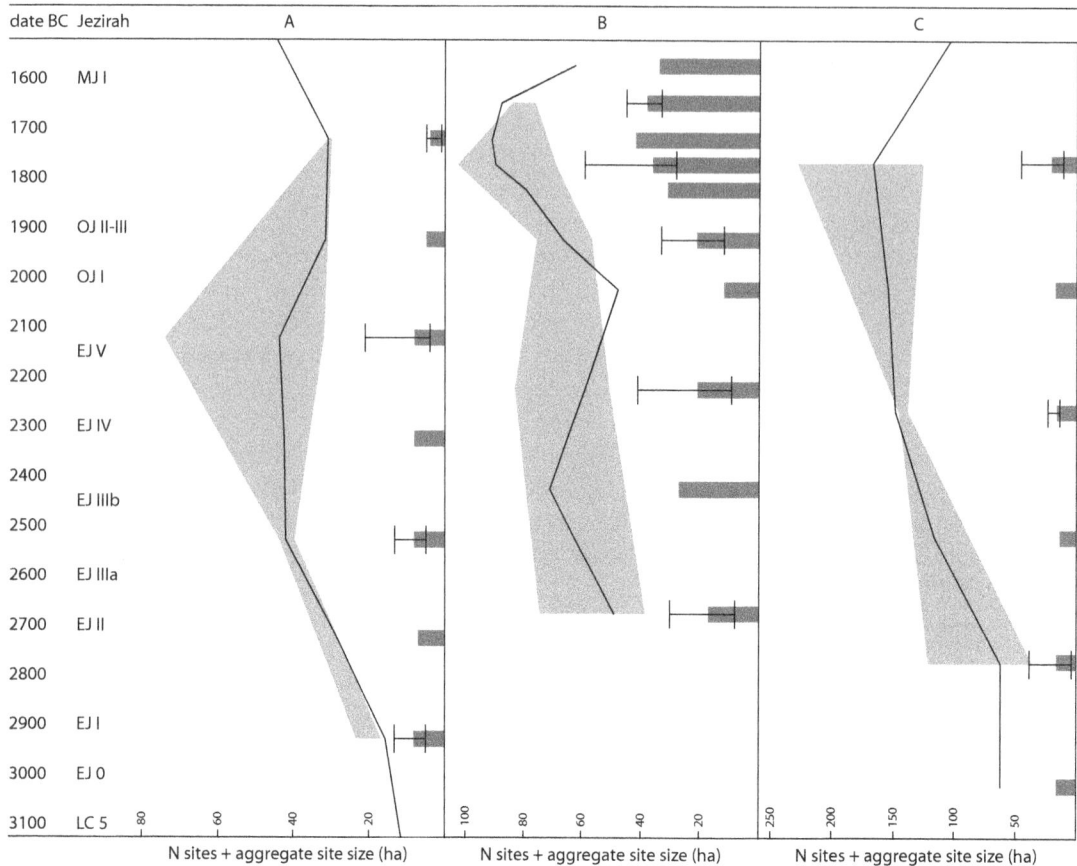

Fig. 5.9: Settlement trends through time. Solid lines indicate the corrected aggregate site size, shaded areas indicate the possible range of aggregate site size, bars indicate the corrected number of contemporaneously occupied sites, and error ranges indicate the possible range of contemporaneously occupied sites. (A) B-EDS region. (B) BS region. (C) NJS region.

the following classes: 0.1–0.3 ha, 0.4–0.9 ha, 1–6 ha, 7–9 ha, 10–30 ha, and 30+ ha. The mean site size for each class is shown in Table 5.8. Within each survey region and within each chronological phase, each site is assigned to one of these classes, and the simulation method is independently performed on each of these classes (see Section 4.4.3). By adding up the site counts and site size estimates per class, one arrives at regional site counts and size estimates for each chronological phase (Fig. 5.9). This figure also displays the potential minimum number of simultaneously occupied sites, that is the number of type *b* sites, and the potential maximum number of simultaneously occupied sites, that is all sites with evidence for occupation within a given phase.

5.3.1 Demographic trends

The total amount of occupied hectares can then be used to reconstruct a diachronic population development, employing the on-site population density value of 150–250 persons/ha, as determined in Section 4.5 (Fig. 5.10). Fig. 5.10 is presented on an absolute timescale rather than per phase to allow comparison of these demographic developments with the climate developments presented in Section 2.3.7.

Between 2900 and 2500 BC, the Birecik-Euphrates area experienced a relatively high demographic growth, followed by a period of stability until 2100 BC (Figs. 5.9A, 5.10A). At that time, population peaked around 6500–8800 persons. After this stable period, the population of the valley declined rapidly to 6200–7800 persons, with only a few small settlements seeing occupation during the Late MBA period. The second population explosion during the second quarter of the second millennium was largely due to the growth of Carchemish. The early third millennium and the middle second millennium population rises were relatively small when expressed in absolute numbers. For the region itself, the population growth may be quite marked, but when compared

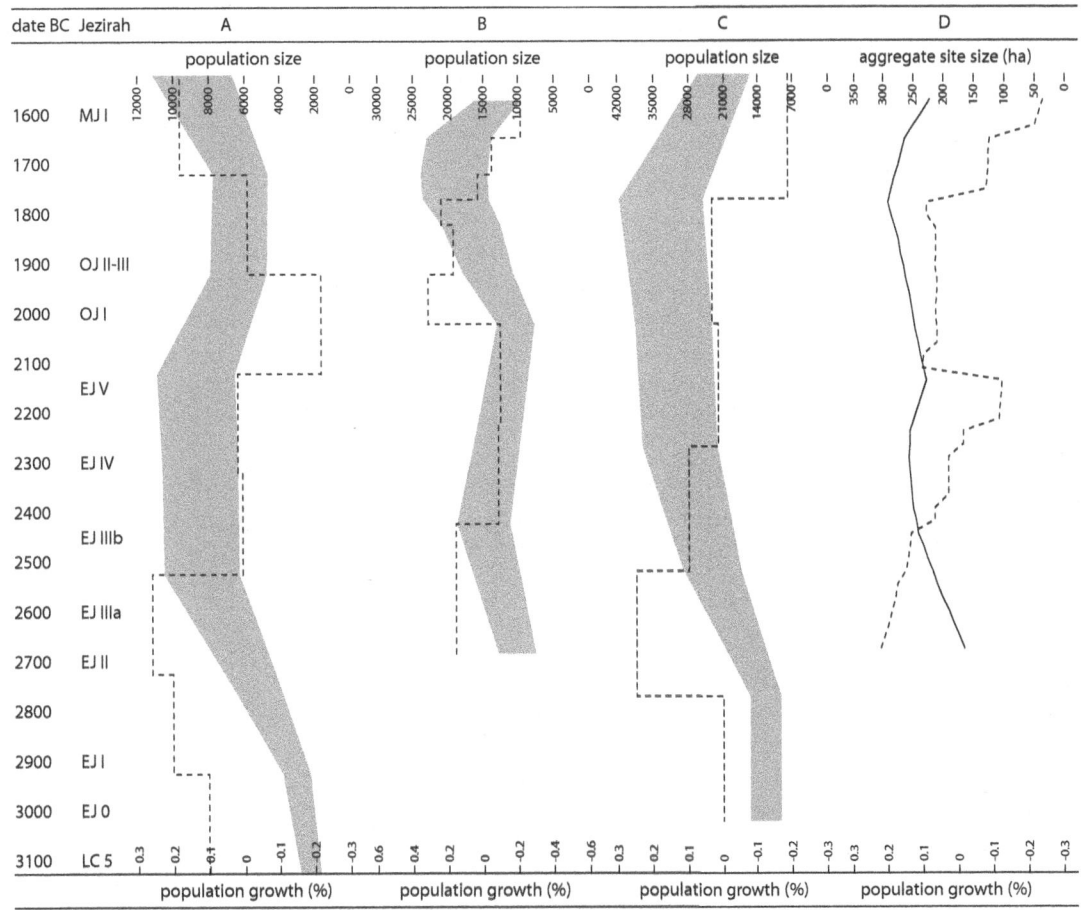

Fig. 5.10: Reconstructed population levels and population growth through time. Shaded area indicates the population size under a population density of 150–250 persons/ha. Dashed line indicates population growth rates through time, based on corrected site counts and site size estimates. Solid line indicates the combined aggregate site size. (A) B-EDS region. (B) BS region. (C) NJS region. (D) The B-EDS, BS and NJS regions combined.

to the developments in the other regions, it is relatively insignificant. This observation may call into question the idea that demographic crashes, such as noted for the LRS or the BS, were compensated by synchronous increases elsewhere (Wilkinson 1999: 53). Apart from the fact that the marked settlement peak that was evident in the reconstruction by Algaze et al. is now much more fluid, it also does not seem to be exactly in phase with the developments in the Balikh Valley. Inspection of Fig. 5.9A furthermore suggests that the increase in settled hectares toward 2100 BC resulted primarily from the growth of already existing settlements. If new people arrived in the region, it might be expected that they founded new settlements, but this seems not to have been the case in this area.

For the Balikh Valley, the following demographic trend can be reconstructed (Figs. 5.9B, 5.10B). From the middle VIab period onwards, and presumably already starting in the Late Chalcolithic, there was a gradual rise in both the number of sites, and the total amount of hectares that was occupied at any moment in time. This results in a rather steep population growth curve for the first half of the third millennium BC. The total population of the Balikh Valley peaked around 2400 BC, when 11000–19000 people lived in the recorded settlements. The Balikh VIcd period witnessed a steadily declining population, with a minimum of 7700–13000 people at the VIcd–VIIa transition. Note that, in tandem with the population decline, the number of occupied sites declined as well, though mean site size increased markedly to almost 4 ha. After this decline there followed a second period of population growth to a level exceeding that of the third millennium. The maximum population of the Early and Middle Bronze Age was reached at the end of the Balikh VIIb period when 14000–24000 persons lived in the Balikh Valley. Period VIIc was, like period VIcd, a time of continuous population decline. This trend continued well into Balikh period VIII and was only halted and reversed during the Middle Assyrian period (Lyon 2000).

In the NJS region, population decreased slightly during the early Ninevite 5 period to a minimum of 9300–15000 people (Figs. 5.9C, 5.10C). This absolute minimum provided the basis for a continuous growth lasting some one thousand years and covering much of the third millennium and the early part of the second millennium. During the Khabur period, an estimated peak population of 25000–42000 persons lived in the settlements of the NJS region. During this period, the large central site of Tell al-Hawa accounted for almost 40 % of the total population living in the area. During the second quarter of the second millennium BC population dwindled, but remained above the low level recorded for the Ninevite 5 period. In this respect, the NJS population curve differs from the BS and the B-EDS curves, which show decreases at the third–second millennium transition. For this period, the NJS growth curve levels off slightly, but there is no evidence for any population decline during this period.

Several researchers have used reconstructed population growth rates to detect periods of anomalous growth and/or decline and to detect periods of population displacement (Wilkinson 2000a; Ristvet 2005: 198–9). The percentual rate of population growth $r_{\%}$ is calculated as (Hassan 1981: 139):

$$(3) \quad r_{\%} = (\ln (N_2 / N_1) / t) \cdot 100$$

where N_2 is the population reached after t years from an initial population N_1. The results of this calculation are presented in Fig. 5.10A–C. It is immediately apparent that population growth fluctuates widely, ranging from very high growth rates to rapid declines. Average growth rates for past societies are estimated at 0.04–0.16 % by Hassan (1981) and at no higher than 0.1 % and 0.13 % by Maisels (1990: 126) and Carneiro and Hilse (1966: 179), respectively. In the surveys analysed here, periods with an average growth rate are rare. These periods include the late VIab, late VIIa, and late VIIb periods of the BS region, the early part of the Middle EBA period of the B-EDS region, and the entire Late third millennium and early part of the Early second millennium periods of the NJS region.

However, closer inspection of the population growth rates suggests that there seems to be a positive correlation between demographic stability and population size. In other words, the region with the largest population, that is, the NJS region, seems to have experienced the highest stability in terms of population growth and decline, and the population growth rate remains closest to the long-term average growth rate. When the population developments in each region are combined, the population growth rate becomes even more stable and generally falls within the assumed range of natural growth (Fig. 5.10D). There was one minor decrease around 2000 BC, and a significant decline only set in around 1700 BC.[50]

At first sight, the observation that the combined demographic graph shows a more stable population development than the regional graphs could indeed lend support to the theory that local population deficits in one region were offset by incoming populations from elsewhere, which would result in the maintenance of a long-term demographic balance (Wilkinson 2000a: 242). However, it should be realized that net population change, as reflected in Fig. 5.10, is the difference between the number of births per year and the number of mortalities per year, which were probably quite high. Relatively minor changes in either the fertility or mortality rate can therefore potentially completely transform the net population growth rate. Such changes can result from changes in inter-birth spacing, child mortality, average age of first pregnancy, and so forth (Hassan 1981: 126). If the birth rate dropped below the mortality rate, the population would decrease without any external interference (Cowgill 1975: 512). This suggests that local demographic trends may have been largely driven by local developments and changes in fertility and mortality rates and that there were no cross-regional processes at play.

---

50   Fig. 5.10D only displays the combined trends for the period for which data from all three surveys is available, i.e. 2650–1550 BC.

5.3.2 Rank-size graphs

Rank-size plots for the three survey regions were constructed according to the method outlined in Section 4.6 (Figs. 5.11–13). Because the exact number of settlements, and more importantly their sizes, is only known for the transition points between periods (see Section 4.4.3), rank-size plots have only been constructed for these specific points in time.

The rank-size plots of the B-EDS settlement system are constructed on the basis of very few sites, making any observations made from them highly suspicious (Fig. 5.11). The primate distribution that is observed for all periods, however, probably indicates that the local settlements always belonged to the *hinterland* of a larger site. The 2200 BC rank-size distribution is slightly reminiscent of a primo-convex distribution (Fig. 5.11B). At least from the mid-third millennium, this primate site was most probably Carchemish.

During the early third millennium, the settlement pattern of the BS region conforms almost perfectly to the log-normal distribution (Fig. 5.12A). This log-normality decreases in favour of a more convex distribution, reaching highest convexity at *c.* 1800 BC (Fig. 5.12B, C). After that point, convexity seems to decrease again (Fig. 5.12D, E), but there remain large error margins, making any more specific observations impossible. In fact, apart from the observation that the 1800 BC and 1700 BC settlement patterns were very probably convex, no other observations can be made with any measure of confidence.

The NJS rank-size plots are again constructed from a reasonably large data set, making conclusions drawn from them more certain (Fig. 5.13). At the beginning of the third millennium, there was a slightly primate or possible primo-convex settlement pattern (Fig. 5.13A). Although the large error margins make any observations problematic, there seems to be a very slight development towards a more log-normal distribution during the late third and early second millennium (Fig. 5.13B–D). The overlapping error ranges for the *A* values of different periods again make any more detailed interpretation hazardous. As expected, the rank-size plots of the NJS region compare reasonably well with those produced by Wilkinson and Tucker (1995: fig. 52) for the uncorrected settlement record of the third and early second millennium BC.

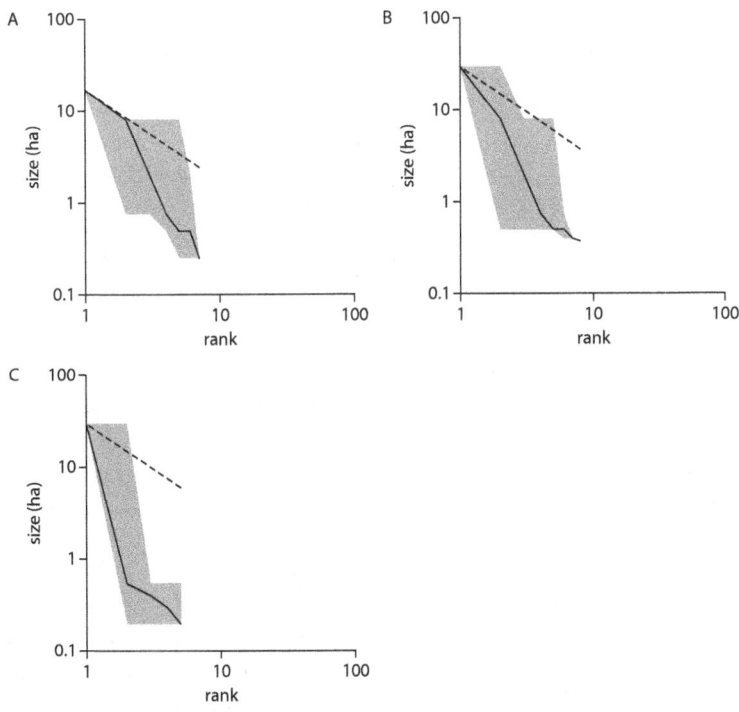

Fig. 5.11: Rank-size plots for the B-EDS region at (A) 2500 BC, (B) 2200 BC, and (C) 1900 BC, based on corrected site counts and site size estimates. Dashed line is log-normal distribution. Grey zone represents 90 % confidence range.

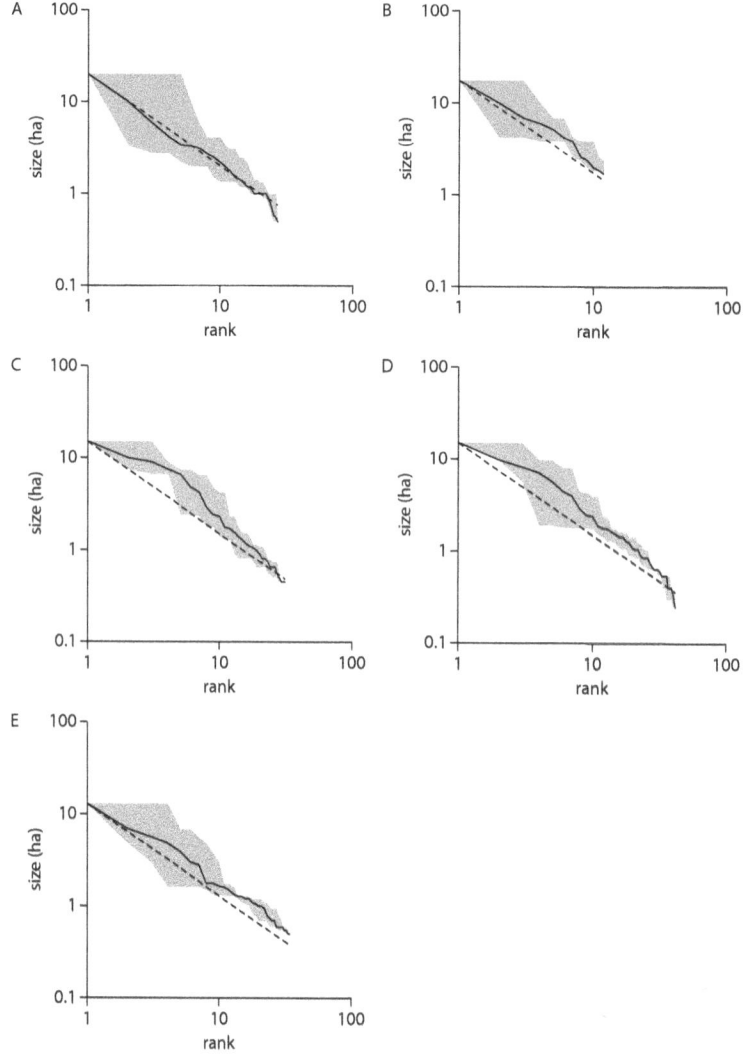

Fig. 5.12: Rank-size plots for the BS region at (A) 2400 BC, (B) 2000 BC, (C) 1800 BC, (D) 1700 BC, and (E) 1550 BC, based on corrected site counts and site size estimates. Dashed line is log-normal distribution. Grey zone represents 90 % confidence range.

As argued in Section 4.6.1, reliable rank-size plots can only be constructed if the limits of the settlement system on which the analysis is performed are correctly identified. Failure to do so may result in a convex distribution because more than one system is analysed, whereas a primate distribution is likely to emerge when only part of a larger settlement system is included in the analysis (Johnson 1977: 498). Comparison of the B-EDS, BS, and NJS suggests that the BS region is least likely to have suffered from these effects, and the NJS most likely. Although the BS focused on the river valley, the survey's western and eastern boundaries were the steppe area where no more sites were encountered (Akkermans 1993: 139). This boundary may have equally applied to the south, where Tell Bi'a and its immediate surroundings, as well as the Euphrates, may have represented a natural boundary prohibiting settlement further south. Only the Syrian–Turkish boundary represents a truly artificial boundary to the survey area (but see Akkermans 1993: 139). The B-EDS was bordered to the south by the Turkish–Syrian border, to the north by Halfeti, and up into the highlands by an artificial line drawn at 400 m above sea level. In fact, the B-EDS covers the northern half of the so-called Carchemish sector (Peltenburg 2007a: 6–7). Finally, the borders of the NJS were dictated by the fact that the area was to become part of a large-scale irrigation project, and therefore apparently did not fully encompass any natural archaeological or landscape zone. The rather arbitrary nature of the borders of the B-EDS and NJS regions make the results of the rank-size analyses for these regions suspect, and any possible effects will be evaluated in Sections 8.2.1 and 8.2.3.

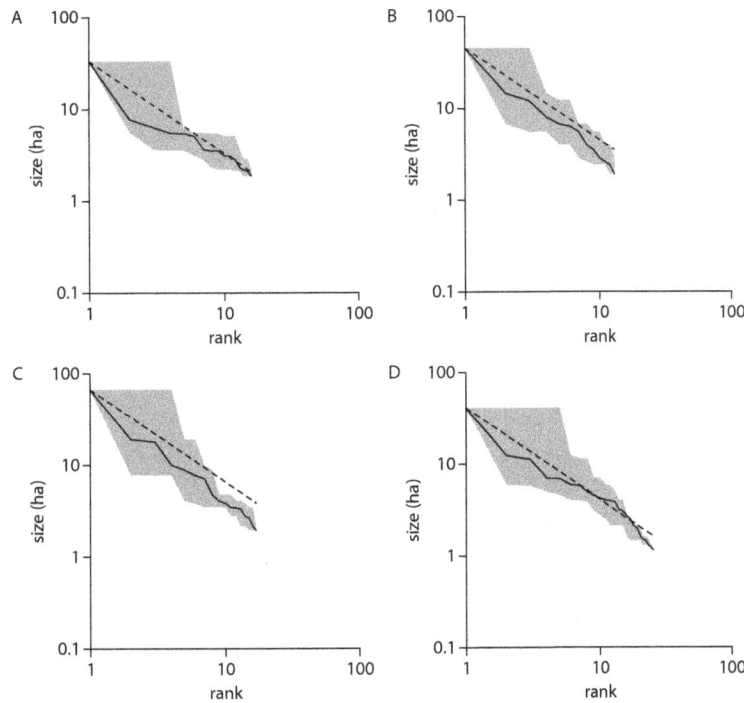

Fig. 5.13: Rank-size plots for the NJS region at (A) 3000 BC, (B) 2500 BC, (C) 2000 BC, and (D) 1500 BC, based on corrected site counts and site size estimates. Dashed line is log-normal distribution. Grey zone represents 90 % confidence range.

5.4 Contextualizing the reconstructions

Section 5.4.1 discusses the evidence from other surveys in order to place the present reconstructions in perspective. The nature of the primary data does not allow the reconstruction of continuous population trends but permits the detection of very general trends, besides touching on issues of scaling and settlement patterning across the landscape. Section 5.4.2 discusses contemporary historical texts that provide information on settlement sizes and organization.

5.4.1 The results from other surveys

Apart from the surveys that have been analyzed here, numerous other surveys have been conducted in the Jezirah (Fig. 5.1). However, many of these surveys have not been published the level of detail necessary for the reconstruction that is attempted here. As a result, the conclusions drawn from these surveys are often more general. Nevertheless, their results can be used to point out large-scale trends and contextualize the current reconstruction. In other words, these surveys can help determine the degree to which the trends identified here are typical for northern Mesopotamia as a whole. The uncorrected settlement trends documented by these surveys are presented at the end of this section in Fig. 5.14.

The Syrian Upper Euphrates settlement system, comprising the extensively researched areas of the Tabqa and Tishrin Dams, has been evaluated by several researchers (Peltenburg 2000; Wilkinson 2004; Cooper 2006; Peltenburg 2007c; Schwartz 2007) (Fig. 5.1B, C). In a sense, it is a continuation of the B-EDS region described above, which has only relatively recently become separated by political borders. This political situation perhaps influenced archaeological research in both areas. Thus, a recent volume on the Upper Euphrates stopped at Carchemish (Cooper 2006), and the present study also uses the large site on the border between Turkey and Syria as an artificial border between two zones that are reviewed in a different context. On the other hand, the intensity of research in both areas is comparable, as the research in both cases derived from dam construction resulting in the permanent loss of cultural heritage. Furthermore, Peltenburg has suggested that the area around Carchemish indeed represented a transitional environmental zone, which may have influenced cultural traditions as well (2007a: 8–9). This gap between two different constructed culture areas is only recently being bridged with the publication of a volume on Carchemish, which focuses on its unique cultural traits and the contacts with both the north and the south (Peltenburg 2007c). The results of the surveys in the Tabqa Dam area were compiled by Reichel (2004), who also included data from the numerous excavations in the region. Wilkinson (2004: table 7.1) estimated an aggregate occupation size of 80–90 ha for the combined Hadidi–Sweyhat settlement system during the middle and late EBA, which was a considerably larger area than that of the early EBA, due mainly to the growth of the two regional centres Tell Hadidi and Tell es-Sweyhat. During the early second millennium, large parts of the Sweyhat plain were abandoned, including Tell es-Sweyhat itself. Schwartz (2007) also documented the apparent wide-spread abandonment of the Tabqa and Tishrim Dam areas, with many sites being at least temporarily abandoned. At other sites, there was evidence for destruction, directly followed by re-occupation (Schwartz 2007: figs. 4, 5).

The first season of the Land of Carchemish survey was carried out in 2006 (Fig. 5.1D).[51] The survey targeted the triangle formed by the Sajur River, the Euphrates, and the Syrian–Turkish border. In 2006, the western part of this area was investigated. Although parts of this triangle had been surveyed before, there still was an incomplete and unbalanced settlement record focusing on tell sites along the Euphrates, and this survey aimed to correct this bias. During the first season, twenty-three sites, typologically divided in small, flat sites, tell-sites, and hilltop sites, were recorded. During the third and early second millennium BC, tells dominated the settlement record. Two sites contained Early EBA material, six sites yielded Middle/Late EBA material, and MBA material was also found at six sites. Four out of six sites with M/L EBA material also had MBA material, suggesting a prolonged settlement preference for these locations (Wilkinson et al. 2007b).

In 1991–2, the Westgazira survey (Fig. 5.1E) targeted the northern half of the upland area between the Euphrates and the Balikh, up to the Syrian–Turkish border.[52] Most sites were located in the Saruj plain in the northwest corner of the survey area, and along the Wadi Qaramogh. EB ceramics were found at numerous sites, including the large sites of Tall Kufaifa, Tall Matin, and Boz Höyük. At Tall Kufaifa and Tall Matin, stone city walls were observed. At Tall Matin, the stone layout of a temple *in antis* was recorded, suggesting a mid–late third millennium BC date for this site, and possibly for the city wall as well. The area also seems to have been relatively densely settled during the MB period, but only a few details on this period are given (Einwag 1993).

The Balikh-Euphrates Uplands survey was carried out in 1996 and 1997 and recorded twenty-eight sites in this area, dating from the early third millennium BC to the early twentieth century AD (Fig. 5.1F).[53] The majority of these sites dated to the Early Bronze Age or to the Roman to early Islamic period. Four sites were dated to the third millennium, these being Tell al-Hassan/Tell Jedi (site 14), Khirbet Taha (site 17), Joub esh-Shayeer (site 22), and Tell Shayeer (site 23). Khirbet Taha was a 4 m high, *c.* 3 ha mound with clear evidence for early third millennium BC occupation. Tell al-Hassan/Tell Jedi was located along a wadi and estimated at 5–10 ha. Stone foundations were dated to the mid to late third millennium BC. Apparently, this was the only period of occupation of this site, although a transitional EB/MB phase may also have been present. Joub esh-Shayeer was also dated to the mid–late third millennium BC, and was again located along a wadi. The site consisted of several small 0.25 ha mounds. Tell Shayeer was dated to the same period. This site consisted of a 5 m high, 2–3 ha conical mound surrounded by a lower tell. In both areas, mid–late third millennium BC material was abundantly present. In so far as any conclusions can be drawn from these four sites, it seems that occupation in this region was more intensive during the mid–late third mil-

---

51 The survey employed a combination of survey by car, transect walking, and recording of sedimentary profiles. Emphasis was placed on locations that were difficult to access by car, i.e. hill tops and wadis.

52 The survey was carried out in three short seasons by a very small group operating from Raqqa. Sites were mapped, and diagnostic ceramics were collected. Only tell-sites seem to have been visited. Several sites along the Wadi Qaramogh were also visited (Einwag 1993).

53 Sites were identified on maps and aerial/satellite imagery. The area between the Euphrates and the Balikh, bordered on the north by the region already investigated by Einwag (1993), was then divided in three different environmental zones, and within these zones, a sample of 5x5 km squares was investigated. Actual fieldwork involved both intensive field walking and driving around by car. In the end, it turned out to be impossible to achieve these objectives, and only squares within two out of three environmental zones were investigated (Danti 2006–7).

lennium than during the early part of the Early Bronze Age, both in terms of the number of sites and their sizes. Sites were preferably located along wadis, indicating a concern for water supplies (Danti 2006–7).

The Euphrates Valley around Tell Bi'a was surveyed by Kohlmeyer in 1983 (Fig. 5.1G).[54] Sites were recorded on both sides of the Euphrates. There were very few large, continuously occupied tell-sites in this area of the valley. This absence may have been caused by the need to occasionally relocate settlements to new locations, as dictated by shifts in the course of the Euphrates. The team found many one-period sites, suggesting that many more have been covered by Euphrates sediments. The EB period may have been among the most densely occupied periods in this area. The large EB site of Tell Aswad was located 15 km southeast of Tell Bi'a. This site had a diameter of *c.* 550 m, and thick ash deposits were found in the sections of a canal cut through the site. A site of similar size was found 12 km west of Tell Bi'a. There are several sites with evidence of stone city walls, especially toward the narrow gap of Halabiye–Zalabiye. A few EB sites show occupation continuity into the second millennium (Kohlmeyer 1984).

A survey of the Harran plain has indicated that the area was intensively occupied during many prehistoric as well as historic periods (Fig. 5.1I). Important sites in this area include Harran itself, occupied from at least the late third millennium BC, as well as Kazane Höyük. Given the proximity to the Balikh Valley, an occupation pattern similar to that recorded there might be expected, but this suggestion remains purely speculative. Unfortunately, apart from a preliminary report (Yardimci 1993), no details of this survey have as yet been published.

The area of the Jebel Bishri up to the Euphrates was recently surveyed by a Finnish team in order to study 'how the pastoral nomadic life has developed in the region in relation to agricultural village life from ancient times to the present' (Lönnqvist 2006: 207) (Fig. 5.1J).[55] The area is of particular interest here because it has long been seen as the homeland of the Amorites (see Section 7.2). Today, the area receives less than 150 mm annually, and farming is only occasionally possible in some wadis. Even during climatically favourable periods, the area remained relatively marginal and for example never developed a continuous tree cover. The area yielded abundant evidence for occupation starting as early as the Middle Palaeolithic period. Some stone enclosures were associated with Early Bronze Age pottery. Cairns/*tumuli* located on hilltops were also associated with Chalcolithic/Early Bronze Age pottery, although it should be noted that these features are still in use today. At one site, an Ubaid sherd, pieces of a holemouth jar, and Early Bronze Age and possibly Middle Bronze Age pottery was found. The terraces bordering the Euphrates Valley yielded evidence of more permanent settlement, including tell-sites. This region also showed more evidence of occupation during the Middle Bronze Age, as attested by sherds with comb-incised straight and wavy lines. It seems that the desert-steppe area was most intensively used during the Early Bronze Age and in the Roman period (Lönnqvist 2006). This conclusion reinforces the picture that had already emerged from the survey of the Euphrates Valley directly north of the Jebel Bishri.

The Wadi Hammar Survey was carried out between 1997 and 2000 within a 10 km radius around the site of Kharab Sayyar, near Tell Chuera, and thus provides contextual data for both the BS and the Upper Khabur Survey (Fig. 5.1K). Sixty-five sites dating to all periods were located by the survey team.[56] There were apparently no Late Chalcolithic/Uruk sites, but a period of apparently low activity was followed by a settlement peak during the third millennium, when eleven sites were occupied. Among these sites was Kharab Sayyar, which now seems to have been occupied during

---

54 Sites were marked on a 1:25000 map, and also mapped at a 1:5000 scale. No aerial imagery was available. Larger sites were divided in sampling units according to site morphology.

55 The survey made extensive use of aerial/satellite imagery to construct a Geographic Information System (GIS) of the region into which all sites and features were plotted using GPS. Small areas were selected for field walking in lines spaced 15 m apart. There was a specific focus on sites, being loci of past human activity, rather than on documenting the continuous sherd and flint blanket covering the landscape, as it was argued that this blanket resulted from erosional activity. The survey also utilized ground penetrating radar in selected areas (Lönnqvist 2006).

56 Details on survey strategy are unavailable, but the survey seems to have been site-based. Of the sixty-five recorded sites, 50 % was found by the survey team, 40 % was found using information provided by the local population, and 10 % was detected on satellite images during the preparation for the survey (http://web.uni-frankfurt.de/fb09/vorderasarch/survey.htm (Accessed 16 November, 2007)).

the entire third millennium (Hempelmann 2008). Only two sites were occupied during the Middle Bronze Age (2000–1600 BC). For the Late Bronze Age (1600–1200 BC), ten sites were recorded (http://web.uni-frankfurt.de/fb09/vorderasarch/survey.htm (Accessed 16 November, 2007)).

The Upper Khabur Survey (UKS) aimed at a *longue durée* population reconstruction of the Upper Khabur Basin (Fig. 5.1L). It was originally envisaged to cover the entire Upper Khabur Basin, but only the western part was surveyed as part of this project. The survey covered a roughly triangular shape that was bordered to the west and southwest by the Khabur, to the north by the Turkish–Syrian border, and to the east by the Wadi Jaghjagh.[57] Much of this material is at the stage of analysis and publication; so far the early periods have been published, as well as the second and early first millennium BC material (Lyonnet 1996, 1998, 2000, 2004; Faivre n.d.). A closer inspection of the adopted survey strategy suggests that, although some measures have been taken to create a representative sample of sites, at the very least small sites that are neither indicated on maps, nor clearly visible in the field, may be severely underrepresented. The focus on potential second millennium sites further suggests that the survey was likely to enhance the picture of second millennium settlement in the region, rather than result in new insights in local settlement patterns. With these caveats in mind, the following settlement development can be sketched (Lyonnet 1996). From the late fourth to early third millennium, the settlement pattern in the western Khabur seems to have been relatively stable, although both the number of sites and the total amount of settled hectares increased considerably. During the Ninevite 5 period, there seems to have been a cultural divide between the northeastern and southwestern part of the survey region. Incised and painted wares were only recognized in the northeastern part, but it is unclear to what degree this cultural divide was also present in non-decorated wares (Lyonnet 1996: 368). During the second part of the third millennium, and more specifically the later part, the number of settlements and their size decreased again, reaching its lowest values during the early second millennium. At that time, the course of the Khabur north of Hassakeh was completely devoid of permanent settlements. Lyonnet interpreted the sites in that region as pastoral nomadic campsites, based on the absence of Khabur ware in this region, and the presence of a certain type of cooking pot (1996: 371). During the Mitannian period, the number of settlements and their aggregate size grew again. Note that this picture is considerably distorted by the fact that several important sites in the wider region of the western Upper Khabur were not surveyed by Lyonnet, including Chagar Bazar, Tell Ailun, and notably Tell Mozan. Tell Mozan alone would have had a larger population than all the other sites in the Upper Khabur Survey region combined, if the size estimate for Tell Mozan of 120 ha during the late third millennium BC is correct (Pfälzner and Wissing 2004: 81). Furthermore, considerable discrepancies exist between Lyonnet's size estimates and those made by the Tell Beydar Survey (Ur 2004). The evidence from the Tell Beydar Survey and from excavated sites, for example Tell Beydar (UKS 15/TBS 1) and Tell Arbid (UKS 42), suggests that Lyonnet's estimates were probably relatively low.[58]

57　The Upper Khabur Survey was carried out in 1989 and 1990. Before the actual fieldwork, sites were located on 1:200000 maps from the French Mandate period. No aerial or satellite photographs were used. Over 300 tells were located, of which 63 were actually surveyed by Lyonnet, and several more by Nishiaki who primarily focused on Neolithic and earlier occupations. The choice for specific sites to be surveyed was made in advance and based on (*a*) site location so as to ensure a representative regional sample that included sites located in a variety of environmental contexts, (*b*) site size, i.e. all large sites were included, (*c*) previous knowledge of sites, i.e. a focus on less well-known sites, and (*d*) a focus on sites where second millennium material could be expected. In the field, some small sites not indicated on the map were included as well. Sites were divided in sampling units according to site morphology. These units were separately sampled in order to allow recognition of shifting on-site settlement and activity areas (Lyonnet 2000).

58　Sites that were surveyed by the Upper Khabur Survey as well as the Tell Beydar Survey included Tell Beydar (UKS 15/ TBS 1), Tell Khatoun (UKS 16/TBS 32), Tell Effendi (UKS 17/TBS 55), Tell Hassek (UKS 28/TBS 43), Tell Berguil Bouz (UKS 30/TBS 22), Tell Rhazal Tahtani (UKS 51/TBS 63), and Tell Boughaz (UKS 52/TBS 65, there called Tell Khazna). The results from the Tell Beydar Survey differed in some important aspects from the Upper Khabur Survey. Notably, the Upper Khabur Survey revealed substantial early third millennium BC occupation at Tell Jamil (UKS 14/TBS 59), Tell Khatoun (UKS 16/TBS 32), Tell Berguil Bouz (UKS 30/TBS 22), and Tell Rhazal Tahtani (UKS 51/TBS 63), identified by a non-decorated pottery assemblage. The Tell Beydar Survey found no evidence at these sites for their period 6, i.e. the early third millennium BC. Note, however, that early third millennium diagnostics are in any case rare in the Tell Beydar Survey, possibly the result of the intensive collection strategy of the earlier Upper

The Tell Beydar Survey (TBS) was carried out in 1997 and 1998 and covered a 452 km² circular area centred on Tell Beydar (Fig. 5.1M). This area included a part of the large basalt plateau to the southwest of Tell Beydar where only very few sites were recorded.[59] Overall, eighty-three sites with a total size of 246 ha were found. Sites were preferably located along or nearby watercourses. Periods 06–08 covered the third and early second millennia BC. During the early third millennium (period 06), there were twelve sites within the Tell Beydar Survey area covering 45.3 ha, of which 17.0 ha was taken up by Tell Beydar itself. Like the Upper Khabur Survey, the Tell Beydar Survey only infrequently found Ninevite 5 sherds. During the mid–late third millennium (period 07), twenty sites were occupied, including all period 06 sites, with a total size of 69.6 ha. Beydar was the largest site at 17.0 ha, but there were also three medium-sized tells covering 7–8 ha each. At sites where period 07 ceramic material was found, it represented the majority of the collected sherds, except for a few sites that were subsequently interpreted as representing seasonal occupations. The close association of period 07 sites with radial lines was also noted. During the early second millennium (period 08), only one site seems to have been permanently occupied, whereas five others may have been used on a seasonal basis (Ur 2004).

The course of the Wadi Jaghjagh south from Tell Barri, and the area around the confluence of the Wadi Jaghjagh and the Wadi Radd yielded abundant evidence for occupation (Eidem and Warburton 1996) (Fig. 5.1O). In this area and around Tell Brak, fifty-six tells were visited of which twenty-nine yielded third or early second millennium material. Of these, fifteen yielded Ninevite 5 material, twenty-five late ED/Akkadian sherds, and nineteen early to mid-second millennium material. It is difficult to draw any conclusions from this material (Eidem and Warburton 1996: 55). Site counts alone are not good indicators of changing human activity, and there are reasons to assume the presence of a significant bias due to the survey strategy adopted.[60] It should further be noted that most sites are located along the Wadi Jaghjagh, which was a perennial stream before the advent of mechanized irrigation works. Fielden's survey of the Jaghjagh between its confluence with the Wadi Radd and Hassakeh yielded the observation that after an Uruk settlement peak, there was a decline in the number of settlements during the third and second millennia BC, and settlement was limited to the larger sites (1978/9: 172).[61]

Several survey seasons were carried out around Tell Leilan between 1984 and 1997 to place the excavation at Tell Leilan in a regional context (Fig. 5.1P). In 1984, a low-intensity site-based survey was carried out within a radius of 15 km from Tell Leilan (Weiss 1986). A more intensive survey in 1987 included an intensive field walking component along the Wadi Jarrah, and controlled site surface survey of selected sites with evidence for Leilan period II and III occupations (Stein and Wattenmaker 1990). In 1995 and 1997, as part of the Leilan Regional Survey (LRS) two further field seasons were carried out that enlarged the survey area to a 30 km wide transect centred on Tell Leilan, and running from the Syrian–Turkish border in the north across the Wadi Radd to the Syrian–Iraqi border in the south. South of the Wadi Radd, survey coverage was limited.[62] The following observations on the local settlement history are mainly drawn from this fieldwork. The Leilan Regional Survey collected material from the Pre-Pottery Neolithic to the late Ottoman period. For the third and second millennia BC, some of this material is now presented in a preliminary fashion by Ristvet (2005, 2008). Ristvet provided a catalogue of third and second millennium sites

---

Khabur Survey. Site size estimates also vary between the Upper Khabur Survey and the Tell Beydar Survey. The UKS estimates Tell Effendi (UKS 17/TBS 55) at 1.4 ha and Tell Hassek (UKS 28/TBS 43) at 1.32 ha, whereas these sites are estimated by the Tell Beydar Survey at 8.5 and 7.4 ha, respectively (Ur 2004).

59 Toponyms, maps, and Corona satellite imagery were used to identify sites before actual fieldwork. On-site collection took place in discrete collection units to allow the recognition of shifts in on-site settlement through time.

60 The fact that all sites are located along wadi courses results from the adopted survey strategy, which was to walk along the wadis and visit all visible tells (Eidem and Warburton 1996: 58). In so doing, many small sites will certainly have been missed.

61 This information will certainly be updated once data from a recent survey in a 15 km radius around Tell Brak under the direction of H. T. Wright (Fig. 5.1N) become available (Oates 2004; Lawler 2006: 1460–1).

62 During this survey, satellite imagery and a wide range of maps were employed. Sites were mapped using GPS and a Total Station. Sites were divided into collection units based on the morphological characteristics of each site. During the 1995 season, so-called hollow ways detected on satellite imagery were investigated, and a geomorphological study of several bulldozer cuts was carried out as well (Ristvet 2005).

with information on the size of the tell and the number of sherds by which chronological phases are represented at each site (2005: appendix 3). In the process, three different classes were distinguished. The class 1.000 indicates the presence of more than four sherds definitely assignable to that phase, 0.001 indicates fewer than four sherds certainly dating to that phase, and 0.0001 indicates 'the presence of sherds that may date to each phase' (Ristvet 2005: 310). Assigning sites to the first two classes seems rather straightforward, but the latter class is rather problematic in that it does not allow distinction between phases to be made. This suggests that if a phase at a site is classified as 0.0001, there are actually no sherds that could be dated to that phase with certainty, unless there are other arguments to do so. The amount of occupied hectares per phase is calculated on the basis of these sherd classes, but how these sherd densities were converted into individual size estimates, which in fact are not given, is unclear.[63]

Ristvet distinguished eight chronological phases in the Leilan Regional Survey material for the period 3000–1500 BC. During phase 1 (Leilan IIIa), seven sites were occupied at a total of 31 ha, almost half of this area being taken up by Tell Leilan. During phase 2 (Leilan IIIb–c), there was a considerable increase in both the number of sites and their aggregate size, although the southern limit of settlement moved slightly north. Ristvet inferred a three-level settlement hierarchy centred on Leilan (LRS 1), Dogir (LRS 16), and Mohammed Diyab (LRS 55). The number of settlements declined to twenty-six during phase 3 (2650–2500 BC). At the same time, Leilan reached a size of 90 ha, and there was strong nucleation into large settlements. The number of sites and their aggregate size increased into phases 4 (2500–2300 BC) and 5 (2300–2200). However, the organization of the settlement system changed slightly, with primary centres except Tell Leilan decreasing in size, interpreted by Ristvet as centralization on Tell Leilan. During phase 5, the area around the Wadi Radd was also settled. Phase 6 (2200–1900 BC) coincided with Leilan IIc and was characterized by a strong reduction of the number of sites and their aggregate size by 69 % and 72 %, respectively. Average settlement size declined as well, and settlement withdrew north of the modelled ancient 400 mm isohyet. During phase 7 (Leilan I), there were 157 sites totalling 767 ha. Over 75 % of these sites were less than 5 ha, in size a significant change from the previous urban-based settlement system. However, Ristvet also observed that correction for the site contemporaneity problem indicated that only twenty-eight sites were occupied at the same time, and this observation severely limits the significance of the Leilan I resettlement. Finally, during phase 8 (1700–1500 BC), the region seems to have been depopulated again, with site numbers and total site size only slightly higher than they were during the phase 6 collapse (Ristvet 2005).

The Tell Hamoukar region was surveyed in 2001 and 2002 (Ur 2004) (Fig. 5.1Q). The survey identified sixty-one sites in a 125 km² area around Tell Hamoukar.[64] Period 06, that is the early third millennium, is divided in a 'pre-urban' and a Late Ninevite 5 phase. Four sites are recorded for each phase, but total site area increases from *c.* 10 ha for the pre-urban phase to *c.* 100 ha for the Late Ninevite 5 phase. However, late excised Ninevite 5 ware is only found at Hamoukar itself, suggesting that the other sites in the THS region had been abandoned by that time (Ur 2004: 158). During period 07, the mid–late third millennium BC, all previously settled sites remained occupied, and two more became settled. Total settled area increased slightly. Apart from these six sites, material diagnostic of period 07 was found at numerous other sites. It may be assumed that the majority of this material resulted from off-site manuring practices (Ur 2004: 159). In a number of cases, sherds were less abraded than might be expected had they resulted from manuring, and these cases could

---

63   The low numbers of sherds involved in classes 1.000 and 0.001, i.e. more or fewer than four sherds, respectively, suggests that only very specific types were considered to be representative for each phase, given the fact that normally hundreds of sherds can be collected from a site. Alternatively, it is possible that the pottery chronology itself is not detailed enough to allow the assigning of large quantities of feature sherds to specific phases. This possibility is hinted at by Ur, who observes that it is presently impossible to accurately assess the validity of the Leilan chronology, because it remains poorly understood (2004: 71). It should also be noted that there are some inconsistencies as regards the number of sites per period between the data published by Ristvet, and those published on the Tell Leilan website (Weiss 2007).

64   Sites were identified from Corona satellite images and maps and then visited in the field by vehicle. It was decided not to identify sites in the field through systematic transect walking, and where fieldwalking was undertaken to document off-site sherd scatters, no new sites were found. Sites were plotted using GPS and each site was sampled in distinct units allowing the reconstruction of diachronic on-site settlement changes (Ur 2004: 127)

represent pastoral encampments or temporary settlements of seasonal labourers who worked at Hamoukar during periods of high agricultural activity, for example during harvests (Ur 2004: 160). Period 08 settlement, the early second millennium BC, was represented at nine sites totalling 25.9 ha. These figures indicate a significant decline of the total population in the Hamoukar area, and also a change in the organization of the settlement pattern. During the late third millennium, there were few large sites, while during the early second millennium there were more sites with a smaller average size.

Meijer (1986, 1990) conducted a survey in the area east of the Wadi Jaghjagh (Fig. 5.1R).[65] To the north and east it was delimited by the borders with Turkey and Iraq, respectively. To the south it was delimited by the road from Tell Brak to 'Ain Diwar. The purpose of this survey was to provide a pottery sequence for the region, and to provide a more solid basis for the interpretation of Old Assyrian textual sources. The survey partly overlapped with other surveys discussed here, that is the LRS and the 1988 survey around Tell Brak. Nevertheless, the figures resulting from this survey are difficult to interpret, due to uncertainties in the ceramic chronology, and the absence of reliable site size estimates.[66] Meijer recorded 16 EB I sites, 12 EB II sites, 55 EB III sites, 50 EB IV sites, 176 MB sites, and 38 LB sites. The assignment of sites to the MB period occurred primarily on the basis of the presence of Khabur ware, which has since been shown to occur already during the final centuries of the third millennium BC. Thus, these MB settlements were spread over a much larger chronological period than any of the other periods distinguished in this survey. Closer inspection of these figures furthermore suggests that the degree of settlement continuity between the MB period and the periods bracketing it is relatively low; indicating that the actual number of contemporaneous sites was also low. Especially the MB sites are primarily one-period sites, indicating a relatively high degree of settlement discontinuity, and an unstable settlement system as a whole. Thus, the MB settlement peak was probably considerably lower, and may not have been as pronounced as the raw data suggest. This observation is in line with the reconstructed and corrected settlement trends of the partly overlapping LRS.

The lower and middle Khabur region shows clear evidence for the waxing and waning of human settlement through the ages. An extensive survey and numerous excavations focusing on third millennium sites provide information on third and second millennium BC human activity in the region (Fig. 5.1T).[67] The following picture emerges from the survey. During the third millennium, settlements were primarily located along the upper stretch of the Khabur, above 36° latitude. There were some larger, potentially urban sites just south and east of the Jebel Abd al-Aziz, that is, outside the valley of the Khabur. The surveyors suggested that some small irrigation canals may have been constructed in this region, but that it is unlikely that the large regional canal indubitably attested for the Late Bronze Age was already in existence.[68] It is, however, observed that EB I assemblages in particular may have been underrepresented in the survey (Röllig and Kühne 1983: 194). During the Middle Bronze Age, the settlements outside the river valley disappeared, and the number of sites along the Khabur clearly diminished in number as well as size. There seems to be no evidence for local irrigation systems during this period (Ergenzinger et al. 1988: 117–20).

---

65 The survey started by locating major sites, followed by visiting smaller sites in the vicinity. No aerial photographs were used. The local population was asked for information. Pottery and other finds were collected in an unsystematic way.

66 The chronological sequence suffered from a lack of well-published and reliable local pottery sequences at that time (Meijer 1986: 32; cf. Ur 2004: 55).

67 The survey of the *Tübinger Atlas des Vorderen Orients* was carried out in 1975 and 1977 and focused on the valley of the Khabur from the Khabur-Euphrates confluence in the south to Hassakeh in the north. The survey focused on tell sites, most of which were already known prior to the survey. Sites were divided in collection units based on their shape, and sherds and artefacts were collected from the surface. In total, 129 sites were mapped and/or visited (Röllig and Kühne 1977/8, 1983).

68 Note, however, that Göyünç and Hütteroth doubt the past existence of a single canal running along the Middle and Lower Khabur, and argue that it is more likely that there were multiple smaller canals that each served a local irrigation system (1997: 75–6).

Directly to the west of the Khabur, the area around the Jebel Abd al-Aziz as far west as a north–south line crossing Ras al-'Ain was surveyed as part of the Yale University Khabur Basin Project (Fig. 5.1U).[69] The area can be considered marginal from the viewpoint of agriculture, which today is carried out with irrigation water supplied by diesel pumps (Hole 1997a: 44). The early third millennium sites, that is those sites dating to EJ I–II, were predominantly small and may have been only seasonally occupied. This settlement system changed dramatically during EJ III, when large villages and walled towns up to 60 ha in size appeared. The number of sites also grew considerably from EJ I–II to EJ III. The large EJ III settlements, typically *Kranzhügel* sites, were located in areas where water was available, typically near karstic sinks and the tails of alluvial fans. Even so, the carrying capacity of the site catchments could not support archaeologically inferred population levels (Kouchoukos 1998: 393). Kouchoukos suggests that there is no evidence for a settlement hierarchy that would support Wilkinson's extraction model, and therefore concludes that these large settlements probably relied on a broad-spectrum economy including agricultural, pastoral, and wild resources (1998: 394). There is apparently no evidence for permanent settlement in the region between 2500 BC, when the *Kranzhügel* settlement system ended, and the Neo-Assyrian period.

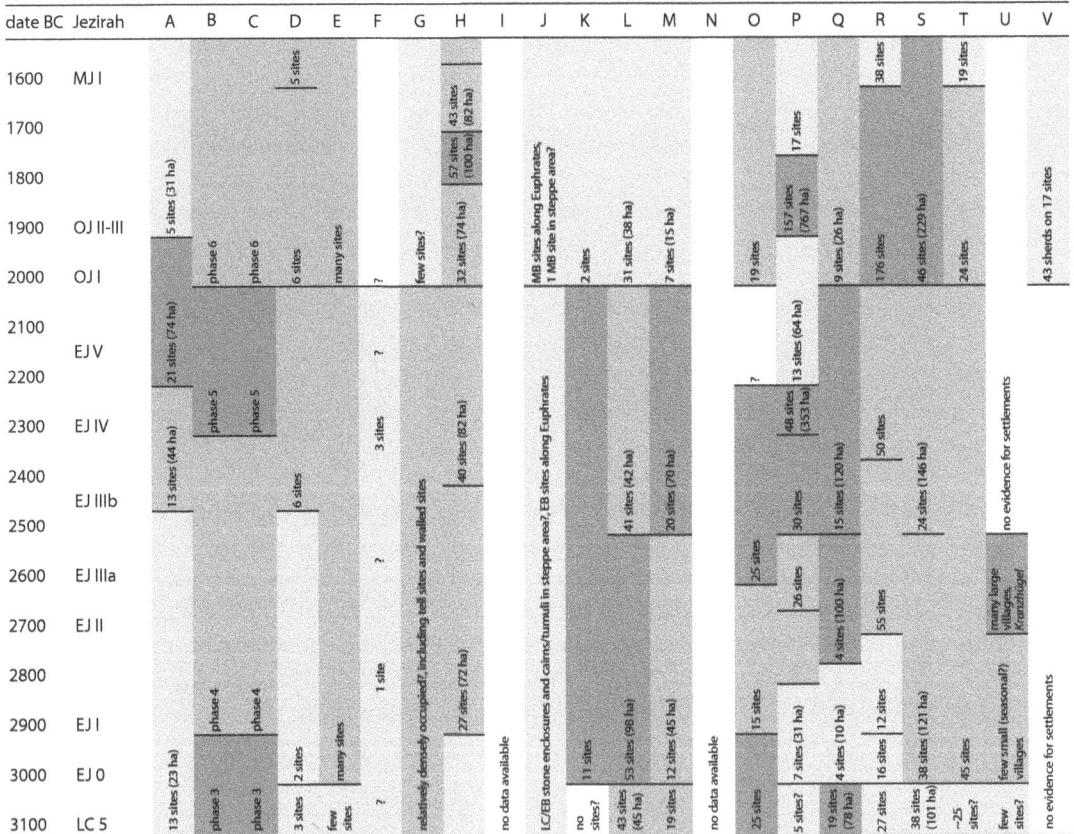

Fig. 5.14: Uncorrected settlement trends in survey regions in northern Mesopotamia for the period 3100–1500 BC. Shading indicates relative intensity of settlement. (A) Birecik-Euphrates Dam Survey (B-EDS). (B) Tishrin Dam surveys. (C) Tabqa Dam surveys. (D) Land of Carchemish survey. (E) Westgazira survey. (F) Balikh-Euphrates uplands survey. (G) Middle Euphrates survey. (H) Balikh Survey (BS). (I) Harran survey. (J) Jebel Bishri survey. (K) Wadi Hammar survey. (L) Upper Khabur Survey. (M) Tell Beydar Survey (TBS). (N) Brak sustaining area survey. (O) 1988 Tell Brak Survey. (P) Leilan Regional Survey (LRS). (Q) Tell Hamoukar Survey (THS). (R) Northeast Syria survey. (S) North Jazira Survey (NJS). (T) Lower and middle Khabur survey. (U) Khabur Basin Project. (V) Wadi Ağiğ survey.

---

69 The survey was carried out in four seasons between 1988 and 1994. The survey was performed by car along systematic transects that were planned on the basis of satellite images. Over 150 sites were recorded (but see Hole 1997b: 44, who mentions 271 sites), of which 50 dated to the third millennium (Kouchoukos 1998: 365–7).

The last survey to be discussed here, that of the Wadi Ağığ, aimed to document traces of permanent occupation in the region east of the lower Khabur (Fig. 5.1V), but west of the Syrian–Iraqi border.[70] Fourth millennium BC material was present at several sites, of which two were probably permanently occupied. The material at other sites may have been transported there, or may have reflected temporary occupations, for example for agricultural activities. Contrary to this settlement pattern, the survey recovered virtually no third millennium material, and only forty-three early second millennium sherds divided over seventeen sites. Given the absence of indications for permanent occupation, Bernbeck suggested that this material reflected temporary use of the sites by mobile groups. The material from the Medieval Islamic period was similarly patterned and could be interpreted along the same lines (Bernbeck 1993: 63).

5.4.2 The historical evidence

Texts from the Mari archive allow the reconstruction of the number of inhabitants for a number of settlements within the core territories of the Mari kingdom under Zimri-Lim (Millet Albà 2004). A number of texts list villages in the districts of Mari, Terqa, and Saggarâtum and their tribal affiliation, that is, whether they were considered to be Sim'alite or Yaminite. At least thirty-nine Yaminite toponyms could be identified, and for twenty-nine of these, data on the number of inhabitants are also available. These texts indicate that settlement sizes varied between as few as 30 to as many as 675 inhabitants. For the district of Mari, seven Yaminite settlements are listed with a total of 1500 occupants. For Terqa, nineteen Yaminite settlements are known; fourteen of which had a combined population of 3875 persons. Finally, for the district of Saggarâtum, thirteen Yaminite settlements are known, and of these eight had a combined population of 2225 persons (Millet Albà 2004: 232). Thus, a total of 7600 persons lived in twenty-nine settlements, and at least ten more Yaminite settlements are known.[71] It is further mentioned that the total population of these three districts may have been 40000 persons, although the origin of this figure is not explained. The district of Suhûm may have housed a further 10000 persons (Millet Albà 2004: 231). Note, however, that the number of known toponyms exceeds the number of sites with a Middle Bronze Age component (Millet Albà 2004: 225). This is perhaps to be expected in the alluvial environment of the Euphrates Valley.

5.5 Conclusions

The present chapter aimed to present a detailed reconstruction of demographic developments in three different survey regions in northern Mesopotamia. Based on the criteria outlined in Section 5.1, the Birecik-Euphrates Dam Survey, the Balikh Survey (BS), and the North Jazira Survey (NJS) were selected for analysis. These surveys, incidentally, also represent three different environmental and cultural zones. The B-EDS region, located upstream of Carchemish along the Euphrates, was probably always, or at least at the end of the third millennium, part of the cultural and economic *hinterland* of Carchemish. The Balikh Valley can be considered a relatively autonomous area that was, as far as can be gleaned from the historical sources, outside the permanent political control of the large late third millennium political entities, such as Ebla, Akkad, and the Ur III state. The area is also interesting because it represents a river valley that crosses the long-term border of dry agriculture. The NJS covers a region without major perennial streams, but with potentially sufficient precipitation for dry agriculture. Today, the area is above the 400 mm isohyet, meaning that even in dry years it is likely that dry agriculture will succeed. Politically, the area was certainly within the

---

70 In an area of 1150 km², sites and sherd scatters were identified by asking local herders, who showed the sites to the survey team. Large sites were divided into collection units based on site morphology, whereas no such distinction was made for small sites. Sherds were unsystematically collected until an amount was reached that was deemed sufficient for chronological and stylistic analysis (Bernbeck 1993: 19).

71 Average settlement size calculated from these numbers would be 262 persons, the median size is 195. In terms of settlement size, and assuming an on-site population density of 150–250 persons/ha, a settlement of 260 persons would be roughly 1.0–1.7 ha.

influence sphere of Akkad and Ur III, but in the absence of local historical sources, it is difficult to assess the impact of these states on the region. The exact implications of these differences will be evaluated in Section 8.2. For the moment, the following observations can be made.

First, each region was characterized by a different population development. The B-EDS region showed a very low overall population and very gradual population developments. The population in the Balikh Valley grew toward the middle third millennium BC, which is in line with the general trend toward urbanization in northern Mesopotamia as a whole. Average settlement size, however, remained at the same level, indicating that the population growth was perhaps more a by-product of urbanization elsewhere than a truly indigenous development. The population of the NJS region grew steadily from the early third to the early second millennium BC. Although the coarse chronology may have masked some minor fluctuations, the overall trend is very clear.

Secondly, there appears to have been no clear correlation between population growth, and settlement system organization as reflected in the various diachronic rank-size plots. In the BS region, there is a trend toward a convex site size distribution from the mid-third to the early second millennium, after which point convexity seems to have decreased again. From 2400 to 1800 BC, there was a population decrease followed by an increase. Potentially, the settlement organization development could be in step with the population development albeit with a 100–200 year time-gap, but this would need much more refinement of the settlement data in order to be tested.

Finally, there are various periods during which population growth exceeded historical natural growth rates, and periods during which populations became smaller. There is some evidence that population growth in one region might be balanced by population decline elsewhere. In fact, the NJS region, which harboured the largest number of sites and the highest absolute population level, demonstrated a consistent population growth that relatively closely approximated the historical natural growth rates.

6 The development of pastoralism

6.1 Introduction

Section 3.5.2 argued that environmental stress is likely to result in the formation of social bonds between groups practising different subsistence strategies. This chapter aims to review the evidence for pastoralism in northern Mesopotamia during the third and early second millennium BC. The central question that will be addressed here is: were there specialized nomadic pastoralists in northern Mesopotamia during the third and early second millennium BC? This question is important, because regional settlement abandonment has been regularly linked to the nomadization of society. This chapter first defines important concepts in the discussion on nomadism and pastoralism (Section 6.2), and outlines the development and organization of the pastoral mode of production up to the beginning of the third millennium BC (Section 6.3). Section 6.4 discusses the zoological and historical evidence for pastoralism during the third and early second millennium BC, and Section 6.5 the development of pastoralism during that period in terms of organization, specialization and mobility. Combined with the evidence presented in Chapter 7, this model will serve as the basis for evaluating the hypothesis that environmental stress should lead to the emergence of cooperative social mechanisms between pastoralists and agriculturalists (Section 8.3).

6.2 Defining pastoralism and nomadism

It is important to have a clear understanding of how the terms nomadism, semi-nomadism, transhumance, and pastoralism should be understood, as they are often incorrectly used interchangeably, but actually represent different yet in many ways connected activities. Here, the definitions provided by Cribb (1991: 18–20) will be followed. Nomadism refers to 'the regular migration of a community *together with much of its productive base* within a single ecological niche'. Semi-nomadism refers to a state of mobility on a scale with untied nomadism, that is, without fixed migration routes, at one end and a fully sedentary existence at the other. Especially transhumance is a confusing term, as it is used to describe a variety of mobile subsistence activities. The definition given by Cribb is as follows: 'a form of livestock management making use of seasonal variations in the availability of pasture'. Pastoralism is a mode of subsistence that exploits herd animals, 'characteristically involving protection of the herd and systematic consumption of its renewable products'. Herding can be understood as a more general term indicating how animals are managed, but not how they are exploited. Thus, nomadism and semi-nomadism refer to communal mobility, whereas pastoralism, transhumance, and herding are principally subsistence activities. Cribb further suggested that there seems to be a strong correlation between mobility and pastoralism on the one hand, and sedentism and agriculture on the other hand, as far as can be gleaned from ethnohistorical case-studies. However, nomadic pastoralism and sedentary agriculture should not be seen as mutually exclusive opposites, but as the extremes on a continuum, with most groups falling somewhere in between (Cribb 1991: 15–17).[72]

---

72 The present discussion will be limited to sheep/goat pastoralism. The camel was not introduced into northern Mesopotamia and the Levant until the late second millennium BC (Köhler-Rollefson 1996). Cattle was used for traction, and some cattle herding may have occurred (Stol 1995: 184), but it is likely that this was confined to river valleys and other environments that could meet the higher water and food requirements of cattle, as compared to sheep/goat. Redding suggested that cattle husbandry was probably more strongly associated with agricultural production than with pure pastoral production (Redding 1993: 86).

In his study on the emergence of nomadic pastoralism in the Central Zagros, Abdi elaborated on these terms by presenting a number of different pastoral production modes, based on degree of mobility, division of labour, and social organization (2003: 400). He distinguished between village-based herding, transhumant pastoralism, semi-nomadic pastoralism, and nomadic pastoralism. Village-based herding can be divided into proximate and distant variants. Proximate village-based herding is organized along the principle that the herds return to the settlement on a daily basis, and is often of only supplementary nature to agricultural practices. Also, it is not a full-time specialized activity, but rather entrusted to younger members of the community who are not necessary for the agricultural production process. The spatial and temporal organization of this type of herding is closely connected with the agricultural cycle, to make sure that the herds do not damage the crops and to ensure the availability of fodder (Abdi 2003: 400–1). Distant village-based herding involves taking the herds to pastures outside the area taken up by agriculture for extended periods of time. However, these periods tend not to exceed several days, after which the herds return to the village. This type of herding is normally carried out by non-specialist herdsmen, such as young adults (Abdi 2003: 401). Under a transhumant pastoral production system, herding is delegated to specialists and the herds are seasonally moved to pastures that are in their most productive season. In mountainous areas, this means that the herds are taken to the mountains in summer and to the lowlands in winter (Abdi 2003: 402–3). In a semi-nomadic pastoral economy, pastoral economic activities are more important than agriculture, which is still practised by the whole group or part of the group who specialize in it. 'Seminomadic pastoralism is characterized by extensive herding and a periodic change of pastures during the course of the entire or greater part of the year' (Abdi 2003: 403). Finally, nomadic pastoralism involves no agriculture at all, not even in a supplementary role. From this perspective, truly nomadic pastoralism occurred only in very specific environments and regions, and semi-nomadic pastoralism with a stronger focus on pastoralism than on agriculture may have been more common (Cribb 1991: 16; Abdi 2003: 404). Other activities that nomads engage in are trade, transport, or hiring themselves out as mercenaries or guides. In such cases, it is probably better to speak of a nomadic multi-resource strategy, rather than pastoralism (Dyson-Hudson and Dyson-Hudson 1980: 18; Abdi 2003: 404–5; Rosen 2003).

6.3 Animal husbandry and pastoralism in pre- and proto-history

The pastoralist mode of production is a fundamental part of the traditional Near Eastern subsistence production system. However, pastoralism as a subsistence strategy exploiting renewable products is a relatively recent development and only arose after a long period of more generalized herding strategies, in which animals were probably predominantly exploited for their meat. Sheep and goat husbandry appeared during the early stages of the Pre-Pottery Neolithic B (8700–6800 cal BC) in the southern piedmont region of the Taurus Mountains, eventually leading to complete domestication. The domestication of cattle probably started somewhat later, and domesticated pig appeared around this time as well, although there is disputed evidence for a much earlier domestication date at the end of the Natufian period. From southeastern Turkey, the practice of animal husbandry seems to have spread south at varying rates, presumably because animals had to adapt to new environmental conditions, and cultural choice cannot be entirely ruled out either (Peters et al. 1999).

There is some evidence for semi-nomadic exploitation of ovicaprine resources during the final stages of the PPNB. The village site of El-Kowm 2 was probably occupied year-round, and the occupants were involved in agriculture as well as animal husbandry. The nearby site of Qdeir was much more ephemeral in terms of architecture, but agriculture was practised there as well, indicating that the inhabitants must have been at least seasonally present at the site (Stordeur 1993). At 'Ain Ghazal (Jordan), reliance on domestic goat increased steadily at the expense of hunted animals from the late PPNB into the Yarmoukian period. It has been suggested that during the PPNC at 'Ain Ghazal, goat husbandry was practised by the site's inhabitants, with herders possibly taking the flocks to seasonal pastures farther removed from the site to relieve pressure on the catchment area. According to this scenario, the site catchment had been depleted during the

Yarmoukian, resulting in the abandonment of the site by its sedentary agricultural population. Herders could, however, continue to frequent the region for pasture, and this may have represented a first step towards a fully specialized domestic animal exploitation, and eventually nomadization (Köhler-Rollefson 1992). Although there is only indirect evidence for this scenario, recent research indicates that a similar spatial separation between the settlement and its pastoral production may also have occurred at Çatalhöyük (Turkey) (Pearson et al. 2007). Pearson et al. conducted an isotope analysis on bone collagen from sheep/goat bones from Pre Level XII D to Level VI from Çatalhöyük East (7300–6400 cal BC). This study indicated that stable carbon ($\delta^{13}C$) and nitrogen ($\delta^{15}N$) isotope levels were relatively uniform in the early period, but that variability between sampled bones increased towards the later periods. Because these isotopes are directly related to the diet of the sampled animals, this change may reflect an increase in diet variability and the consumption of a wide variety of plants from different ecological zones. Pearson et al. suggested that this increased variability reflected 'an increase in the significance of pastoralism with herders moving separate flocks over more extensive territories around villages thus becoming more likely to encounter multiple isotopically distinctive plant biomasses' (2007: 2178). As at 'Ain Ghazal, it is possible that this shift in the spatial organization of pastoral production was linked to the need for the population of Çatalhöyük to increase agricultural production, either through intensification, or through an increase in the amount of land that was under production. Summarizing, the evidence presented here suggests that some limited form of transhumant herding may have emerged at the end of the PPNB or during the early Pottery Neolithic, although the degree of specialization and the distances that were covered remain difficult to assess.

During the Late Neolithic in Syria (6800–5300 cal BC), ovicaprines were the dominant species in faunal assemblages, although there were clear differences between larger, sedentary sites and smaller, presumably short-lived sites where wild animals were also important. There is indirect evidence for high mobility that can be associated with herding, again indicating the importance of this subsistence strategy. At Late Neolithic Tell Sabi Abyad in the Balikh Valley, sealing practices have been linked with the periodic absence of a part of the settlement's population, which may have been involved in herding. Early Chalcolithic subsistence in northern Mesopotamia is much more poorly understood. During the Ubaid period in northern Mesopotamia, the importance of ovicaprines increased to the detriment of cattle and pig. Ovicaprines were probably herded in small flocks, and hunting remained important as well (Akkermans and Schwartz 2003).

During these early periods, ovicaprines must have been mainly exploited for their meat and hides. Pastoralism as a subsistence strategy relying on renewable products could have only emerged after the secondary products revolution. The secondary products revolution would have greatly enhanced the value of sheep, that not only produced meat but also milk and wool. The selection of wool-bearing and milk-producing qualities in animal breeds is, however, a long process that would have yielded results only long after the initial domestication. For that reason, the secondary products revolution has traditionally been placed somewhere in the fourth millennium (Sherratt 1981), although it is possible that Neolithic people used secondary products on a more modest and unsystematic scale (Akkermans and Schwartz 2003: 74). In fact, the use of milk has recently been attested for the middle seventh millennium BC in Anatolia and northern Mesopotamia (Evershed et al. 2008).

The faunal assemblage of late fourth millennium Tell Qraya, south of where the Khabur joins the Euphrates, indicated a 'generalized, multipurpose form of herding' including production for dairy as well as meat, but probably not wool (Galvin 1987: 123). At other sites, however, wool was certainly exploited during the late fourth millennium BC. Texts from Uruk indicated the existence of wool-bearing sheep breeds, and archaeozoological evidence indicates that herding strategies favouring wool production were employed (McCorriston 1997: 521). Although it has been suggested for northern Mesopotamia that specialization in sheep/goat pastoralism was primarily associated with Uruk colony sites, it seems that this development also occurred as an indigenous development before the occurrence of Uruk influence. Even though faunal material from many sites indicated an increasing importance of sheep/goat, there were notable exceptions where wild resources remained important (Akkermans and Schwartz 2003: 206).

Abdi (2003) presented a model for the emergence of nomadic pastoralism in the Islamabad Plain in the Central Zagros during the Middle/Late Chalcolithic. Based on archaeological survey evidence, he suggested that from the Early Neolithic to the Middle Chalcolithic period, there was a spectacular increase in the number and size of sedentary occupations, followed by a similarly dramatic decline during the Late Chalcolithic. The Middle Chalcolithic also yielded the strongest evidence for irrigation. Synchronously with the Late Chalcolithic sedentary settlement decline, the number of temporary campsites increased, as did the distance between these campsites and the permanent occupations. Excavations at selected permanent and temporary occupations revealed significant differences relating to the activities that were carried out at each site. For example, the faunal material from Chogha Gavaneh, a permanent occupation, indicated a diversified animal economy, whereas the material from the temporary campsite of Tuwah Khoskeh indicated seasonal use focusing on sheep/goat animal husbandry. Pottery and flint tools were also indicative of a rather limited array of activities, and large storage jars for example were absent. According to Abdi, these developments highlighted an increasing distance between increasingly specialized subsistence strategies. The evidence for intensive agriculture is contemporaneous with a settlement peak and would have forced pastoralists to move their flocks over increasingly greater distances. This development would in turn have provided to an incentive for pastoralists to specialize in herding because the movement and management of flocks demanded greater skill and knowledge.

Late fourth/early third millennium BC texts from Uruk attest to the highly developed and diversified pastoral economy that existed at that time. The texts list various breeds of sheep and goat, as well as the products for which they were kept, including wool and various dairy products (Szarzynska 2002). Based on the relative frequencies of attestations of sheep and goat, and the fact that the signs for sheep breeds were much more diverse, it has been suggested that sheep represented the dominant animal in this economy (Szarzynska 2002: 7).[73] These texts may have been herding contracts in which a flock was placed in the care of a shepherd for a period of one year. The texts probably described the condition of the herd at the end of that year, so that the performance of the shepherd and the herd could be monitored (Green 1980: 15).

6.4 Pastoralism during the third and early second millennium BC

Extensive evidence is available for pastoral production in northern Mesopotamia during the third and early second millennium BC. The faunal material recovered from numerous excavations, textual evidence, and secondary evidence on settlement patterns and settlement types allow at least a partial reconstruction of northern Mesopotamian pastoral production. This section will present an overview of the available evidence, and aims to reconstruct changes through time in the organization of pastoral production.

Numerous sites in northern Mesopotamia have yielded archaeozoological data, allowing inter-regional comparison of local production strategies, and elucidation of trends through time (Fig. 6.1). Recent evaluations of pastoral production throughout northern Mesopotamia indicated that there were clear differences between regions bordering the Taurus foothills and areas further north on the one hand, and the drier regions to the south on the other hand (Weber 1997; Clason and Buitenhuis 1998; Zeder 1998). Both along the Euphrates and along the Khabur Rivers, the importance of sheep/goat increased from north to south at the expense of other domesticated animals. The percentage of sheep/goat at fourth to early second millennium sites along the Turkish Euphrates ranged between 40–70 % (Fig. 6.2A), whereas for sites in the Tabqa–Tishrin Dam area, this percentage varied between 50–90 % (Fig. 6.2B). Similar percentages were observed at Hammam et-Turkman (Fig. 6.2C), the only site in the Balikh Valley to have yielded an Early/Middle Bronze Age faunal assemblage (Buitenhuis forthcoming). At Tell Chuera (Fig. 6.2C), 70–90 % of the faunal assemblage consisted of sheep/goat, whereas wild animals were virtually absent (Boessneck 1988; Vila 1995). At Umm el-Marra, on the other hand, sheep/goat were much less dominant and accounted for only 40–60 % of the total assemblage in favour of gazelle and

---

73 Szarzynska (2002: 5) listed fourteen signs for sheep and five for goats, including such variants as 'breeding adult male sheep' (udu utua), 'fat-tailed sheep' (gukkal), and 'wool sheep' (udu sig).

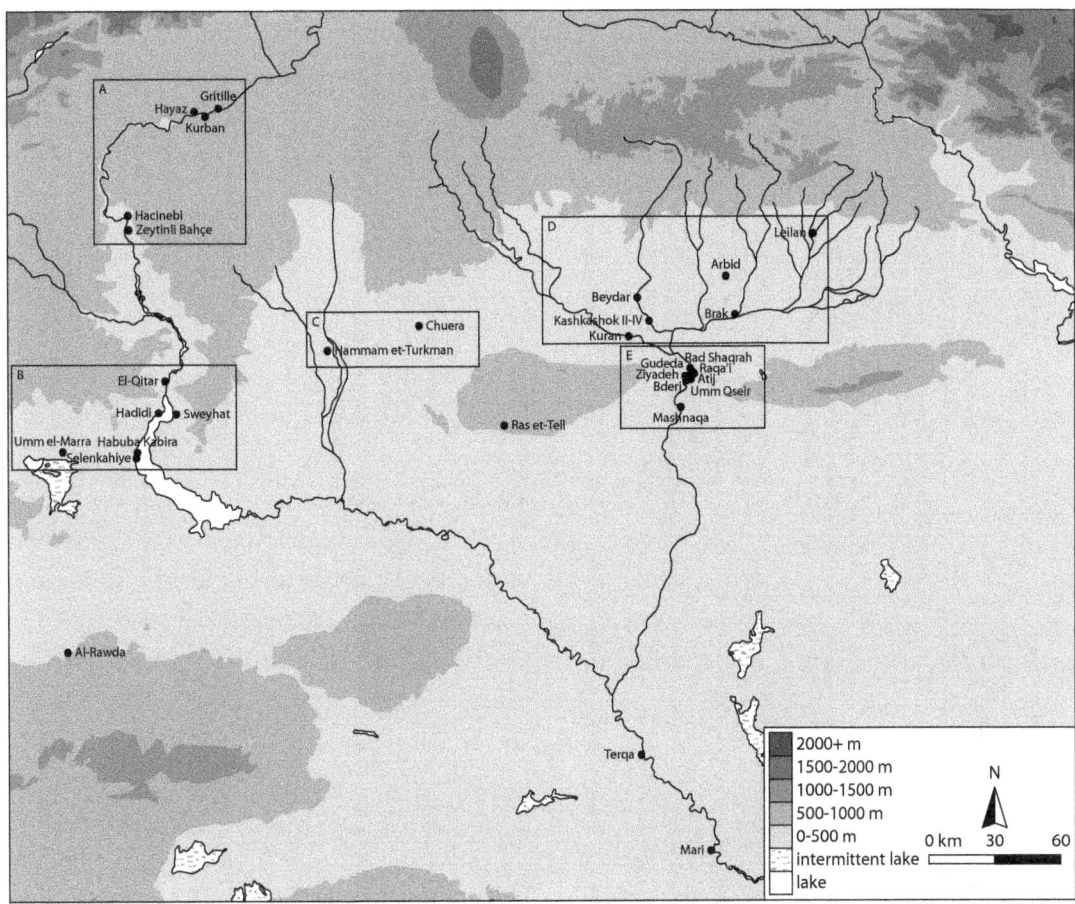

Fig. 6.1: Map of northern Mesopotamia indicating sites with faunal assemblages discussed in this study. The rectangles indicate the site clusters used in Figs. 6.2 and 6.4. (A) Turkish Euphrates. (B) Syrian Upper Euphrates. (C) Balikh/Wadi Hammar. (D) Upper Khabur. (E) Middle Khabur.

onager (Schwartz et al. 2000: table 2). Similarly, sheep/goat represented between 40–80 % of faunal assemblages from sites in the Upper Khabur (Fig. 6.2D). Along the Middle Khabur percentages of sheep/goat ranged between 30–90 %, with a considerable increase in the percentage of sheep/goat taking place since the late third millennium (Fig. 6.2E) (Zeder 1998). Both in the Euphrates and Khabur assemblages, the importance of sheep/goat increased over time.[74]

These differences between northern and southern sites in northern Mesopotamia may be partially explained by environmental conditions. However, other determinants may have played a role as well. At various sites, differences in the faunal assemblage through time as well as between excavation areas have allowed the reconstruction of diversified local subsistence strategies.

Wattenmaker argued that at Kurban Höyük socioeconomic differences accounted for the variability in the faunal assemblages recovered from different contexts. She argued that households of lower socioeconomic status relied more heavily on pigs, whereas elite households apparently focused on sheep/goat. The sheep/goat assemblage from period IV.2/3 (2400–2200 BC) suggested that herding was an activity that was carried out by individual households. The absence of prime-aged sheep/goat indicated tribute payments to a higher-order settlement or elite. Both elite and non-elite households were involved in sheep/goat herding, but elite households may have been able to focus primarily on sheep/goat whereas non-elite households had to rear other species as well, that is, cattle and pigs, to reduce risks associated with specialized subsistence strate-

---

74 The number of identified specimens (NISP) is used as indicator of species distribution, rather than the minimum number of individuals (MNI) (see Renfrew and Bahn 1996 (1991): 272–3, for the potential biases resulting from this choice). It should further be noted that only those data were used where bones could be identified to species level. If animal remains could only be assigned to a particular size, or could not be determined at all, these data were excluded from the present study.

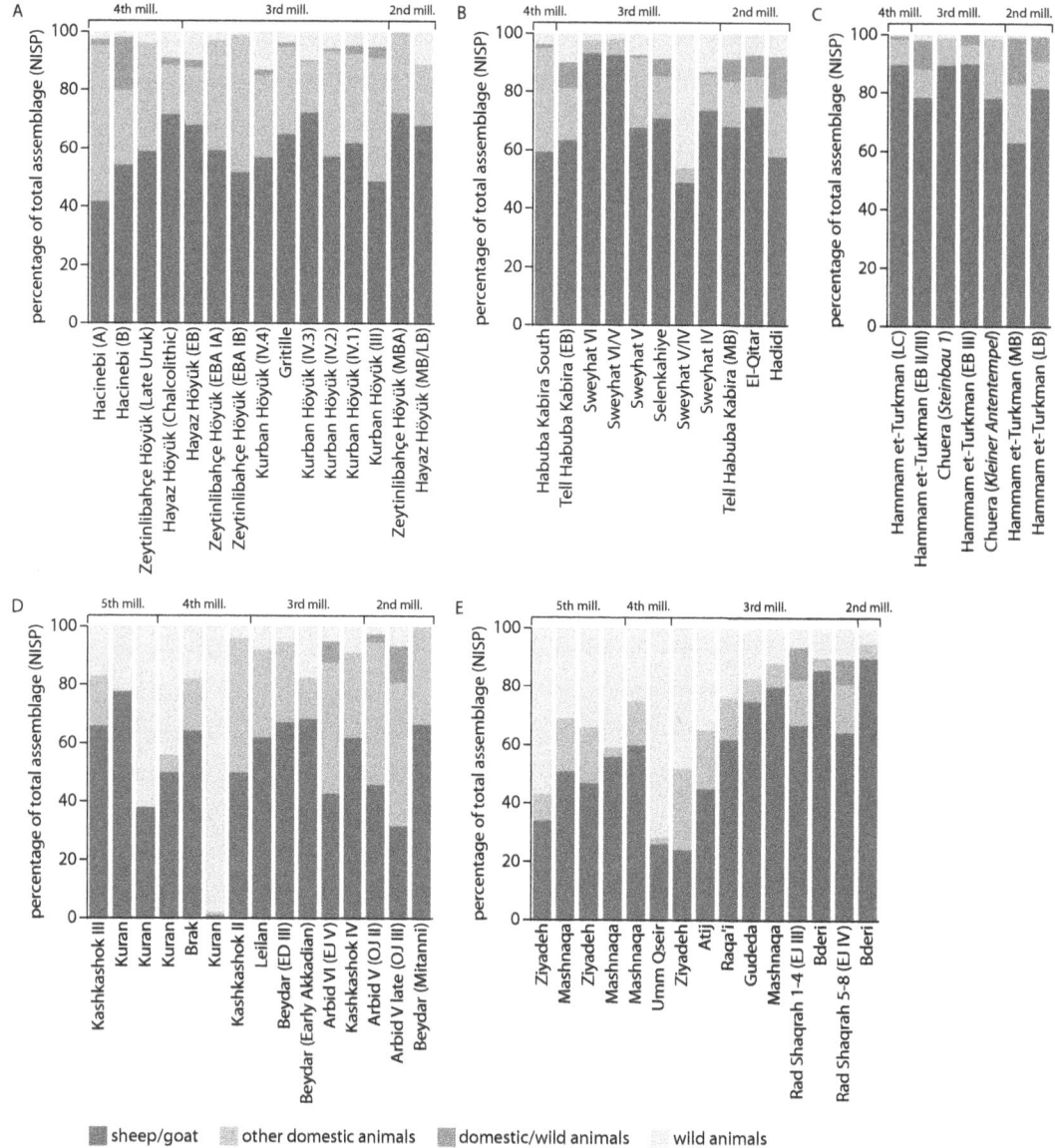

Fig. 6.2: Presence of sheep/goat, other domestic, domestic/wild, and wild animals at fifth to second millennium BC sites in northern Mesopotamia (after Clason and Buitenhuis 1978; Stein 1987; Wattenmaker 1987; Boessneck 1988; Buitenhuis 1988; von den Driesch 1993; Vila 1995; Stein et al. 1996; Rothman et al. 1998; Zeder 1998; van Neer and Decupere 2000; IJzereef 2001; Frangipane et al. 2004; Koliński and Piątkowska-Malecka 2008; Buitenhuis forthcoming). Note that in some cases, the class domestic/wild could not be distinguished on the basis of the available data. (A) Turkish Euphrates. (B) Syrian Upper Euphrates. (C) Balikh/Wadi Hammar. (D) Upper Khabur. (E) Middle Khabur.

gies. A shift in the slaughter-pattern towards younger animals probably occurred during period IV.1 (2200–2100 BC), as did an increase in the importance of sheep/goat in the total assemblage, suggesting that sheep/goat herding may have become a specialized activity carried out by full-time herders, from whom the sedentary households at Kurban procured their animal products. During period III (2100–2000 BC), household production seems to have been the rule again, as evidenced by data on age of death, and the higher diversity of species recorded in the assemblage (Wattenmaker 1987, 1998).

Evidence for a dual pastoral economy consisting of generalized pastoral production along-side specialized pastoral production came from Tell es-Sweyhat. Wilkinson (2004: 168) noted the marked differences between analyses by Buitenhuis of faunal material from the excavations by Holland, and by Weber of material from the more recent excavations by Zettler. Buitenhuis' analy-sis of mid–late third millennium material (1988) indicated that ovicaprines made up 50–90 %

of the total assemblage, with especially high percentages recorded for the early periods. Weber's analysis of the material from nearby and contemporary Hajji Ibrahim demonstrated the existence of a much more diversified animal economy, with greater percentages of pig and cattle (1997). According to Wilkinson, this difference probably reflected the existence of two subsistence strategies. The wide range of animals identified at Hajji Ibrahim was incorporated in a village-based production system, in which agriculture was combined with small-scale pastoral production. The higher sheep/goat percentages of Tell es-Sweyhat resulted from a more specialized pastoralism, involving seasonal migration or fully nomadic movement to pasture areas beyond the immediate catchment area of Tell es-Sweyhat (Wilkinson 2004: 169).

The agro-pastoral economy of Tell Beydar during the EJ IIIb period is relatively well-known due to the recovery of numerous texts from the Official Block and Chantier B, dealing with various aspects of the centralized institutional economy (Sallaberger 1996, 2004b). Although inconsistencies and uncertainties remain, these texts allow the following reconstruction of institutionalized pastoral production. All aspects of agro-pastoral production were overseen by five officials, who each headed a public household. These households consisted of workmen who were assigned various tasks, and who received monthly rations of grain. The largest group of workmen was involved in agricultural production and designated by lu_2-giš-DU apin. The second-largest group consisted of the ba-ri_2 udu, probably to be translated as 'the one overseeing the sheep', and usually associated with some form of sheep/goat husbandry. There were between 19 and 28 ba-ri_2 udu associated with each household overseer, indicating that there may have been 100–40 ba-ri_2 udu at the same time. However, the exact function of the ba-ri_2 udu is uncertain. If a sealing with the name Ma-ti belonged to a ba-ri_2 udu with the same name, it is possible that ba-ri_2 udu was not a shepherd, but some higher official. This is not necessarily the case, since at least three persons that go by the name of Ma-ti are known from the Beydar texts. Furthermore, it is possible that the term ba-ri_2 udu indicated not only shepherd in the strictest sense, but was more widely applied to a 'social group living on animal husbandry'. These narrow and wide uses of the designation ba-ri_2 udu could partly explain why there are only twenty-eight ba-ri_2 udu known by name to have been actually engaged in herding activities (Sallaberger 2004b: 18).[75]

Several texts described the size and composition of herds of sheep or goat belonging to the central institution of Beydar. Sheep and goat were herded in separate flocks. Only two out of twenty-eight ba-ri_2 udu involved in herding were associated with both sheep and goat, whereas sixteen were exclusively associated with sheep, and ten with goat. Each flock was entrusted to a ba-ri_2 udu, who probably had to take care of the flock for the duration of one year. These texts all dated to the Month of the Sun-God, which probably fell somewhere in spring-time. It is likely, therefore, that the sheep were returned to Beydar during spring to be counted and plucked. The average size of sheep herds was 210 animals, whereas goat herds consisted of 300 animals. The sheep were designated as being destined 'for plucking' (ur). Sheep herds consisted of relatively large numbers of female sheep, given the fact that rams gave more wool and would be expected to have been the dominant sex if wool-production was maximized. Based on the available evidence, it has been suggested that the central institution of Beydar may have administered 7400 sheep and goat, divided over thirty flocks (Sallaberger 2004b: 20).

Finally, the Beydar texts also recorded the rationing of grain to sheep (Sallaberger and Talon 1996: texts 7, 17). These sheep were described as 'grain-fed sheep' (udu niga) and udu nita HAR (male ... sheep) and were apparently different from the sheep that were designated by ur. The number of grain-fattened sheep was much lower than the number of sheep that were herded in flocks. Sallaberger suggested that grain-fattened sheep and grazing sheep were usually kept separate in Mesopotamia. Grain-fattened sheep may have been permanently kept at the site, whereas grazing sheep were taken out to pasture areas (Sallaberger 2004b: 21). This division could reflect the situation at Tell es-Sweyhat, where the faunal assemblage also indicated two different pastoral production strategies.

---

75 Usually, shepherds are designated by the term sipa, but this word occurred only once in the Beydar tablets (Sallaberger 2004b: 18).

The faunal assemblages from the Middle Khabur sites clearly indicate that these settlements were involved in specialized pastoral production during the early third millennium. It has been observed that many early third millennium (EJ I–II) sites along the Middle Khabur were relatively small, that is, below 1 ha in size, and had facilities capable of storing much more grain than would be needed by the hypothesized populations. These sites clustered around the modern 250 mm isohyet, and included Atij, Kerma, Raqa'i, Ziyadeh, and possibly several others (Hole 1999: 274). There has been considerable debate on the function of these storage facilities. Schwartz (1994) suggested that these sites provided agricultural surplus production for Mari, but this explanation failed to account for the relatively marginal location of these sites. As an elaboration of this argument, it was suggested that these sites were way stations for grain or trade transports coming from the north and going south along the Khabur (Margueron 1991: 99). The main argument for this thesis was that the Khabur may have been navigable in the third millennium, and that anchor stones were found at Tell Atij (Fortin 2000: 122). However, the evidence for higher water levels of the Khabur during the third millennium is inconclusive, and Hole has calculated that the surplus that could be produced by the population of Raqa'i was much smaller than suggested by Schwartz (Hole 1999: 277–8). It is more likely, therefore, that any surplus was used locally, that some storage volume was not devoted to grain, but to other products such as wool, or that not all storage facilities were in use at the same time. According to Hole, the Middle Khabur storage facilities were used by small-scale communities that were involved in both agriculture and pastoralism, with agriculture probably being of secondary importance to provide fodder for flocks (1999: 280).[76]

Many researchers have associated the *Kranzhügel* phenomenon with pastoral activity. Most *Kranzhügel* are found in a broad zone extending from the Wadi Hammar region in the northwest to the Jebel Abd al-Aziz in the southeast (Fig. 6.3). However, in the absence of excavations of *Kranzhügel*, with the exceptions of Beydar and Chuera, many of these associations remain rather hypothetical (e.g. van Liere 1963: 114; Lyonnet 1998; Meijer 2000). Recently, Kouchoukos (1998) carried out an extensive analysis of the environment of the *Kranzhügel* and contemporary settlements around the Jebel Abd al-Aziz, evaluating their agricultural potential. He first reconstructed potential population levels. The excavations and geophysical research at sites like Chuera and Beydar have shown that both the inner and outer towns of *Kranzhügel* could be relatively densely built up, suggesting that *Kranzhügel* may have housed considerable populations (Meyer 2006). Oppenheim's observations on the presence of significant stone architecture at other *Kranzhügel* suggest that their layout probably resembled those of Chuera and Beydar (Moortgat-Correns 1972: 26ff.). Kouchoukos then compared these figures with the regional carrying capacity for barley agriculture. Based on the conclusion that the catchment areas of EJ III settlements could not support the reconstructed populations, he argued that the inhabitants of these settlements must have relied heavily on pastoral production in order to supplement their diet.

Indirect evidence suggests a similar strong reliance on pastoralism at the site of Al-Rawda (2400–2100/2000 BC). Geophysical research at the site has revealed a dense urban layout within the city wall. During the later occupation phases of the site, houses were also built outside the city wall. Thus, as with the *Kranzhügel* further north, Al-Rawda may have housed a considerable urban population (Castel 2007: 173). The site is located well below the 250 mm isohyet, suggesting that agricultural production without irrigation was unreliable. Al-Rawda is however favourably located on a *fayda*, or natural depression, taking full advantage of any precipitation and runoff. This advantage was further enhanced by the construction of hydraulic installations in the vicinity of Al-Rawda (Barge and Moulin 2008). Analysis of botanical remains from the site has shown that agriculture was an important subsistence strategy. Next to barley, crops such as vines and olives were also cultivated, probably requiring some form of irrigation or water management (Herveux 2004). Despite the fact that agriculture was important for the inhabitants of Al-Rawda, pastoralism may also have been an essential subsistence strategy. The presence of numerous large desert-kites

---

76 Recent research indicated that third millennium inhabitants of Raqa'i, or at least persons that were buried there, consumed relatively small amounts of barley and wheat. Analysis of dental calculus indicated that plant microfossils identifiable as wheat or barley made up between 7–16 % of the total amount of starches from grass seeds. The staple plant food could unfortunately not be identified (Henry and Piperno 2008).

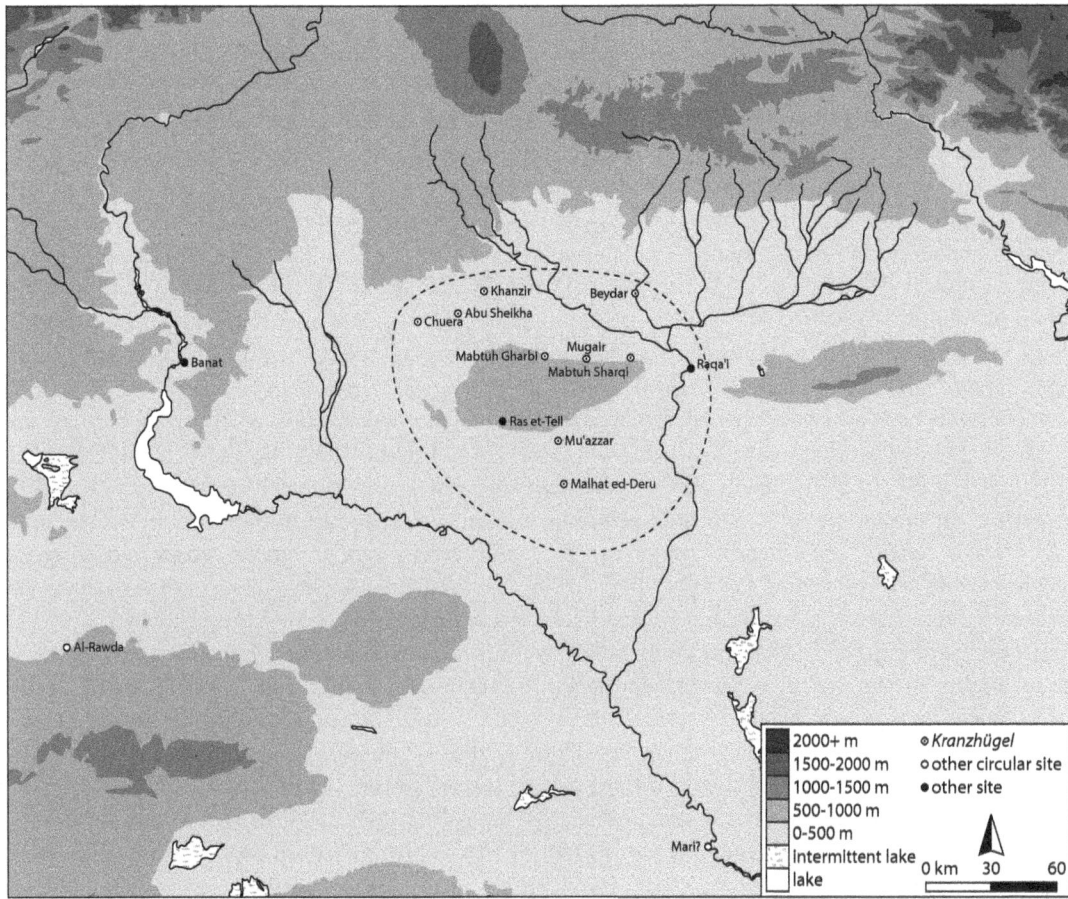

Fig. 6.3: Map of northern Mesopotamia indicating the zone where *Kranzhügel* are found (based on Moortgat-Correns 1972: map 2; Lyonnet 1998: fig. 4).

in the wide surroundings of Al-Rawda makes a convincing case for the existence of a large-scale pastoral economy in this area, in which Al-Rawda may have played an important role (Barge and Moulin 2008: 20–1).[77]

For the Ur III period, there are numerous texts dealing with agricultural and pastoral production from the core area of the Ur III state. Although the conditions there were dissimilar to those of northern Mesopotamia, some comments here may serve to illustrate the development and importance of pastoral production in wider Mesopotamia, especially because textual and zoological sources from northern Mesopotamia are relatively rare for the late third–early second millennium BC. The Drehem texts mainly documented the centralized state-economy, although a significant private economy probably existed as well. Bureaucratic control of pastoral production is particularly well-known thanks to the Drehem archives. Drehem functioned as a redistribution centre, mainly for animal products. Animals, including large numbers of sheep/goat were brought in and processed at or around the site, and secondary products, as well as the animals themselves, were redistributed to various officials. Sheep/goat were brought in during the entire year, but it seems that the number of incoming animals increased during the fall and winter months. During this period, 175 animals were delivered daily, whereas during summer and spring, 10 animals were brought in per day. This discrepancy between winter and summer figures could reflect the presence of herders around Drehem who were normally away on seasonal movements. Indirect evidence from Umma, however, suggests that the distances covered by herders may have been relatively small. Herders

---

77 On the proposed function of so-called desert kites in a pastoral economy, rather than as structures facilitating (gazelle) hunting, see Échallier and Braemer 1995.

were contracted to care for flocks of sheep/goat and deliver production to the central administration. It is possible that herding was not carried out by a single herder, but that he was assisted by several others or that herding was carried out on a family basis (Zeder 1994: 179; Adams 2006).[78]

The evidence for specialized pastoral production in northern Mesopotamia is much less clear during the early second millennium. The number of excavations that yielded faunal data is small, but it seems that the percentage of ovicaprines in all regions analysed here (Figs. 6.2, 6.4) is slightly larger than during the third millennium. Evidence from Terqa for the period 1750–1500 BC suggests that sheep and goat dominated the faunal assemblage at 37 % and 40 %, respectively. Cattle were much less dominant at 11 % and pig was virtually absent. Sheep seem to have been kept for wool production and some minor meat production as well, whereas goat may have been primarily kept for meat production. However, the degree to which wool and meat production were specialized seems to have been limited (Galvin 1987).[79]

The Mari texts allow a fragmented reconstruction of the organization of the early second millennium pastoral economy (e.g. Kupper 1957; Luke 1965; Rowton 1974; Matthews 1978; Buccellati 1990; Streck 2002). Many details are not recorded, and especially practical issues such as breeding practices, or the composition of herds, remain completely unknown. On the other hand, the texts do provide detailed information on many socio-cultural aspects, including migration patterns, pastures, and the social and economic position of pastoralists. Nevertheless, other social aspects are not covered at all, and can only be reconstructed after careful analysis. Any reconstruction of pastoralism during the early second millennium BC must therefore take into account the rather one-sided nature of the texts that form the basis for the reconstruction.

The central concept in Mari pastoralism was the *nawûm*. The term *nawûm* had several related meanings, summarized as 'the total pastoral group and its surroundings' (Matthews 1978: 59). More specifically, *nawûm* could be used to designate the encampments of *hana* pastoralists, groups of *hana* pastoralists, sheep/goat flocks, and pasture areas (Matthews 1978: 60–2; Anbar 1991: 161). Where *nawûm* was used to designate a pastoral group, it was probably used on different scales, from relatively small numbers to groups of thousands of people. The term *nawûm*, then, referred to various aspects of a type of pastoralism that was carried out outside settlements (see Section 6.5.2).

The terms *hasiratum* and *rubsatim*, respectively translated by Matthews as 'camps' and 'pens', (Matthews 1978: 52), possibly referred to small-scale pastoral production in settlements. These terms perhaps reflected the continued dichotomy between small-scale pastoral production in settlements, and large-scale specialized pastoral production in areas beyond the agricultural catchments of these settlements.

Several Mari texts describe migration routes and pasture areas, thus providing insight into the degree of mobility practised by pastoralists. The pastoralists' movements were monitored by the state when they were within the core territory of the Mari kingdom, that is, Ah Purattim or the 'Banks of the Euphrates'. There are texts describing herders with their flocks crossing the Euphrates, and describing the grazing of flocks on agricultural fields. Other texts described how sheep/goat flocks were moved in response to (un)favourable grazing conditions (Matthews 1978: 55, ARM II 102, 9–16): 'The men who were living at Sagarātum have turned their faces toward the Upper Country only, and with their flocks which were grazing on *laskim* (?), they have set out for the Upper Country. When he (the informant sent to investigate the situation by Yaqim-Addu) questioned them, they answered in this manner: "There is no pasture (here). Therefore, we are setting out for the Upper Country."' Another text described how sheep/goat flocks of Yasmah-Addu were moved to the territory of Qatna, because the pasture was more abundant there (Matthews 1978: 54). It has recently been proposed that the term *nighum* may have served to designate the *hana* summer pastures, whereby different areas in northern Mesopotamia were assigned to either the Sim'alites or the Yaminites. More specifically, Ida-Marash, in the Upper Khabur, was described as Sim'alite territory, whereas Yamhad and Qatna were Yaminite pasturages (Durand 2004a: 391–2, 2004b: 118–21). Thus, it is likely that seasonal migration patterns existed, suggesting that some

---

78 According to Russell (1988: 83, 85), flocks of up to 400 sheep/goat can be herded by a single herder, although assistance of another full-time herder and a part-time herder is needed during the breeding season.

79 Galvin used MNI instead of NISP, and the figures for Terqa are, therefore, not included in Figs. 6.2 and 6.4.

	archaeozoology	administrative texts	literary texts
early third millennium	+	-	-
mid-third millennium	+	+/-	-
late third millennium	+/-	+	-
early second millennium	-	-	+

Table 6.1: Comparison of available evidence for third and early second millennium BC pastoral subsistence strategies (+: strong evidence; +/-: some evidence; -: very little/no evidence).

sort of transhumance was practised over relatively large distances in comparison to earlier periods. As part of these seasonal migrations, flocks could graze on agricultural fields, presumably after the harvest when grain stubble provided abundant food for sheep/goat.

Looking more closely at the data presented in this section, it becomes clear that there is a disparity in the nature of the evidence of the third millennium BC, as opposed to that of the early second millennium BC (Table 6.1). It is possible that this discrepancy creates differences in the observed organization of the pastoral economy. This possibility is particularly true for the zoological and textual evidence. The number of second millennium sites with faunal assemblages is relatively small compared to the number of third millennium sites. This discrepancy and the fact that there were relatively large inter-site differences between third millennium BC faunal assemblages suggest that the confidence with which second millennium regional animal husbandry patterns can be reconstructed is relatively low. In the textual evidence, there is a shift from administrative toward literary texts as the main source on pastoral subsistence strategies at the third–second millennium transition. These disparities must be carefully considered in the evaluation of third–second millennium pastoralism that is the subject of the following section.

6.5 The emergence of specialized pastoralism in northern Mesopotamia

The present review has shown that by the end of the fourth millennium BC, a fully developed pastoral production system had emerged. At least from that time onwards, a large variety of breeds was exploited for meat as well as for a full array of secondary products. Before the third millennium, there had been periods when economic and environmental circumstances necessitated a removal of pastoral production from the direct environment of a settlement. Thus, at 'Ain Ghazal and Çatalhöyük, environmental degradation, growing population, or a combination of both possibly led to the development of some form of distant village-based herding, and the same development has been reconstructed for the Islamabad Plain during the Chalcolithic period. It seems, then, that spatial separation between agricultural and pastoral production emerged at various places and various points in time as an adaptive strategy to the demands of agricultural intensification and/or population pressure. The present section discusses the development of specialized pastoralism during the third and early second millennium BC.

6.5.1 Specialized pastoralism during the third millennium BC

A similar scenario of increasing spatial separation of specialized pastoralism and agriculture can be postulated for the Upper and Middle Khabur, and possibly to a lesser extent for the Syrian Upper Euphrates during the mid-third millennium. Zeder has suggested that the faunal patterns from third millennium sites in the Upper and Middle Khabur not only reflected environmental differences, but also differences in the dominant pastoral production strategy (1998). The faunal assemblages from sites along the Middle Khabur such as Raqa'i and Ziyadeh were characterized by ever greater percentages of sheep/goat, at the expense of other domesticates such as pig and cattle, and wild animals. At the same time, the importance of sheep/goat at northern sites decreased, with other domesticates becoming more important. According to Zeder, these developments resulted from the emergence of large urban settlements during the early to mid third millennium BC. She suggested that the local economy of the Middle Khabur, characterized by exchange between sedentary agro-pastoral communities along the Khabur, and more mobile fully pastoralist groups, was eventually incorporated into the urban economy of the Upper Khabur. In fact, the highly specialized Middle Khabur economy may have played a key role in the formation of urban society further to the north (Zeder 1998: 65).

However, Zeder did not fully explain why specialized pastoral production emerged along the Middle Khabur, and not in areas located closer to the urban centres of the Upper Khabur. Wilkinson (2003) suggested that there may not have been sufficient land in the Upper Khabur to sustain large flocks of sheep/goat. He used Corona satellite images to identify tells, and plotted sustaining areas of 3 and 5 km around each of them. Even if agriculture was highly productive, resulting in relatively small cultivated areas around each site, only a limited amount of land would have been available as pasture (Wilkinson 2003: 121–2). Thus, urbanization in the Upper Khabur may have provided both the incentive for specialized pastoral production, as well as the necessity for relocating (parts of) this production into the less densely inhabited regions of the western Upper Khabur and the Middle Khabur. The survey evidence for occupation in the area between the Euphrates and the Balikh, and around Tell Chuera, also points to increased activity during the mid-third millennium (Fig. 5.14). This could indicate that a similar process may have been operating in the western Jezirah as well, although perhaps on a less massive scale than in the Khabur region. In fact, the evidence from Tell es-Sweyhat suggests that specialized pastoral and generalized agro-pastoral production strategies may have been carried out in relatively close proximity. One explanation may be that pasture lands unsuitable for agriculture could be found in close proximity to Tell es-Sweyhat, that is, outside the Euphrates Valley, and that there was less incentive to keep the two economies strictly separated. Alternatively, the assemblage at Tell es-Sweyhat reflects the fact that urban communities consumed pastoral products produced mainly under a specialized system, which was not necessarily carried out near the site.

Hole (1999) suggested that by the mid-third millennium (EJ III), the small-scale riverine agro-pastoral communities along the Middle Khabur were gradually replaced by larger and more specialized pastoral communities around the Jebel Abd al-Aziz. He argued that the sequence of small storage sites along the Khabur and the slightly later *Kranzhügel* and other large sites near the Jebel Abd al-Aziz reflect a development whereby pastoralists became increasingly specialized and relocated their activities from the Middle Khabur toward the west. This development would have occurred at a time when urban settlements in the north such as Mozan, Leilan, and Tell al-Hawa were flourishing, and demand for pastoral products would be high. Thus, although it is perhaps premature to point to urbanism as a causal mechanism in the emergence of specialized pastoralism in this region, the synchronous development is at least indicative of such a correlation. It is likely that this development reflected a monetization of the pastoral and agricultural economies, as is possibly indicated by texts from Beydar that record the hiring of herders to take flocks to the pasture areas.

No third millennium site around the Jebel Abd al-Aziz has been excavated to test the importance of pastoralism and its degree of specialization. However, there is circumstantial evidence to suggest that the region was visited by mobile communities during the mid-third millennium. The site of Ras et-Tell on top of the Jebelet el-Bēdā, a small hilly outcrop southwest of the Jebel Abd al-Aziz, may have functioned as a central place for surrounding (semi-)nomadic communities. Ras et-Tell was located at 775 m above sea-level, providing excellent views in most directions. Several basalt sculpture fragments were found on the slopes surrounding Ras et-Tell, and it is likely that the sculptures, from which these fragments originated, originally stood on the top of the hill. There must have been at least four sculptures at the site, all of which showed men in fleece skirts. At and around Ras et-Tell, some 20 stone cist graves were found. No grave gifts were associated with these graves, and they were marked on the surface by stone circles filled with smaller pebbles (Moortgat-Correns 1972). This site has usually been interpreted as a victory monument based on the sculptures' stylistic links with southern Mesopotamia (Moortgat-Correns 1972: 21–4). However, it was recently proposed that the site functioned as a regional burial and ancestor cult centre (Meyer 1997). According to Meyer, the grave groups may have reflected separate burials by clans or lineages of (semi-)nomadic groups living in the area of the Jebel Abd al-Aziz. The apparent absence of settlements in the immediate vicinity of Ras et-Tell might support this

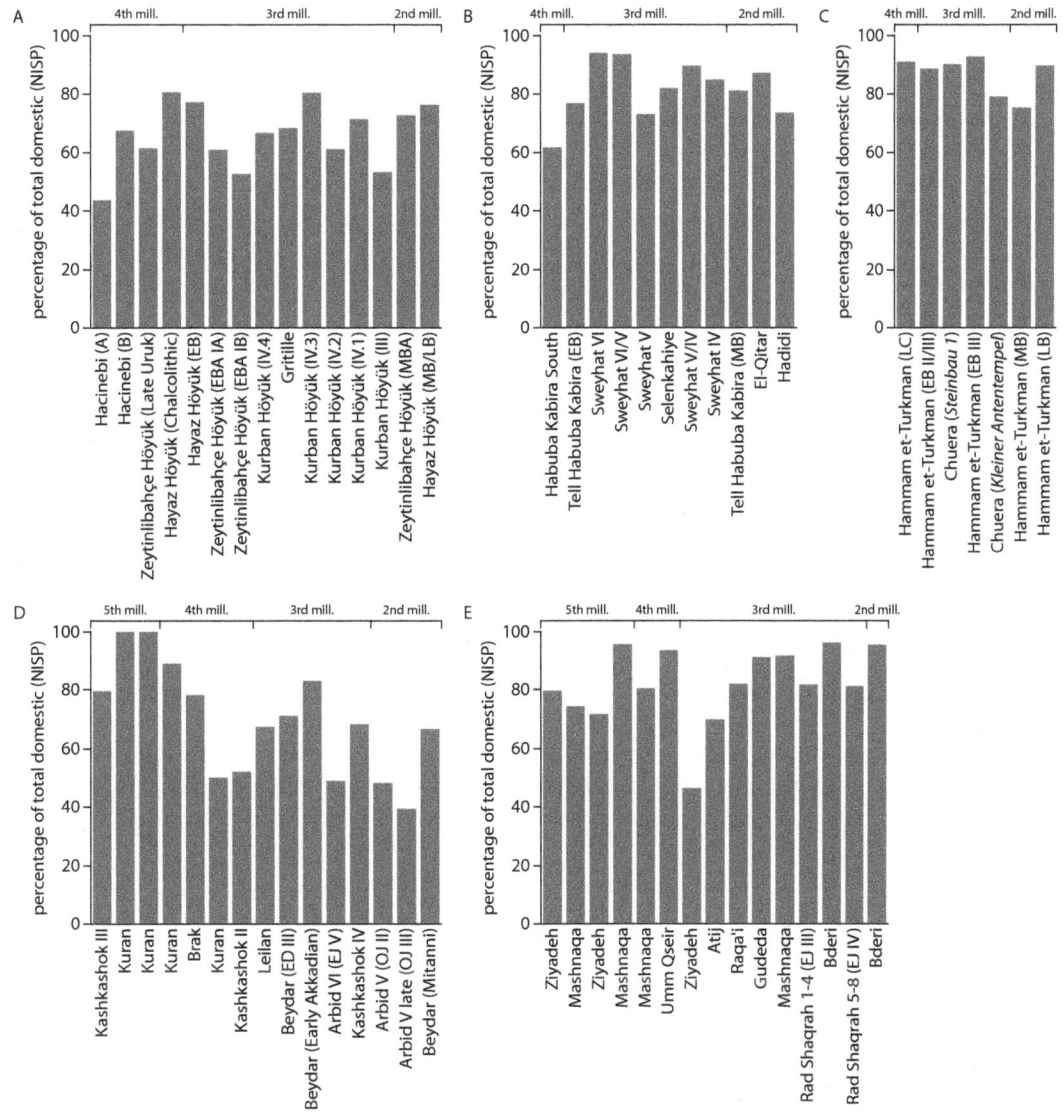

Fig. 6.4: Proportion of sheep/goat in the total assemblage of domestic animals at fifth to second millennium BC sites in northern Mesopotamia. Sites are chronologically ordered. (A) Turkish Euphrates. (B) Syrian Upper Euphrates. (C) Balikh/Wadi Hammar. (D) Upper Khabur. (E) Middle Khabur.

interpretation. Although the date of the monumental sculptures remains problematic, it is likely that they date to EJ II–III, making them contemporaneous with the emergence of *Kranzhügel* in the Jebel Abd al-Aziz region.[80]

The so-called White Monument of Tell Banat North may have represented a similar attestation of tribal presence in northern Mesopotamia. This striking landmark was located along the Upper Syrian Euphrates and consisted of a conical mound that was reconstructed and modified several times during its lifespan from 2600 to 2300 BC. The White Monument was part of a larger complex that at one time included not only the monument itself, but also a complex of partly public, partly private buildings at Tell Banat. The activities carried out at the Tell Banat complex changed repeatedly, and the White Monument was constructed in three stages, indicating shifts in the function of the complex as a whole. Both the building complex and the White Monument must have played

---

80 Ristvet argued that the cultic complex of Jebelet el-Bēdā, as well as the monumental sites of Banat and Hazna 1 reflected a 'landscape of pilgrimage' (2005: 105) in which city-based kings travelled to the countryside to legitimize their authority and help 'construct a new form of political landscape–that of kingdoms, not isolated villages' (2005: 112).

a key role in funerary traditions, as the activity area of Tell Banat contained several stone-lined tombs, and the White Monument yielded burials as well. Based on these considerations, Porter has demonstrated that, throughout its entire history, the Banat complex functioned as a focal point for pastoral communities enacting rituals and burying their dead (Porter 2002).[81]

The relocation of specialized pastoral production to non-urban areas is furthermore suggested by the importance of the sheep/goat faunal assemblage of sites in the Khabur region. At sites in the northern Khabur area, the proportion of sheep/goat decreased in favour of other domestic animals, including pig, suggesting that production was carried out at a small-scale household level. At the Middle Khabur sites, however, the proportion of sheep/goat not only increased in the total faunal assemblage, but also at the expense of other domestic animals. The contemporaneity and inverse correlation of these developments fits nicely in the scenario presented above (Fig. 6.4D, E). This development seems to be absent along the Euphrates and the Balikh, again suggesting that the spatial separation between pastoral and agricultural production may have been less severe (Fig. 6.4A–C).

Thus, by the mid to late third millennium BC a fully specialized pastoral economy had emerged, centring on settlements located in areas where no large-scale occupation had existed previously, and spatially separated from the urban core-areas. Although pastoralism was the main subsistence strategy, agriculture was carried out on a small scale as well. Kouchoukos (1998) has shown that the large settlements to the west of the Middle Khabur were located in areas that received runoff from the Jebel Abd al-Aziz. It seems likely, then, that these communities practised some form of distant village-based herding, whereby herders took the sheep/goat flocks to pastures around the settlements, whereas other community members tended the small-scale agricultural fields, taking full advantage of rainfall, runoff, and soil moisture. The large sedentary communities in the river valleys of the Euphrates, Khabur, and Balikh provided the necessary outlets for the pastoral production of the pastoral communities in the upland areas, notably textiles (McCorriston 1997; Kouchoukos 1998).

A similar scenario has been envisaged for the site of Al-Rawda in western Syria. The site's favourable location with respect to agricultural productivity has already been pointed out, as well as the central function the site may have fulfilled in the region's pastoral economy (Section 6.4). The circular layout of Al-Rawda makes it likely that the site was a planned settlement constructed within a short period of time. It has been suggested that Al-Rawda was founded by people from Ebla, or that the site at least had close economic ties with that important centre to the west. If so, Al-Rawda may have fulfilled a key role in pastoralism and textile production. Additionally, Al-Rawda was strategically located along the trade route between the urban centres of the Orontes Valley, and those of the Euphrates Valley (Castel and Peltenburg 2007).

6.5.2 Specialized pastoralism during the early second millennium BC

The transition from the third to the second millennium is much more poorly understood. There are no texts for this period that shed light on the organization of the subsistence economy, making reconstructions of this economy rather hypothetical. The few faunal assemblages dating to this period suggest that little changed with respect to the mid-third millennium in terms of exploited animals. Thus, a dual pastoral economy with a specialized and a small-scale component probably continued to exist. By the time of the Mari archives, however, a shift in the organization of the specialized pastoral economy must have occurred. This is suggested by the abandonment of sites in the more arid upland zones, as indicated by the Wadi Hammar survey and the survey around the Jebel Abd al-Aziz (Fig. 5.14), and by the appearance of a surface assemblage in the Wadi Ağiğ that displays many similarities with more recent assemblages that can be associated with nomadic pastoralists (see Section 5.4.1). It seems, then, that at least in some regions the mid-third millennium village-based herding system was replaced by (semi-)nomadic pastoralism.

---

81   Note that there are several mounds throughout northern Mesopotamia that, based on their size and shape, are hypothesized to contain structures similar to the White Monument of Tell Banat (Porter 2002: n. 31).

It has recently been suggested that the term *hana* or *hanûm* occurring in the Mari archives should be translated as 'tent-dweller' and that it cannot be taken to represent a tribal or ethnic identity similar to Sim'alite or Yaminite. According to Durand, the term *hana* derived from the root *hn'*, 'to live in a tent'. Because there are several texts describing Yaminites and Sim'alites as *hana*, it is likely that *hana* was a designation for a particular subgroup of Yaminites or Sim'alites, rather than a tribal or ethnic group (Durand 1998: 417–18; Streck 2002: 175; Fleming 2004: 47–9). Previous studies interpreted *hana* or Hanean as a tribal or ethnic group, partly on the grounds that the designation occurred very regularly in the Mari texts. It is, however, plausible that the term *hana* without ethnic or tribal determinative was very specifically used to designate Sim'alite tent-dwellers. This may have been the case because the Lim-dynasty was of Sim'alite affiliation and writers may not have felt the need to mention this affiliation because it so obviously dominated the political scene of Mari.[82]

The equation of *hana* with mobile pastoralist is further strengthened by the fact that in some cases *hana* was opposed to the town (*ālum*) or townspeople (Fleming 2004: 48, n. 63). Thus, in *ARM* 8 11, a distinction is made between members of the tribal elite living in the town of Appan, and those living with the *nawûm*. Another text from the reign of Yahdun-Lim listed soldiers according to their descent. This list clearly distinguished between soldiers classified as lu₂-meš *hana* and dumu-meš *mâtim*, that is, *hana* people and sons of the land, respectively. The dumu-meš *mâtim* were furthermore identified by district and town, making clear that this list distinguished between (semi-)nomadic *hana* and sedentary townspeople (Charpin 2004). The mobility of the *hana* is also apparent in a letter from Ibâl-El, in which he writes that 'My lord knows that I govern the *hana*, and that the *hana* travel through (lands in) war and peace as merchants do, and that during their movements they learn what (the people of) the lands say' (translated from Durand 1997: t. 333).

Despite the epigraphical arguments in favour of the interpretation of *hana* as tent-dweller, archaeological evidence for the use of tents during the early second millennium BC is currently lacking. It has been argued that the tent was a relatively late phenomenon and that clear evidence for its use only appeared during the first millennium BC. Based on circumstantial evidence, it has for example been suggested that large woven tents such as those traditionally used by Bedouin only appeared during the first millennium BC (Rosen 2008: 125; but see Hole 2009: 264). Although such tents facilitate higher mobility, they are not essential for a mobile way of life. A diversity of temporary structures can be observed among (pre-)modern Near Eastern nomadic pastoralists, including structures made of reed, and tents that were roofed with multiple pieces of woven cloth as well as felt and animal skins (Cribb 1991: 105–8). Many of these structures would not significantly impede mobility. Horne described temporary camps in Iran that often included permanent features such as walls, whereby only valuable and transportable materials were brought in by the occupants when they came to the site (1993: 45–7). Mobile groups travelling along fixed routes or within a circumscribed region may have maintained several of such stations at preferred camping locations. The absence of clear evidence for tents during the early second millennium BC, then, could suggest that the translation of the term *hana* as tent-dweller may not be entirely appropriate, but this absence cannot be seized upon to disregard the strong focus on mobility that is evident in the *hana* way of life.[83]

In order to understand the position of the *hana* in the Mari state, it is necessary to look in detail at the organization of the kingdom's power structure. In fact, under the king there were two separate power structures. The first of these was concerned with the rule of the 'districts' (*halsum*)

---

82 Durand's translation of the term *hana* as Bedouin is extremely unfortunate, as the term Bedouin is associated with certain cultural and social features that may not have been part of Mari society, and that suggest direct continuity between *hana* and Bedouin. Fleming wrote on this issue: 'Dominique Charpin brings to my attention the fact that French does not offer the possibility of a compound word that follows the proposed etymology, as in the English "tent-dweller." This means that "bédouin" must serve to designate the same general category, without any forced anachronism by association with peoples from the modern Middle East' (2004: 252, n. 67).

83 In fact, one text from Mari mentions that the *hana* are in the steppe with their livestock and tents (Durand 2004a: 385).

of the 'Banks of the Euphrates' (Ah Purattim). These districts were headed by governors (*šāpitum*) that were identified by town and ruled from a local palace. Individual towns in the Ah Purattim and beyond were led by 'leaders' (*sugāgum*) (Fleming 2004: 25).

The second power structure focused on the *hana*. Sim'alite *hana* were organized in a number of 'divisions' (*gayum*). The term *gayum* was generally only applied to *hana* Sim'alites, although some Sim'alites living in towns were also identified by *gayum* (Fleming 2004: 43–4). It seems that the term *gayum* was applied to several types and levels of divisions, so that it is difficult to define a more precise meaning of the term. Furthermore, no political functions seem to have been associated with the *gayum*, as these functions were carried out by officials called *merhûm* and *sugāgum* (Fleming 2004: 43). The *sugāgum* were lower-level officials or 'leaders' who were not identified by towns, but by tribal affiliation. Above the *hana sugāgum* stood two 'chiefs of pasture' (*merhûm*). These *merhûm* reported directly to the king.

The position of *sugāgum* existed in the power structure of the Ah Purattim, as well as the structure of the *hana*. It seems that a further distinction existed between *sugāgum* based on their tribal affiliation. Thus, Sim'alite *sugāgum* were primarily identified by their membership of a *gayum*, although they were not the leaders of these divisions. Yaminite *sugāgum* on the other hand were often identified by town. Rather than assuming that this dichotomy reflected a difference in the degree of sedentarization of Sim'alites and Yaminites, it is better to assume that there were organizational differences between the two confederacies. Furthermore, it is possible that *sugāgum* ruled towns, even when they were not identified by these towns (Fleming 2004: 51).

The relation between *hana*, *sugāgum*, and *merhûm* is clarified by a letter about a deserter from Zimri-Lim's army, quoted by Fleming (2004: 50, FM III 136): 'If [(my lord?) does not arrest(?)] the one who takes leave (of his troop), the life of the Hana will take leave (as well). He will break the fortune(?) of the Hana. Now then, Lawasum, the Yumahammû, has taken leave. His *sugāgum* is Dadi-[Lim], and his property consists of 200 sheep and five donkeys. My lord should confiscate (this).' The ordering of the political structures of which the deserter was a part reveals that he was first of all a Sim'alite *hana*, thereby excluding Sim'alite townspeople, that he was a Yumahammû, which is known from other texts to be a *gayum*, that his *sugāgum* was Dadi-Lim, and that he owned sheep, rather than land or a house. Thus, there was a clear distinction in how the different groups that constituted the Mari state were governed, that is, sedentary townspeople on the one hand, and (semi)nomadic pastoralists or *hana* on the other hand (Fleming 2004: 25).[84]

Despite the strong evidence in the Mari archives for a highly mobile population component that is involved in specialized sheep/goat pastoralism, the composition of this group is much less clear. It is clear that the term *hana* designates a specific group within larger tribal divisions, and that the term is primarily defined through its association with pastoralist activities. The inter-tribal relation between mobile *hana* and their sedentary counterparts is however much more difficult to define. Due to the large variety in pastoralist modes of existence, and in the associated degree of mobility, it is difficult to determine exactly whether the term *hana* designated entire households that were fully mobile, or fully or semi-mobile parts of households.

Nevertheless, it is suggested here that the term *hana* designated entire households that were primarily engaged in pastoralist subsistence activities, and that pursued a (semi-)nomadic way of life. Specialization in pastoralism thus occurred at the level of the household, rather than on the basis of traits such as gender or age within households. The strongest evidence for this interpretation is the marked duality in the power structure of the Mari state, in which there are different officials for the sedentary and the *hana* population components. If members of a particular household split off to become nomadic for part of the year, they would be governed under different systems, depending on the time of the year. The duality is also and perhaps especially evident in the military organization of Mari. It has long been recognized that *hana* played an important role in the army, and that they were often separately counted and listed (Matthews 1978: 95; Charpin 2004: 87–9).

---

84 The term *sugāgum* that can be roughly translated as leader seems to have designated a number of different positions within the power structure of the Mari kingdom. These positions varied not only according to their duties, but also to their tribal affiliations (Fleming 2004: 63–8). According to Fleming (2004: 53), *sugāgum* of towns paid levies in silver, whereas *hana sugāgum* paid with sheep.

This would have been much more difficult if the distinction between *hana* and townsmen was blurred or based on a seasonal cycle. Finally, a letter from Ibâl-El to Zimri-Lim describes the nature of the exchanges taking place between the *hana* and more sedentary agriculturalist. The letter describes the symbiotic relationship between several kings in the Upper Khabur area and the *hana* as follows: 'The *hana* have always been your pastoralists, whereas you [the kings] are their farmers' (translated from Guichard 2002: 160). Apparently, the kings had not given the *hana* the grain deliveries to which they were entitled, because Ibâl-El settled the matter by giving grain and oil rations to them (Guichard 2002: 160). This letter again emphasizes the distinct nature of the different population components, and also suggests that transactions taking place between them may have occurred on the basis of generalized reciprocity (Guichard 2002: 164; Durand 2004b: 191).

A shift in the nature of the evidence for pastoralism has been observed to occur from the mid–late third millennium to the early second millennium (see Section 6.4). For the earlier period, the faunal material is dominant, whereas for the second millennium, the number of sites with faunal assemblages is rather limited, and textual evidence is much more abundant. It is important, therefore, to consider the possibility that the reconstructed late third–early second millennium BC shift in the organization of specialized pastoralism does not represent a historical development, but is recognized as result of changes in the nature of the available evidence. Although the early second millennium faunal material is relatively scarce compared to the earlier periods, no significant changes seem to have occurred in the composition of the range of domestic animals that were kept at sites. It is clear, however, that epigraphical as well as settlement evidence indicate changes in the organization of pastoral production during the late third–early second millennium BC. The changes documented by the respective types of evidence cannot presently be definitively linked to each other, but the Mari evidence clearly documents the result of this transition.

6.6 Conclusions

The present chapter has highlighted the evidence for pastoralism in the Near East in general, and in northern Mesopotamia in particular. Animal husbandry has been practised since the late PPNB, but the organization of this subsistence strategy has proved much more difficult to reconstruct. It seems that at various points in time, and as a response to different environmental and socio-cultural conditions, pastoralism was spatially separated from areas where agriculture was practised in order to meet higher subsistence requirements.

During the mid-third millennium BC, this process seems to have taken place in the Upper Khabur area, whereby the Middle Khabur and Jebel Abd al-Aziz regions functioned as overflow regions where large-scale pastoralism could be practised without encroaching upon agricultural production. This specialized pastoral subsistence strategy was practised out of relatively large settlements around the Jebel Abd al-Aziz, that took advantage of local favourable environmental conditions and where agriculture was practised only in a subsidiary fashion. At the same time, however, small-scale mixed animal husbandry was carried out at urban sites, possibly by individual households and aimed at household subsistence production. For the late third millennium, the faunal assemblages from sites and textual evidence suggest that this dual economy continued to exist, although evidence on the exact nature and localization of specialized pastoralism is lacking.

During the early second millennium, the Mari evidence suggests the existence of a highly mobile population component involved in pastoral production. Although second millennium bone assemblages from northern Mesopotamia are relatively scarce, it seems that sheep/goat remained the dominant domesticated animal. Sheep/goat flocks were moved over relatively large distances, and the texts indicate that pastoralists were recognized as a distinct group within tribal confederacies. The absence of settlement evidence comparable to the mid-third millennium *Kranzhügel* suggests that if the *hana* maintained settlements or camps in the *nawûm* areas, these settlements were much more ephemeral than before, again suggesting a higher mobility than existed in earlier periods. Alongside this mobile pastoralism, village-based herding continued to be practised, as possibly evidenced by the occurrence of the terms *hasiratum* and *rubsatim*, referring to 'camps' and 'pens'. Thus, it is possible to reconstruct for the early second millennium the dual pastoral economy that existed previously, but with a change in the organization of the fully specialized subsistence strategy. The archaeological settlement evidence from the Wadi Ağiğ region seems to support this reconstruction (see Section 5.4.1). There, third millennium BC material was conspicuously absent from the region, whereas second millennium BC material was relatively scarce. The spatial patterning of this material resembled that of the Medieval Islamic period, when nomadic pastoralists certainly visited the region. Thus, in this area that was certainly within the influence sphere of Mari, evidence for mobile occupation in the early second millennium appeared after an absence of some thousand years.

The present chapter thus argues for the existence of two separate pastoral economies, distinguished by the degree of specialization. It is unlikely that there was ever a fully specialized agriculturalist or pastoralist way of life. Rather, each strategy was implemented as dictated by environmental and social conditions. In each subsistence system, agriculture as well as pastoralism was practised, but the degree of specialization differed, as well as the degree to which the subsistence strategy could meet the demands of the social group. At least from the mid-third millennium onwards, a dual pastoral economy consisting of a highly specialized component and a component representing pastoralism practised in a subsidiary way can be reconstructed. This dual economy fluctuated not only in terms of the relation between the two components, but also in the organization of each component. Whereas the pastoral production carried out by agriculturalists was geared toward local consumption, the specialized pastoral production was geared toward the production of exchangeable products, notably wool, though the potential importance of dairy products should not be underestimated. Whereas this distinction can be identified for the mid to late third millennium, in the absence of economic administrative archives of the Mari state, it remains to be determined whether wool production was also the economic rationale for the *hana* pastoralists.

7 The development of social networks

7.1 Introduction

This chapter will focus on exchange networks as a mechanism against environmental stress. It has already been argued that the geographical configuration of such networks is not random. More specifically, exchange networks will preferentially develop between groups and/or regions relying on different resources, or experiencing different environmental conditions (Section 3.5.2). In other words, exchange networks can only effectively buffer against environmental stress if their geographical expansion exceeds the area where stress is experienced, or if they forge ties between groups practising different subsistence strategies. These exchange networks can potentially be reconstructed through their social component, that is, the identity or components thereof likely to be shared by the network participants (Section 3.6). Historical records provide details on the constellation of northern Mesopotamian society in the form of the names of various social groups that lived there during the third and early second millennium BC. By analyzing the social nature of these groups, it can be determined whether or not they represented identities functioning in the way that was hypothesized in Section 3.5.3.

There is one Mesopotamian identity that has the greatest potential for functioning as the social expression of an exchange network, namely the Amorite identity. The historical records make clear that the group called the Amorites existed in numerous regions of the Near East and was involved in various activities and subsistence activities, thereby fulfilling the two main criteria for functioning as a social expression of an exchange network. The present chapter will start with an evaluation of the main historical sources on the Amorites (Section 7.2), followed by a discussion of how the Amorite identity should be understood, and whether or not this identity actually functioned as the social expression of an exchange network (Section 7.3). This chapter does not aim to give a complete overview of the evidence for the existence of Amorites in greater Mesopotamia. Rather, the focus will be on those texts that highlight the nature of Amorite identity, their perception of themselves and how they were perceived by others.[85]

7.2 The Amorite identity in the textual record

The term Amorite is the English translation of several loosely connected terms in ancient Near Eastern languages. It is thought to derive from the Sumerian martu through the Akkadian *amurru* and the Hebrew `*ēmōrī*. These terms had several interconnected meanings, complicating interpretations of texts in which the terms were used. At different times and in different contexts, the term martu/*amurru* could be used to designate the compass direction west, a specific area, and/or a people. How each of these uses evolved, and which meaning led to the others, is not entirely clear, as the different meanings are related and probably evolved out of each other (Buccellati 1966: 361–2; Whiting 1995: 1231–2). Any differences in meaning that may have existed between martu and *amurru*, and between different periods, are even further conflated by the translation of these terms into a single English term Amorite. The issue is complicated further by the fact that Amorite does not survive as a written language. What is known of the Amorite language primarily stems from names that are thought to be Amorite, and from rare words and expressions in Akkadian texts (Whiting 1995: 1233). Despite these uncertainties, the term Amorite will be retained here, partly because no adequate alternative is currently available, and partly because the present study deals

---

85 See Buccellati 1966, Gelb 1980, and Streck 2000, 2004 for statistical analyses of the occurrences of Amorite names and words in historical texts.

date BC	period/reign	description
1792–1750	Hammurabi	Amorite genealogy, claimed Amorite affiliation
1810–1761	Mari	Mari as Amorite state under Yahdun-Lim, Shamshi-Adad/Yasmah-Addu, and Zimri-Lim
1830–1776	Shamshi-Adad	Amorite genealogy
1834–1823	Warad-Sîn	Warad-Sîn's father Kudur-mabuk was called 'father of the Amorite land'
1941–1933	Zabāia	ruler of Larsa claimed Amorite affiliation
2019–1987	Išbi-Erra	described as 'a man from Mari, with a dog's intelligence'
2026–2003	Ibbi-Sin	correspondence between Ibbi-Sin and Išbi-Erra concerning grain transports and Martu activities
2032–2031	Šu-Sin 4–5	year name mentioning construction of the *muriq-tidnum*
2032	Šu-Sin 4 (*terminus post quem*)	correspondence between Šu-Sin and Šarrum-bani concerning the *muriq-tidnum*
2056–2055	Šulgi 37–38	year names mentioning construction of the 'wall of the land'
2056	Šulgi 37 (*terminus post quem*)	correspondence between Šulgi and Puzur-Šulgi concerning the 'wall of the land'
2056	Šulgi 37 (*terminus post quem*)	correspondence between Šulgi and Arad-ĝu concerning the 'wall of the land'
2064	Šulgi 29 (*terminus post quem*)	Amorite names in administrative records
c. 2100	Gutian/Ur III?	Curse of Agade: Amorites are described as 'people ignorant of agriculture'
c. 2100	Gudea	transport of building materials from the mountain range of the Martu
c. 2160	Šar-kali-šarri	year name mentioning victory over Martu at Bašar
	Naram-Sin	victory over Amar-Girid of Uruk at Bašar; capture of Martu leaders
c. 2350	Tell Beydar	Amorite names (?)
c. 2350	Ebla palace G	various interactions between Ebla, Martu, and Tidnum (?)
c. 2500	Lagash (ED III)	Amorite toponyms
c. 2500	Shuruppak (ED IIIa)	Amorite personal names

Table 7.1: Selected textual references to Amorites in chronological order. See main text for references.

primarily with texts from the third and early second millennium BC, thereby circumventing some ambiguities that were introduced by the slightly later use of *amurru* to refer to a northern Levantine state, rather than to a group of people (Klengel 1992: 160; Whiting 1995: 1236).

Due to the uneven spatial and chronological distribution of the available evidence, the present review includes historical records from greater Mesopotamia (Table 7.1). For the purpose of the present overview, the mid–late third millennium and early second millennium will be divided in three periods, for which clear differences in the nature and availability of the evidence can be demonstrated. The first period includes all evidence predating the Ur III period. This evidence is meagre and sometimes rather ambiguous. The middle period coincides with the Ur III period, when there is much more evidence on Amorite presence and identity, although most evidence comes from southern Mesopotamia. The final period covers the early second millennium and roughly corresponds to the Isin-Larsa/Old Babylonian periods. The degree to which differences in the available evidence influence the interpretation of the texts will be addressed below (Section 7.3.1).

7.2.1 Amorites before the Ur III period

Steinkeller has suggested that the term Tidnum, which is used nearly synonymously for Amorite during the Ur III period (see Section 7.2.2), possibly already occurs in archaic ED I texts from Ur (1992: 265), but this suggestion cannot be further substantiated at the moment (Marchesi 2006: 22–3). The first reliable references to people of supposedly Amorite affiliation date to the ED IIIa period and come from texts found in Shuruppak. There are a few names of supposedly Amorite origin, despite the fact that the determinant Martu is lacking (Bauer et al. 1998: 265, n. 268). However, this qualification does appear at this site in combination with personal names, for example in a list of barley rations: '900 litres of barley for Esur'agga, the Amorite' (Marchesi 2006: n. 95). Another text reads '45 men receiving 1 loaf of bread (each), (and) 28 women receiving one loaf of bread (each) – (they are) Amorites' (Marchesi 2006: n. 95). In addition, a 'land of the Amorites' is mentioned in a literary composition (Marchesi 2006: n. 95). In ED III texts from Lagash, several toponyms occur in association with Martu. An inscription from Ur-Nanshe mentions the digging of a 'canal of the Amorites' (Marchesi 2006: 23–4; Frayne 2008: E1.9.1.31).[86] References to a 'bridge of the Amorites' occur several times in these texts (Marchesi 2006: 24). A pre-Sargonic literary text from Mari, possibly based on an earlier but otherwise unknown text from Lagash, mentions an 'Amorite shepherd' (Marchesi 2006: n. 99 quoting Bonechi and Durand 1992).

The Ebla palace G archives contained several references to Amorites and an Amorite region (Archi 1985). An inhabitant of Tuttul (Tell Bi'a) was called a resident of Martu (Archi 1985: text 30). Inhabitants of Emar were in some way involved in a victory over Amorites: '(fabrics) for Aršum of Emar, for a votive offering (when/because?) Martu was destroyed, and for one of (Aršum's) sheep for the victory. (Fabrics) for Baluzú of Emar, when Martu was destroyed' (Archi 1985: text 9). There were Amorite rulers and councils of elders who each received gifts of cloth (Archi 1985: text 14). Several texts recorded the delivery of sheep by people from Martu. Finally, there was a list including personnel from Martu. Note that individuals designated as coming from Martu(m) never had Amorite names. A possibly older variant of the term Tidnum has also been attested in the Ebla archives (Marchesi 2006: 11).[87]

There is discussion about the occurrence of Amorite personal names in the roughly contemporaneous texts from Tell Beydar. According to Talon, there were no Amorite names or Amorite elements in the Beydar text corpus (1996: 75). However, Streck (2002: 191) suggested that several names may have been Amorite.

Amorites were mentioned by Naram-Sin and his successor Šar-kali-šarri. Naram-Sin's account of the Great Revolt described how he crossed the Euphrates and fought the decisive battle against Amar-Girid of Uruk at Bašar, the mountain of Martu. After the battle, Naram-Sin captured Amar-Girid and, among many others, two persons named Belili and Kindagal, who were described as leaders of the Martu people (Sommerfeld 2000: 423–6). Naram-Sin's successor Šar-kali-šarri was the second Akkadian king to mention the Martu. One of Šar-kali-šarri's year names recorded a victory over the Martu at Bašar: 'The year in which Šar-kali-šarri defeated the MAR.TU in Bašar' (Sommerfeld 2000: 435).

The Curse of Agade, probably to be dated within 100–200 years after the Akkadian period (Cooper 1983: 11–12), described how the Amorites submitted to Naram-Sin and offered to the goddess Inanna: 'The highland Martu, people ignorant of agriculture, brought spirited cattle and kids for her' (Black et al. 1998–2006: t.2.1.5).

Gudea statue B described, among other things, the building of the Ningirsu temple, and paid close attention to where building materials came from: 'From Umanum, in the mountain range of Menu'a, and from Pusala, in the mountain range of the Amorites, he (=Gude'a) brought down big stone slabs, … From Tidanum, in the mountain range of the Amorites, he brought alabaster blocks' (Marchesi 2006: 16). According to Marchesi, Pusala and Tidanum were two different places in the mountain range of the Amorites, rather than being two synonyms for the mountain range of the Amorites, as had been suggested earlier.[88]

---

86 Frayne remarks that it is not entirely clear whether Martu here indicates the Amorite identity or the compass direction west. He suggests that the Amorite interpretation is plausible, based on the mentioning of LUM-ma as Eanatum of Lagash' Tidnum name. However, Marchesi recently argued that the link between LUM-ma, Tidnum, and Eanatum is based on a false reading of the text, and suggested that LUM-ma should be understood as an as yet unknown ruler of Lagash (2006: 124-5).

87 Note that mar-TUM[ki] has recently been identified as a cult place near Ebla, possibly necessitating a revision of some of the conclusions drawn by Archi (Sallaberger 2007: 445, n. 138).

88 Marchesi's interpretation also contrasts with earlier readings in its reading of Pusala, which had previously been read as Bašar, for example by Edzard 'He brought down big stone slabs from Umānum, the mountain range of Menua, and from Basar, the mountain range of the Martu, … He brought alabaster blocks from Tidanum, the mountain range of the Martu' (1997: E3/1.1.7.StB, vi 3-14). It also contrasts with Michalowski (1976: 111) who argued that Bašar and Tidnum were two distinct places that were each called 'Mardu highland'. See Sallaberger (2007: n. 79) for further discussion of the problematic reading of Gudea Statue B.

7.2.2 Amorites in the Ur III period

There are numerous historical records from the Ur III period (2110-2003 BC) dealing with or mentioning Amorites. These records generally fall into one of two classes: administrative records mentioning Amorite names or literary texts including year names, royal correspondence and other public literature describing activities of Amorite groups, usually without mentioning individual names. As will be seen below, these two classes of historical records present the Amorite identity in fundamentally different ways.

The occurrence of Amorite names in Ur III administrative records has been studied by Buccellati (1966). His study of Amorites during the Ur III and early Isin-Larsa periods provided a picture of Ur III society in which Amorites played an active role and were involved in various economic activities. His main sources of information were administrative texts from archives found in Puzrish-Dagan, Isin, and Lagash. He was able to identify over 300 names with the determinant mar-tu, or with West-Semitic elements that are thought to be Amorite. He then studied the economic and social contexts in which these names occurred, and the spatial and chronological distribution of these names.

The greater share of the Amorite names occurred in the Puzrish-Dagan texts, which dealt with the circulation of animals by the central administration. Amorites were recorded as delivering and receiving animals for a variety of purposes. As contributors, Amorites delivered relatively large numbers of fat-tail sheep. Although Buccellati found no seasonal pattern in the presence of Amorites at Puzrish-Dagan (1966: 346), it has been shown that there were seasonal fluctuations in the number of animals that were delivered (see Section 6.4). The texts from Isin described transactions of leather products. The majority of Amorites apparently received these products, and, with a few exceptions, were involved neither in the production of these products, nor in the contribution of the raw materials. Finally, the Lagash administration primarily documented the distribution of food rations to various officials, several of whom were Amorites. The Amorite officials primarily held lower positions within the bureaucracy. Amorites also held positions outside the state administration, including such diverse functions as solder, priest, brewer, or farmer (Buccellati 1966: 340-2).

Buccellati cursorily referred to the textual evidence from other sites (Buccellati 1966: 316). Thus, at Eshnunna, there were numerous references to Amorites in texts from both the Ur III and Isin-Larsa periods. At Nippur, Amorites were virtually absent from the textual record. The texts from Umma suggested to Buccellati that relatively few Amorites lived there. At Larsa, evidence for an Amorite presence came from the first two names on the Larsa king list, which occurred as Amorite names elsewhere. At Ur, attestations of Amorites were also few and far between, and they primarily occurred in administrative texts recording the distribution of goods.

The onomastic evidence suggests that Amorites were well-integrated into the Ur III society, and that they functioned in the state bureaucracy, as well as in other parts of society. The references to Amorites in literary sources are of a different nature. Literary sources, mainly consisting of year-names and the letters from the Royal Correspondence of Ur, usually provide a rather one-sided picture of the Amorites, using expressions and descriptions that clearly aimed to portray Amorites as enemies of the state. The Amorites were first mentioned in year names from Šulgi's reign. The year names of Šulgi 37 and Šulgi 38, mentioned the construction of the 'wall of the land' (bad$_3$-ma-da) or the 'wall facing the highlands' (bad-igi-hursaga).

This 'wall of the land' also played a significant role in the so-called Royal Correspondence of Ur, a body of letters between several Ur III kings and high officials within the state (Michalowski 1976). It should be noted, however, that several researchers have questioned the historicity of the so-called Royal Correspondence of Ur, of which the letters from or to the Ur III rulers quoted here are a part. These letters are known only from Old Babylonian copies, in other words, they were made at a time when fundamental changes had occurred in the socio-political constellation of southern Mesopotamia as compared to the Ur III period. The problem is that it is difficult, if not impossible, to distinguish historical details from parts that were added or invented when the texts were composed or copied. Michalowski, however, has shown that personal names in the royal correspondence of Ur also occurred in genuinely Ur III texts, suggesting that the texts at least incorporated some historical details (Michalowski 1976: 65, 67). Therefore, it is perhaps best to treat these texts as secondary historical sources (Streck 2000: 36; Charpin et al. 2004: 59, n. 145).

A letter from the high official Aradĝu to Šulgi referred to the construction of the wall: 'As to the fortification which my lord sent me back to, the work on it has been put into effect. The approach of the enemy is kept at a distance from the Land. My lord continues to maintain his sublime reputation in the south and the uplands, from the rising to the setting sun, as far as the borders of the entire Land. {The rebellious (?) Martu have turned back ......}' (Black et al. 1998–2006: t.3.1.06). The correspondence between Šulgi and another official, Puzur-Šulgi, was also concerned with the construction and restoration works of the 'wall facing the highlands'. After boasting that he had built the wall and brought peace and prosperity to the land, Šulgi wrote: 'When the master-builder (?) has taken up the work concerned, he is to re-establish securely any place where the fortification has fallen into ruins. Let him reinforce and also rebuild it. The neglected work load is to be completed within one month; I shall be questioning him about this work.' (Black et al. 1998–2006: t.3.1.08, version A, segment B). The urgency of the task was again stressed at the end of the letter: 'The orders are rigorous: you should not neglect your work load. They are to proceed with the building work by night and in the heat of noon. You will not be sleeping during the night or in the heat of noon! You should know this! It is urgent!' (Black et al. 1998–2006: t.3.1.08, version A, segment B). The same letter also mentioned that the '...... the Tidnum have returned (?) from their mountain land' (Black et al. 1998–2006). In another letter, Puzur-Šulgi wrote to Šulgi: 'I am also well-informed about the oracular signs concerning the enemy: the enemy has replenished his strength for battle. However, my strength is limited. I cannot strengthen the fortress further or {guard it}' (Black et al. 1998–2006: t.3.1.07). He furthermore suggested that Šulgi send him 7200 workers for the construction of the wall (Black et al. 1998–2006: t.3.1.07).

The mentioning of the Tidnum in close association with the 'wall of the land' is important, because it establishes a link between the Tidnum and the Amorites as groups that are apparently in some way related. In different copies of texts belonging to the Royal Correspondence of Ur, Tidnum and Martu are used synonymously for both a group of people, and a geographical place name, suggesting at least a considerable overlap in meaning. This suggestion is further enhanced by several lexical lists that equate Tidnum with Martu (Marchesi 2006: n. 20, 23). At the same time, a text from Šu-sin describing a battle indicates that Tidnum could also be used in a more restricted way to describe a particular group of Amorites: 'Hostile Amorites – Tidnum and Yamatiyum – came forth (to do battle): their kings confronted him (=Šusîn) in battle, (but) by the strength of Enlil and Ninlil he was victorious in battle' (Frayne 1993b: E3/2.1.4.1; Marchesi 2006: 12).

During the reign of Šu-sin (2035-2027 BC), there was again work on a wall as a defence against Amorite attacks. Šu-sin's fourth and fifth years were named after the construction of this wall. The official Šarrum-bani, who had been ordered to build or reconstruct the wall, wrote to Šu-sin: 'You sent me a message ordering me to work on the construction of the great fortification Murīq-Tidnim. {You presented yourself before me} [...], announcing: "The Martu have invaded the land." {You instructed me} [...] to build the fortification, so as to cut off their route; also, that no breaches of the Tigris or the Euphrates should cover the fields with water' (Black et al. 1998–2006: t.3.1.15). This letter also indicated that Šarrum-bani had been building and fighting at the same time. Šu-sin replied that Šarrum-bani had not been instructed to fight, but only to reconstruct the wall that Šulgi had built (Black et al. 1998–2006: t.3.1.16), indicating that the 'wall facing the highlands' (bad$_3$ igi-hursaga) and the 'wall which repels the Tidnum' (bad$_3$ mu-ri-iq-tidnim) were probably the same, or at least located in the same region.

In the absence of archaeological evidence for the 'wall of the land', its location can only be reconstructed from the textual evidence. Michalowski has shown that the 'wall of the land' was usually mentioned in association with place names that can be located in or near the Diyala region. More specifically, the wall is described as running between the Abgal waterway and Zimudar, and having a length of 26 danna (Black et al. 1998–2006: t.3.1.15).[89] The Abgal waterway has been equated with the western branch of the Euphrates at Sippar, whereas the location of Zimudar is uncertain but probably lies somewhere in the Diyala region (Michalowski 1976: 114). This location in the Diyala region and the fact that the 'wall which repels the Tidnum' was also called 'wall facing the highlands' suggest that the wall was built toward the Jebel Hamrin (Michalowski 1976: 115). The nature of the wall remains unclear. It is unlikely that it was a large continuous wall, and a chain of small fortifications or watchtowers has been proposed as a more likely alternative (Buccellati 1997: 109).[90]

A letter by Išbi-Erra, the later ruler of Isin (2019-1987 BC), to the last Ur III king Ibbi-Sin (2026-2003 BC) indicated that Išbi-Erra was instructed to purchase grain, but was forced to abandon this expedition because of the presence of Martu: 'I heard news that the hostile Martu have entered inside your territories. {I entered with 72,000 gur of grain} [...] – the entire amount of grain – inside Isin. Now I have let the Martu, all of them, penetrate inside the Land, and one by one I have seized all the fortifications therein. Because of the Martu, {I am unable to hand over} [...] this grain for threshing. They are stronger than me, while I am condemned to sitting around.' (Black et al. 1998–2006: t.3.1.17). The letter furthermore indicated that Elam, with which the Ur III state was also in conflict, was likewise suffering from grain shortages (Black et al. 1998–2006: t.3.1.17).[91]

In another letter from the Royal Correspondence of Ur, Ibbi-Sin described Išbi-Erra as 'not of Sumerian origin' and as a 'man from Mari, with the understanding of a dog' (Black et al. 1998–2006: t.3.1.20). Cooper noted that the qualification 'with a dog's intelligence' was usually applied to Amorites and Gutians, but not to foreigners in general because Elamites were almost never described as such, leading him to doubt the Babylonian origin of Išbi-Erra (1983: 33). The Mari origins of Išbi-Erra were also highlighted in a hymn known from Old Babylonian copies, but probably originating from the time of Išbi-Erra, that described how the gods Anu and Enlil elevated Išbi-Erra to kingship (Michalowski 2005).

Amorites also appeared in several literary compositions. In the well-known literary composition *The marriage of Martu*, the god Martu is described as follows (Black et al. 1998–2006: t.1.7.1): 'Now listen, their hands are destructive and their features are those of monkeys; he is one who eats what Nanna forbids and does not show reverence. They never stop roaming about ......, they are an abomination to the gods' dwellings. Their ideas are confused; they cause only disturbance. He is clothed in sack-leather ......, lives in a tent, exposed to wind and rain, and cannot properly recite prayers. He lives in the mountains and ignores the places of gods, digs up truffles in the foothills, does not know how to bend the knee, and eats raw flesh. He has no house during his life, and when he dies he will not be carried to a burial-place'. Note, however, that this composition does not describe the Amorites as such but the god Martu/Amurru, necessitating an investigation of the link between this god and the Amorites. It has been observed that Martu/Amurru does not occur as a theophoric element in Amorite names, but that it does occur in Akkadian names. Beaulieu therefore suggested that the god Martu/Amurru was a Sumerian/Akkadian construct 'on the basis of their perception of a separate ethnic identity' (Beaulieu 2005: 35). As such, the god Martu/Amurru may have served as a stereotypical image of the Amorites in various aspects of Mesopotamian society.

---

89 The exact length of 1 danna is not entirely clear but may range between 7.3 and 11.5 km, with 10.8 km being most probable (Powell 1990: 467), making the wall between 190 and 300 km long, and 280 km if 1 danna≈10.8 km is accepted.

90 Sallaberger rightly points to the low wall that has recently been found near Al-Rawda and that may eventually shed new light on the 'wall facing the highlands' (2007: n. 136)

91 During the Ur III period, 1 gur=300 sila, and 1 sila≈1 litre (Powell 1990: 493), so that 72000 gur would be 21600000 litre. It should be noted however that this letter is not an administrative, but a literary text, making the accuracy of these figures doubtful.

There is evidence from the Ur III period that Amorite existed as a spoken language. A literary text mentioned an 'Amorite interpreter', and Šulgi claimed to be able to speak Sumerian, Elamite, and Amorite. Letters from Old Babylonian Mari recognized Akkadian, Sumerian, Subarean, and Amorite as separate languages (Streck 2000: 76). Nevertheless, Sumerian and later Akkadian remained the dominant written languages.

For northern Mesopotamia, there is some evidence for Amorite names in texts from Mari and Tell Bi'a. Several of the latest Mari *šakkanakkus* bore Amorite names and words like *gayyum* (clan) were also already present (Streck 2004: 340).

7.2.3 Amorites in the early second millennium BC

The Ur III period and the *šakkanakku* period in Mari ended shortly before 2000 BC, after which documentation on the presence of Amorites in Upper Mesopotamia ceased, only to start again some 200 years later. This is unfortunate, since this period must have witnessed significant changes, including the appearance of Amorite kingdoms across Mesopotamia. Thus, Charpin and Ziegler (2003: 29) called this period 'la grande vague' and argued that it witnessed large population displacements or 'invasions amorrites'. By the time this epigraphic silence had been broken, the political landscape of greater Mesopotamia had changed considerably. At the same time, the Sumerian determinative mar-tu to designate Amorite persons disappeared almost completely. Buccellati observed that this development had already commenced during the Ur III period (1966: 358). After this time, Amorites could only be distinguished through their names, or if they specifically stated their Amorite affinity.[92]

During the Old Babylonian period, numerous city-states were now ruled by rulers with Amorite names, or who claimed Amorite descent. Amorite dynasties became established at Babylon, Eshnunna, Isin, Larsa, Mari, Qatna, and Yamhad. Several of these kings specifically mentioned their Amorite affiliation. Thus, Zabāia, fourth king of Larsa (1941-1933 BC), wrote: 'Zabāia, Amorite chief, son of Sāmium, built the Ebabbar' (Frayne 1990: E4.2.4.1). The title 'Amorite chief' or 'commander' was also used by Abi-Sare of Larsa (1905-1895 BC) and Sin-gamil of Diniktum (Michalowski 1976: 121). Kudur-mabuk, the father of Warad-Sîn of Larsa (1834-1823 BC), was called 'father of the Amorite land' (Frayne 1990: E4.2.13.13, l. 14-15). Note the kinship element present in this title. Hammurabi of Babylon called himself 'king of all the Amorite land' (Frayne 1990: E4.3.6.8) and 'king of the Amorites' (Frayne 1990: E4.3.6.2001).

Some kings apparently even claimed or constructed a shared Amorite ancestry, that is, Shamshi-Adad of Ekallatum and Hammurabi of Babylon. Comparison of their genealogies, as recorded by the Assyrian King List (AKL) and the Genealogy of the Hammurabi Dynasty (GHD), reveals some interesting patterns (Finkelstein 1966). Several names of rulers preceding the first dynasty of Babylon on the GHD also appear on the AKL, although in a different order. These names are *Heana/Hanū*, *Ditānu/Didānu*, and *Zummabu/Zu'abu*. In the AKL, these kings were among the seventeen kings who lived in tents. *Ditānu/Didānu* is known as an Amorite group, whereas *Heana/Hanū* is a term that is closely associated with pastoralism is general (see Section 6.5.2). Two later entries in the GHD are very similar to the names of Amorite groups living in Sippar. Finkelstein furthermore suggested that several personal names in the AKL appeared in a fused and/or distorted way in the GHD, and vice versa, eventually concluding that 'the genealogical traditions of the Hammurapi dynasty and those of the Assyrian King List – the first two sections of which must almost certainly be identified as the "Ahnentafel" of Šamši-Adad I – are one and the same insofar as they represent a consciousness of tribal origins' (1966: 99).

The structure of the emerging Amorite kingdoms is most clearly visible in the archives from Mari, supported by several smaller archives from other northern Mesopotamian sites, including Tell Leilan, Chagar Bazar, and Tell al-Rimah. These archives primarily document a relatively short time-period centred on the reigns of Shamshi-Adad and Zimri-Lim. The Mari archives, being the largest among these text corpuses and containing *c.* 20000 complete and fragmented tablets,

---

92   The appearance of Amorite dynasties across Mesopotamia and western Syria led Charpin to suggest that the Old Babylonian period could more appropriately be named the Amorite period (2004: 31).

consist predominantly of administrative documents relating to the functioning of the palace, and letters dealing with a variety of subjects, from which state affairs can be reconstructed in unprecedented detail. Precise data on the archaeological contexts are lacking for the texts found in the early excavations by Parrot, but some patterns have emerged from the more recent excavations at Mari. These indicate that the majority of texts pre-dating the reign of Yasmah-Addu were probably discarded or buried in the past, possibly indicating a lack of interest on Yasmah-Addu's part for keeping alive the remembrance of earlier rulers. The archives from the reign of Zimri-Lim, on the other hand, contained letters from the reign of his predecessor, perhaps indicating a stronger historical consciousness (Durand 1997: 29). Nevertheless, it should always be kept in mind that these texts were part of the palace archives, and written by palace officials. The views expressed in them therefore were the views of the state, and the matters discussed in them were of importance to the state.

Before the structure of the Mari state is more fully explored, it may be useful to clarify the use of the terms tribe and tribal, and whether these are applicable to the society that is documented in the Mari archives. This study will follow the proposals made by Fleming (2004: 26), building on earlier work by Salzman (1978a) and Khoury and Kostiner (1990a). In this approach, the term tribal is used to describe a type of social organization usually characterized by real or constructed kinship ties, and by shared cultural traits such as language and customs (Khoury and Kostiner 1990b: 5). A tribal confederacy represents a coalition of tribes that is politically united, although this does not necessarily involve a centralized government system (Fleming 2004: 27). Within this approach, the term tribe is devoid of the evolutionary connotations evident in the works of Sahlins and Service. Thus, a tribal organization is not *a priori* more advanced or primitive than other archetypical forms of political and social organizations such as the state or the chiefdom (Parkinson 2002: 9).

It appears that Mari society was organized along lines of tribal affiliation. Rowton already realized the importance of tribal affiliations in Mari and used the terms 'dimorphic chiefdom' or 'dimorphic state' to describe a polity uniting tribal (semi-)nomadic groups and non-tribal sedentary people living in towns (Rowton 1973a).[93] In Rowton's model of the Mari state, the state was essentially non-tribal in nature, headed by a non-tribal ruler who was assisted by primarily non-tribal officials. The state continuously tried to control the tribal elements that were present within its jurisdiction. These tribal groups, however, actively resisted and, because they were (semi-)nomadic, could and often tried to evade state control by moving into the steppe beyond the state boundaries. However, due to the cyclical nature of resources, (semi-)nomadic pastoralists always had to return to the more fertile areas controlled by the state, and this cycle of intrusion and retreat was central to Rowton's notion of enclosed nomadism (Rowton 1973b, 1974). Rowton built his model of the structure of the Mari state on observations of pre-modern and historical tribe-state interaction in the wider Near East. Central to his model of the dimorphic state was the idea that town and tribe were both governed by a single ruler or state administration. Thus, according to Rowton (1973a: 202), 'The hallmark of dimorphic structure is an autonomous chiefdom centered on a town in tribal territory. From this base a local dynasty exerts a varying blend of rule and influence over the nomadic and sedentary tribes in the countryside. The population of the chiefdom includes both a nontribal and a tribal element.' Within this structure, the power of tribal leaders fluctuated, but these leaders were eventually absorbed into or replaced by the essentially non-tribal state organization (Rowton 1974: 22).

However, recent research has indicated that the tribes or tribal confederacies were even more strongly interwoven with the Mari state than assumed by Rowton or Matthews (1978). The Mari state was not above or on a par with the tribes, but was itself entirely tribal. Thus, rather than assuming that the Mari rulers dealt with the tribal confederacies through intermediate (non-)tribal officials such as the *merhûm*, the assumption now is that the rulers themselves were members of one of these confederacies, the Sim'alites (Charpin and Durand 1986: 150). Within this view, the absence of indications for a tribal identity should not be understood as an absence of tribal affiliations, but as an indication that these affiliations were so obvious that they were not always

---

93   Rowton has written extensively on pastoralism, nomadism, and tribe-state interaction in recent times, as well as during the Old Babylonian period (see Rowton 1980: n. 2, for an extensive list of Rowton's articles on these subjects).

mentioned. In other words, in many cases where all indication of tribal identity is absent, the text probably related to the Sim'alite confederacy, because that confederacy dominated Mari politics (see Section 6.5.2).[94]

Several Amorite tribal confederacies can be recognized, of which the Sim'alites 'of the left hand' and Yaminites 'of the right hand' were the most important (Streck 2002: 175).[95] These tribal confederacies were ruled according to different power structures. This difference is most clearly visible in the fact that the Sim'alites were governed by a single ruler, that is, Yahdun-Lim and later Zimri-Lim, whereas the Yaminites had five rulers. Under their single ruler, the Sim'alites were organized in *gayum*, of which at least twelve are known by name (Durand 2004b: 179-80). The term *gayum* is often translated as 'clan' or 'tribe', whereas Fleming suggested the more neutral 'division' (2004: 43). No clear official position can be assigned to the *gayum*, and they may not have had any political functions (Fleming 2004: 58). Apparently a very strong social relation between *hana* and the *gayum* existed, in the sense that virtually only *hana* were identified as members of a *gayum*. Nevertheless, sedentary individuals could identify themselves as members of a *gayum* as well, indicating that a *gayum* was not a strictly pastoralist social unit. (Fleming 2004: 54-5).

Some evidence suggests that there was an overarching division of the *gayum* into *hana* of Yabasu and *hana* of Ašarugayum, to which most of the known *gayum* could be assigned. The Yabasu may have been located in the western part of the Jezirah, whereas the Ašarugayum were found around the Jebel Sinjar and along the Euphrates south of Mari (Durand 2004b: 180-3). This overarching division seems to have primarily functioned in military contexts, and Sim'alites were otherwise always identified by their *gayum* (Fleming 2004: 57).

The Yaminites were divided into *li'mum*, five of which are known by name: the Amnanû, Rabbû, Uprapû, Yahrurû, and Yarihû (Durand 2004b: 158). In fact, Yaminites were generally identified as members of one of these *li'mum*, rather than as Yaminites. Each Yaminite *li'mum* was governed by a *šarrum* or king (Durand 2004b: 158). Despite these organizational differences between the Sim'alite and the Yaminite tribal confederacies, the *gayum* and the *li'mum* were very similar in terms of membership, as they both represented the first tribal level by which individual Sim'alites and Yaminites were identified. The Yaminite *li'mum* similarly consisted of mobile pastoralists and sedentary people (Fleming 2004: 61-3). Among the Yaminites, the term *hibrum* seems to have denoted a mobile pastoralist population component in much the same way as *hana* and *gayum* did among the Sim'alites (Matthews 1978: 65; Durand 2004a: 390-1; Fleming 2004: 97).[96]

In his discussion on enclosed nomadism, Rowton was similarly concerned with the tribal organization as attested in the Mari archives. He drew attention to the fact that the Mari tribal confederacies could include sedentary as well as mobile population components. He used the term 'integrated tribe' to describe social groups consisting of sedentary as well as (semi-) nomadic households (1974: 2). For example, in a letter to Zimri-Lim by Hammî-ištamar, a ruler of the Uprapû and therefore a Yaminite, one reads: 'My lord knows that the leaders and the tent-dwellers are with me at Samânum. Because they had not met with their brothers, residents of the city, for a long time, they spent several days without moving' (translated from Durand 1998: text 733). The same distinction is even more strongly expressed in another letter about the division of land. According to this text, the 'house of Awin', part of the Yaminite *li'mum* Rabbû, is divided in five sons of Awin who live in Appan, and eight sons of Awin of the *hibrum* of the *nawûm*, presumably in the meaning of pasture area (Matthews 1978: 90). These texts not only clearly indicate that *hana* could be used within a Yaminite social context, but also, and more importantly, that mobile *hana* and *hibrum* on the one hand, and sedentary townsmen on the other were part of the same tribe and that their affiliation was described along kinship ties.

---

94 Consequently, as Streck rightly observed, the tribal-non-tribal dichotomy is not an apt characterization of a dimorphic chiefdom or state (Streck 2002: n. 145).

95 Early scientific literature on Mari nomadic pastoralism recognized the *hana* or Haneans as a further tribal group on an equal organizational level as the Sim'alites and Yaminites (e.g. Kupper 1957; Anbar 1991), but it has recently been shown that this view is erroneous (see Section 6.5.2).

96 Some social groups apparently stood outside the tribal system outlined here, i.e. the *habiru* (see Matthews 1978: 159-62).

Although there were areas with mixed tribal populations, the available evidence nevertheless indicates that attested individual tribal identities were more strongly associated with particular geographical regions than with others. This evidence points to the existence of settlements with a specific tribal affiliation, as well as to the location of pasture areas of specific tribes. According to Anbar (1991: 116), the Yaminite settlements were primarily located in the districts of Mari, Terqa, and Saggaratum. Their pasture areas were widely dispersed across northern Mesopotamia, including Yamhad, Zalmaqum, and the Jebel Bishri. Anbar distinguished the 'Haneans' from other tribes, but as has become clear, references to them should be understood as references to Sim'alites. Thus, the observations by Anbar that the 'Haneans' had a special connection with Ida-Marash, and that many Sim'alite settlements could also be found there actually reinforce each other (Anbar 1991: 116-17). The smaller tribal groups were primarily found along the periphery of Mari's influence. Thus, the Ya'ilānum were found east of the Tigris, whereas the Numha and Yamutbal were located in Terqa, Saggaratum, and south of the Jebel Sinjar.

It is clear, then, that there was not only a dichotomy in Mari between a mobile pastoralist population component and a sedentary population component (see Section 6.5.2), but also a tribal dichotomy. The first dichotomy opposed *hana* against those people associated with the *ālum*, whereas the second pitted the tribal confederacy of the Sim'alites against that of the Yaminites. Although these tribal confederacies were characterized by very different power structures, they were social units containing and therefore uniting smaller units from both the *hana* and *ālum* population components. An important consequence of this interpretation of Mari society is that conflicts between the Sim'alite rulers of Mari and the Yaminite confederacy were not conflicts between the desert and the sown, but between two tribal confederacies, each consisting of households practising pastoralist or agricultural subsistence strategies (Fleming 2004: 46).

Two such conflicts are particularly well known. The first of these is recorded in Yahdun-Lim's foundation inscription of the Shamash-temple. In this inscription, Yahdun-Lim described how he defeated a Yaminite coalition consisting of Amnanû, Rabbû, and Uprapû (Matthews 1978: 134): 'La'um, king of Samānum and the land of the Uprabum, Bahlukulim, king of Tuttul and the land of the Amnanum, Aiālum, king of Abattum and the land of the Rabbium, these kings attacked him (Yahdun-Lim) and to aid them (also) came the army of Sumu-Ebuh of the land of Yamhad. At the town, Samānum, the troops of the Yaminites assembled to him (Sumu-Ebuh) as a single entity and in a mighty battle these three kings of the Yaminites he (Yahdun-Lim) put in chains.' According to Matthews, this conflict represented a war between the urban ruler of Mari and the tribal Yaminites. However, because the Sim'alites were ruled by an individual ruler, whereas the Yaminites had several rulers, this text does not indicate a conflict between tribe and state, but between two tribal confederacies. The same situation seems to have arisen when Zimri-Lim ascended the throne of Mari and had to quell a rebellion of several Yaminite *li'mum* (cf. Durand 1998: 420).

Streck (2004) analysed the chronological and spatial development in the occurrence of Amorites names in greater Mesopotamia during the Old Babylonian period (2000-1600 BC), thereby supplementing Buccellati's work on the Ur III period. Using a database of Amorite names appearing in texts from numerous sites across greater Mesopotamia, he was able to document the relative importance of Amorite names in the total corpus. Because textual data from different periods were available at several sites, it was also possible to document changes through time. Lower Mesopotamia, the region east of the Tigris, and the Jezirah were each characterized by different diachronic developments. In lower Mesopotamia, the percentage of Amorite names in the total corpus decreased from an average of 15 % for early sites to 4 % for late sites. In the Diyala and Hamrin regions, the percentage of Amorite names was 23 %. The number of sites and their chronological spread is insufficient to point to any diachronic changes. Amorite names were particularly strongly represented in the Mari region and the Euphrates upstream to Emar. Archives from this region contained on average 78 % Amorite names. Further west, at sites like Carchemish, Aleppo, Mishrife, and Hazor, 75 % of the names was Amorite. At sites further east, like Chagar Bazar, Rimah, and Barri, the importance of Amorite names was much lower, ranging between 11-36 % (Streck 2004). Both Anbar (1991: 89-90) and Streck (2004: 334) pointed to the marked differences in the occurrence of Amorite names in tribal and non-tribal contexts in the Mari archives. Although the absence of tribal affiliation should not be taken as an indicator of non-tribal ascent, the results are nevertheless interesting. Amorite names made up only *c.* 40 % of the total number of names known from Mari. That percentage, however, increased to over 75-80 % when only names of tribal members were taken into account.

Streck was also able to study the occurrence of Amorite names within single families and over several generations in the archives of Mari and Sippar. He collected the names of parents and children whenever one or all of them bore an Amorite name. In 70 % of these cases, both the parent and the child had an Amorite name, whereas in 17 % of all cases, the parent had an Amorite name, but the child had an Akkadian name. In the remaining 13 %, the names of either the parent or the child were Akkadian, Hurrian, or Sumerian. The texts from Sippar that predate the reign of Hammurabi indicated that in almost half of the cases both the parent and the child had Amorite names. Akkadian-Amorite and Amorite-Akkadian parent-child combinations made up 30 % and 21 % of the total number of cases. In the later texts from Sippar, Amorite-Amorite name combinations were less common (11 %), whereas Amorite-Akkadian and Akkadian-Amorite parent-child name combinations had risen to 53 % and 33 %, respectively (Streck 2004: 342).

7.3 Determining the nature of the Amorite identity

The previous sections have presented historical evidence for the presence of Amorites in greater Mesopotamia. Due to the distribution of the data, this review has emphasized southern Mesopotamia over the north, and this bias will be addressed below. The previous sections have shown that the label Amorite could be applied in a number of ways, and that it may have changed its meaning over time. It is therefore important to discuss how the term Amorite may be interpreted. Previous researchers argued that the Amorites were nomadic pastoralists, possibly in the process of becoming sedentary (Buccellati 1966; Kenyon 1966; Lönnqvist 2008), that they represented a rural population, to be juxtaposed against an Akkadian urban population (Buccellati 1990: 102), or that they came from northern Mesopotamia, and were already well-accustomed to urban life (Porter 2007). However, the often proposed direct equation of Amorites with (semi-)nomadic pastoralists, or a more general rural class, is not supported by the data presented above. Instead, many researchers have described the Amorite identity as an ethnic label, allowing for a more fuzzy distinction between pastoralists and agriculturalists, or nomads and city-dwellers (Kamp and Yoffee 1980). However, this labelling is only useful if there is a shared understanding of the term ethnicity. Therefore, the application of ethnicity to archaeological case-studies and the Amorite case in particular will be discussed below.

7.3.1 Understanding ethnicity

There has been considerable debate on the nature of ethnicity, and on the application of ethnicity in archaeology (e.g. Barth 1969a; Bentley 1987; Emberling 1997; Jones 1997; Stone 2003; Bahrani 2006), severely limiting the usefulness of the concept of ethnicity in the debate on the Amorites in Mesopotamia. This debate suffered from (*a*) disagreement on a definition of ethnicity, (*b*) a historical burden resulting from an abusive association of ethnicity, race, and archaeology by modern ideologies and states, and (*c*) the inability to identify archaeological correlates of ethnicity. Principally, there are two approaches to ethnicity: the primordialist perspective that explains ethnicity through *habitus*, arguing that ethnicity is part of the unconscious values shared by a group; and the instrumentalist perspective that views ethnicity as an active identity that is maintained by an individual or group in the interaction with others (Stone 2003: 32). The second approach is inspired by the work of Barth (1969a), and has emerged as the dominant paradigm in the study of ethnicity (Emberling 1997: 295).[97]

---

97 Jones recently attempted to reconcile the two approaches by arguing that 'ethnic identification involves an objectification of cultural practices (which otherwise constitute subliminal modes of behaviour) in the recognition and significance of difference in opposition to others' (1997: 128).

The implications of recent anthropological research for the concept of ethnicity have not yet been fully considered in all studies of ethnicity in the ancient Near East. For example, van Driel's recent evaluation of the concept of ethnicity (2005) is a large step backwards in this discussion, completely neglecting Kamp and Yoffee's (1980) much more thoughtful consideration of the subject. According to van Driel, ethnicity and its signalling primarily relate to problems caused by conflicting land-use strategies (2005: 3). He further assumed that ethnic change occurred through contact, *Landnahme*, and acculturation (van Driel 2005: 3-4). Thus, writing about the Amorites, van Driel suggested that at the end of the Ur III period a *Landnahme* had occurred whereby land was given to military colonists (2005: 7). Through this argumentation, ethnic change is reduced to a process of migration and replacement of one social group by another, without consideration of the complex interactions between ethnic groups that are documented for the more recent past. Furthermore, the supposed Amorite take-over of power only seems to have occurred effortlessly because there is a relative lack of historical evidence on exactly the period during which this take-over allegedly took place, that is, the period immediately following upon the Ur III period.

Fundamental for the modern conception of ethnicity has been Barth's observation that ethnic identity is not only acknowledged by others but also self-ascribed (1969a: 11). Within this approach to ethnicity, it is not important to define an ethnic culture-core, but to define how, why, and when ethnic groups distinguish themselves from other groups. This is logical, since any social division can only be meaningfully articulated if both its members and any outsiders have a clear understanding of where and how the boundary or difference is drawn. Thus, according to Emberling (1997: 304), 'an ethnic group is most essentially a group whose members view themselves as having common ancestry, therefore as being kin. As kin units larger than any others, they must include members of more than one lineage or extended family. Members of an ethnic group usually possess some common language. Ethnic groups often are unified by constructions of their past, by perception of injustice in the past or in the present, and often by hopes of a future reunification.' In this perspective, ethnicity is separated from other social categories including race, language, and culture. Ethnic identity is the result of the dynamic social process between the actors themselves. Rather than looking for what constitutes and is included in an ethnic identity, the differences between ethnicities should be the focus of attention. In archaeology, this translates into identifying differences in material culture that are meaningful for the expression of ethnic differences (Stone 2003: 43). However, the situational nature of ethnic expression, that is, whether or not an individual or group decides to signal its ethnic identity, and through which channels that signalling is achieved, makes it impossible for archaeologists to clearly identify ethnicity from archaeological or historical records. No clear one-to-one relationships exist between ethnicity and particular cultural traits (Stone 2003: 43). Rather, depending on the context, different features can be used to signal ethnicity. Additionally, ethnicity is usually expressed through a combination of features (Horowitz 1985: 47). Because ethnicity is one of many identities an individual or group can assume, it must further be considered whether particular features signal ethnicity at all, or not some other aspect of identity (Emberling 1997: 299, 313; Stone 2003: 43).

Archaeologists thus face specific problems when attempting to identify the emergence and expression of ethnic identity. This is not only the case when ethnicity must be reconstructed from material culture alone, but also when historical records naming groups and documenting interactions between them are available. For example, there has been a long-standing debate on how the occurrence of non-Sumerian words in the earliest texts and the replacement of Sumerian by Akkadian as the most important written language should be interpreted. This debate has often been phrased in terms of migration. However, languages cannot be equated with ethnic identities, and other cultural and/or social processes may be invoked to explain changes in scribal traditions, which is the actual change observed by archaeologists and philologists (Yoffee 1995: 290; Emberling 1997: 314). As regards the reconstruction of ethnic identities from material culture, this can be most fruitfully pursued by using as many lines of evidence as possible. More specifically, important social boundaries will be more strongly marked than less important ones (Emberling 1997: 318; Stone 2003: 43), although this correlation does not indicate if in a given context ethnicity is an important social structure.

	self-identification	ascribed identity	shared traits
pre-Ur III	-	+/-	-
Ur III	-	+	-
early second millennium	+/-	+/-	+/-

Table 7.2: Comparison of available evidence for ethnic properties of Amorite identity through time (+: strong evidence; +/-: some evidence; -: very little/no evidence).

There are numerous examples of ethnic change, that is, cases where an individual or group changes its ethnic identity, and adopts a new identity. Because ethnicity is a social category, the mechanisms through which it is changed are also of a social nature, that is, they involve the acceptance or adoption of new socio-cultural processes and/or ideas. Ethnic change first and foremost assumes the acceptance of newcomers by members of the receiving ethnic group, and it assumes that there are socio-cultural processes through which inclusion into the group can be achieved. If such processes are in place, several motivations can lead to individuals or groups changing their ethnic identities. On a very general level, such motivations entail a clear economical, cultural, or social advantage of some sort to the group (Barth 1969a: 22; Abruzzi 1982: 21). Barth (1969b) for example clearly showed in his analysis of Pathan ethnic identity that Pathans assumed new identities when the situation forced them to do so, or when it was advantageous to them, despite a clear reluctance to abandon the socio-cultural concepts associated with being Pathan. This ethnic change was, however, only possible because the receiving group also saw advantages in increasing their numbers, and in adopting new members into their social system (cf. Horowitz 1985: 48).

Having reviewed some important aspects of ethnicity as identity, it is now possible to discuss whether or not these aspects apply to what is known about the Amorites. In other words, can the Amorite identity indeed be considered an ethnic identity? Table 7.2 indicates the relative strength of the evidence for ethnic properties of the Amorite identity for the three chronological periods discussed earlier. As is clear from this table, the overall evidence is not very strong, at first sight suggesting that the Amorite identity can only be considered an ethnicity during the Old Babylonian period. However, there are reasons to believe that the weak ethnic markers of the Amorite identity in the earlier periods are only partly due to the nature of the evidence, and that the Amorite identity may also be considered an ethnicity in those periods for which insufficient evidence is available. Below, it will be argued that the absence of these markers can to a substantial degree be attributed to the cultural and environmental contexts in which the Amorite identity developed.

7.3.2 The development of the Amorite identity

The development of the Amorite identity can be traced from *c.* 2500 BC, when it was first mentioned in historical sources. It is important to realize, however, that this date does not necessarily signify the emergence of that identity, but rather its appearance in the historical record. This appearance indicates either that the writer(s) had become aware of the existence of an Amorite identity, or that they considered that identity to be important enough to be mentioned. The scattered references to Amorites before the Ur III period are difficult to assess. Whereas the Amorites fought by Naram-Sin were part of the army of Amar-Girid and may have been mercenaries, the context of the conflict between Šar-kali-šarri and the Amorites was less clear. The Amorites may have been the main opponents, but the reason for the battle remains unknown. On the other hand, the earlier references to Amorites from Shuruppak and Lagash illustrate that already at that time Amorites were integrated in Mesopotamian urban society. More importantly, they are mentioned in very different contexts, that is, in a toponym, as recipients of barley rations, and as shepherds. Although it is dangerous to rely on one or two textual references, these references nevertheless suggest, together with what is known from the Ur III period, that Amorites did not enter lower Mesopotamia as foreigners at the end of the third millennium BC, but were there well before that time.

During the Ur III period, the historical sources point to two fundamentally different perceptions of Amorite individuals and groups (Buccellati 1966: 336). Administrative documents clearly illustrate the high degree of integration of Amorite individuals in all levels of Ur III society.

More interestingly, Amorites could reach high positions in the state bureaucracy, as demonstrated by Išbi-Erra, the later ruler of Isin. This observation contrasts the perception of Amorites as emerges from the literary sources and the Royal Correspondence of Ur. These sources focus on the unnatural and non-human aspects of the Amorite identity, thereby creating a clear dichotomy between friends, that is, the Ur III state and its allies, and foes, that is, Amorite outsiders, that only partially reflects reality. The official documents are generally concerned with the usually hostile interactions between the state and Amorite groups. It seems therefore logical to assume that these documents purposefully created an image of these Amorite groups as non-human enemies as a kind of state propaganda (cf. Meijer forthcoming). The fact that the Ur III state considered these groups as enemies derived not from their Amorite ethnicity *per se*, but from their raids into land that was controlled or claimed by the state. Thus, some Amorite groups acted in a way that the Ur III state deemed hostile, and it is specifically against these groups that the state employed its anti-Amorite rhetoric, rather than against Amorites in general.

It is important to note in this context that the term Tidnum in Ur III sources occurred frequently as a synonym for Martu. Tidnum was mentioned in texts from Ebla and on Gudea Statue B, but only as a geographical place name. The Tidnum in the Ebla texts cannot be precisely located, but the Tid(a)num in the 'mountain range of the Amorites' mentioned by Gudea should be located in the Jebel Bishri. This location is based on Akkadian sources that clearly demonstrate that the 'mountain range of the Amorites' was another name for Bašar. However, in the Royal Correspondence of Ur, Tidnum is used to designate a group of people, and more specifically as the people against which the 'wall of the land' was built. The Šu-sin text, in which he describes his victory over the Amorites, suggests that the Tidnum were one among several Amorite subgroups, thus again reinforcing the view that not all Amorites were hostile toward the Ur III state, but only specific subgroups. The combined evidence on the location of the 'wall of the land', as well as some references to Tidnum in combination with geographical places that can be located in the northeastern Zagros suggests that the Tidnum probably lived north and northeast of the Ur III state (Sallaberger 2007: 448-9). Thus, Tidnum as a place name can be located in the Jebel Bishri, whereas the group identified as such must be sought in the Zagros. The references to Tidnum in Ebla and possibly Ur cannot be interpreted, but may point to additional regions where the name Tidnum was known. Concluding, the use of the same name for different regions, or for Amorite people living there, may be an early reflection of the mirror toponymy that is clearly attested for the early second millennium BC (see below).

The early second millennium BC represents a third stage in the development of the Amorite identity. Administrative texts no longer labelled individuals as Amorites, but Amorites were still recognizable as such by their distinctive names. Furthermore, as Streck's analysis has shown, there was a proliferation of Amorite personal names across greater Mesopotamia and western Syria, with Amorite names occurring as far south as Hazor. At the same time, royal inscriptions appeared in which rulers from cities across greater Mesopotamia pointed out their Amorite ancestry, and described themselves as being Amorites. These claims represented a fundamental change from the royal historical sources of the Ur III period, which were generally concerned with portraying Amorites as enemies of the state, and which did not consider them to be human at all. Apparently, early second millennium rulers found it advantageous to emphasize their Amorite ancestry. Many researchers have interpreted this shift as a result of some sort of Amorite population displacement at the end of the third millennium, leading to the replacement of the old urban elites by new Amorite rulers (e.g. Kenyon 1966: 34; Klengel 1972: 44; Weiss et al. 1993: 1002; Charpin et al. 2004: 57-60; Buccellati 2008: 143). However, the historical and archaeological evidence does not uniformly support this interpretation (Yoffee 1988: 50). Furthermore, this interpretation is based on flawed analogies between Amorite-Ur III interactions and ethnohistorically documented nomad-sedentary interactions.

First, the physical passage of Amorites through the wider Near East in the early second millennium BC has been inferred from the occurrence of identical place names from western Syria to northern and southern Mesopotamia. Charpin called this phenomenon a 'toponymie en mirroir' or mirror toponymy (Charpin 2003). This phenomenon is most clearly seen in letters that actually state that there are two places with the same name, but can also be inferred from place names where a qualification is added, that is, whether the city is located near a river, or whether it is associated with a particular deity (Charpin 2003: 7-10). Charpin suggested four possible mechanisms for the establishment of double place names: *(a)* chance, that is, place names with obvious meanings or roots like 'encampment' or 'ruin' are likely to occur multiple times without any further link existing between them, *(b)* deportations whereby the deportees introduced new names into the region to which they were relocated, *(c)* the place names were clearly meaningfully linked in some way, as may possibly be inferred from the double occurrence of groups of place names between region, and *(d)* the place names were established in the course of a migration of Amorite groups across the Near East (Charpin 2003: 12-15). Whereas the first three mechanisms seem to have operated only in a limited number of cases, Charpin argues that the fourth mechanism must have been very important. The main argument in favour of migration is that none of the early second millennium BC toponyms is found in third millennium texts, and that a systematic renaming would have occurred, whereby ancient names were dropped in favour of new, proper Amorite names (Charpin 2003).

There are several arguments against this interpretation. First, it is unclear whether many toponyms in northern Mesopotamia actually were of Amorite origin (Eidem 2000: 262). Secondly, Porter (2009: 204-5) recently argued that, in the absence of information on when each place name came into use, it is equally possible that the duplication of geographical names was intended to reinforce a common social identity over large distances. In this perspective, the adoption of identical place names served to actively strengthen social ties between possibly mobile groups that met only irregularly. In other words, the place names themselves became part of social structure. Based on the historical evidence alone, neither the explanation favouring migration nor favouring changes in social structures can be ruled out. At the same time, the mirror toponymy can in and of itself no longer be used to infer migration, unless other evidence can be produced to support that hypothesis. Thirdly, it should be noted that the mirror toponymy of the early second millennium BC is primarily known from the Mari texts and contemporary archives elsewhere. The period covered by these archives, as well as smaller contemporary text corpuses, is exceptionally well-known in terms of historical geography. The somewhat sudden occurrence of a mirror toponymy might therefore at least partially be the result of an increase in the amount of information that is available for this period with respect to previous periods, and of an increase in the geographical range that is covered by the Mari texts. Finally, the possible occurrence of a mirror toponymy prior to and during the Ur III period suggests that the early second millennium BC can no longer be regarded as the only period during which this phenomenon occurred, thereby further weakening the link between the mirror toponymy and a unique episode of large-scale migration during the early second millennium BC.

A second argument against an Amorite migration comes from Streck's analysis of the Amorite onomasticon. As described earlier (Section 7.2.3), Streck has demonstrated that Amorite names can appear together with Akkadian and Hurrian names within the same family, and that parents with non-Amorite names could give their children Amorite names. in so much as ethnic affiliation can be deduced from personal names only, these data indicate that ethnicity may have been a relatively fluid concept, and that naming conventions may have changed over one or two generations. If there existed proper naming conventions, and if names were a defining trait of the Amorite ethnicity, this evidence does not support the replacement of previous populations by a new Amorite population. This evidence rather suggests that becoming Amorite was a conscious choice that was open to at least a part of Mesopotamian society.

The ethnic fluidity that is implied by the onomastic evidence is also apparent in a letter from Mari from the high official Sammetar to Zimri-Lim, describing a group's transition from the Yaminite into the Sim'alite confederacy: 'Another thing. Uranum and the elders of Dabiš came, saying: "By extraction, we are among the Yahrurû, but never (as) *yarrādum*; also, in the back country, we have neither *hibrum* nor *kadûm*." We are native (?) to the Yahrurû. Let us now come into the midst of the Sim'alite(s) as (part of) the Nihadû (tribe), so that we may slay the (treaty) donkey' (Durand 1992: 118; Fleming 2004: 63). In the remainder of the letter, Sammetar asks the elders of Dabiš three times whether they want him to write to the king about this matter. In the final paragraph, Sammetar requests the king's permission to carry out the ritual that is necessary to turn the Sim'alites into Yaminites. The Yahrurû are one of the five Yaminite *li'mum*, and the Nihadû are a known Sim'alite *gayum*. This letter makes clear, then, that social mobility between the two tribal confederacies was possible if there was consensus among all involved parties.

Thirdly, the idea that Amorites forcefully made forays into regions controlled by sedentary states is a direct analogy of the situation in the Near East as it existed during the late Ottoman period, and more particularly in the areas that today make up Jordan and Syria. If this analogy is examined more closely, it becomes apparent that many aspects of nomadic pastoralist society during the Ottoman period were absent or different during the third and early second millennium BC (cf. van der Steen 2009: 114). The Ottoman government paid Bedouin sheikhs to protect villages from raids by other Bedouin. This system was, however, not very effective, and sedentary farming villages often had to pay tribute or *khuwah* to more than one Bedouin tribe (Lewis 1987: 12). There is currently no textual evidence to suggest that a similar system functioned during the Ur III or Old Babylonian periods. The success of Bedouin raids depended not only on the fact that Ottoman protection of these villages was insufficient, but also on their ability to quickly disappear into the steppe when Ottoman soldiers did appear. Bedouin were able to do so because they were nomadic camel-pastoralists. In this respect it should be mentioned that third/early second millennium BC (semi-)nomadic groups were probably much less mobile than later groups because horse and camel had not yet been introduced as riding animals (Khazanov 2009: 123). Khazanov furthermore suggested that the absence of the camel may have severely limited the number of people that could be sustained by steppe areas, simply because these areas could not be exploited to the fullest by sheep/goat pastoralists. Thus, any third to early second millennium BC equivalents of the Bedouin would be severely limited in their mobility, their numbers, and consequently their military advantage against more sedentary enemies (Khazanov 2009: 123).

The Ur III did not end because of attacks by Amorites, but Amorite groups, as well as others, were able to raid into Ur III territories because the state was already weakened by internal processes. Išbi-Erra's rise to power is a case in point. According to the Royal Correspondence of Ur, Išbi-Erra was an important official in the Ur III state before he became the ruler of Isin. Not only does this case indicate that Amorites were well-integrated into Ur III society, but also that, at least in the case of Isin, the emergence of Amorite rulers was not related to, or caused by, any population displacements. Išbi-Erra's usurpation is more indicative of internal power struggles during the final years of the Ur III state, in which Amorites actively participated (cf. Wilcke 1969: 15; Yoffee 1995: 297).

Finally, another striking aspect of the emergence of an Amorite identity is the fluidity with which Amorite rulers embedded themselves in a pre-existing urban culture. Thus, in terms of material culture and the expression of kingship, Amorite rulers and supposedly Amorite towns were not very different from what preceded them. Traditionally, this continuity has been interpreted either as an attempt by Amorite rulers to imitate or adopt pre-existing structures to legitimize Amorite rule, or as evidence for a complete absorption of the Amorite newcomers into the supposedly more sophisticated urbanized Mesopotamian cultural traditions (Buccellati 1966: 355-6).

The problem that is posed by this fluid transition was correctly realized by Porter, who therefore recently suggested that the Amorites were 'part and parcel of the mainstream urban world' (2007: 70). Instead of arguing how and why nomadic pastoralists could have overthrown the Ur III state, only to be rapidly assimilated into the urbanized Mesopotamian culture, she suggested that the Amorites originated from northern Mesopotamia, where they were already familiar with an urbanized environment. Thus, Porter explained the impact of the Amorites on southern Mesopotamia primarily in terms of the heterarchical system of city-states that replaced the Ur III state (Porter 2007: 107). Porter rightly drew attention to a number of important issues, some of which were also emphasized by other researchers: (*a*) the possibility that the Amorites originated not from marginal steppe areas but from more fertile regions in the northern Jezirah (Meijer 2000: 206), (*b*) the possibility that there may have been no assimilation of the Amorites into southern Mesopotamian culture because they already possessed many similar cultural features (Kamp and Yoffee 1980: 98-9; Yoffee 1988: 50), and (*c*) the fact that the end of the Ur III state represented primarily a transition towards a scattered political landscape of independent city-states retaining much of their previous socio-cultural traditions (Yoffee 1995: 297).

However, whereas Porter successfully disassociated herself from the traditional view of Amorites as fulltime nomads and pastoralists, she, as well as many other researchers, failed to completely move away from the idea that the emergence of an Amorite identity must have entailed some sort of population displacement. As was argued above, neither the idea that there was a large-scale Amorite population displacement, nor the idea that Amorite nomadic pastoralists constantly pressed into southern Mesopotamia from the more marginal steppe regions, fully explains the available evidence. Especially the observation that at least some Amorites were obviously well-integrated into southern Mesopotamian society already before the final centuries of the third millennium BC remains unexplained.

7.3.3 The Amorite identity as ethnicity

In order to address the issues that have been raised in the previous section, and more importantly to explain the particular development of the Amorite identity throughout the third and early second millennium, the following model will be put forward as an alternative to previous explanations. Rather than focussing on population displacement as the primary cause of the emergence of Amorite states in greater Mesopotamia during the early second millennium BC, this model emphasizes ethnicity as a flexible identity that can emerge and disappear without necessarily involving population displacement.

The model that is presented here argues that Amorite is an ethnic identity that was at least partly constructed around a shared past and ancestry. The use of the determinative Martu in Ur III records, as well as the references to Amorite ancestors by early second millennium rulers, reflect a general concern for maintaining an identity that is firmly rooted in a perceived or constructed past. From this perspective, Shamshi-Adad's and Hammurabi's claims to the same ancestral roots need not necessarily reflect real kinship. In fact, it is possible that these genealogies were built around the social networks in which Shamshi-Adad, Hammurabi and other early second millennium rulers participated, rather than that these networks emerged out of true historical kinship ties. This loose use of kinship to legitimize and strengthen existing social relations has been observed among Rwala Bedouin. They deliberately lost track of their genealogies after four or five generations in order to maintain a flexibility of membership to particular lineages that could be modified when necessary (Barfield 1993: 74-6). The system of constructed kinship among the Rwala was characterized by the persistence of a limited number of named ancestors, even though these should have changed if the system reflected true kinship (Barfield 1993: 75). The persistence of the term Tidnum over several centuries and its slightly different meanings, that is, as tribal and ancestral name, might reflect a similar intentionally fuzzy system in which true kinship was of secondary importance to the maintenance of a common ethnic identity.[98]

The evidence for a separate Amorite language possibly further strengthens the interpretation of the Amorite identity as a distinct ethnicity. However, it is difficult to determine whether language was considered a defining trait of the Amorite identity. The frequent occurrence at Sippar and Mari of Amorite names in families previously dominated by Akkadian names might suggest at least a minimal knowledge of or familiarity with Amorite. The continuing dominance of Akkadian as written language does not necessarily reflect the linguistic composition of greater Mesopotamia, as is clear from the references to people who spoke multiple languages. It would in this regard be interesting to know the degree to which speakers of Amorite and Akkadian could understand each other. Although Buccellati has suggested that the languages may initially have been 'socio-lects' with a significant degree of overlap (1992: 98), the mentioning of an Amorite interpreter in the Ur III texts (Section 7.2.2) suggests that this mutual understanding was rather low.

The strong association with pastoralism evident in texts from the Ur III period and the early second millennium BC represents an additional common feature of the Amorite identity. This association does not necessarily mean that all Amorites were pastoralists, which they clearly were not, but rather that many Amorites were pastoralists, and, more importantly, that many pastoral-

---

98 The names, or parts thereof, of several Sim'alite *gayum* during the reign of Zimri-Lim were structured as personal names (Durand 2004b: 183). This could indicate a similar concern with ancestry as is evident in the genealogies of Hammurabi and Shamshi-Adad.

ists were Amorites. It is here suggested that practising pastoralism may have represented an ideal life-way by which ultimately all Amorites were defined, including those that were not pastoralists themselves. In other words, being pastoralist was at the core of being Amorite, but being Amorite did not require being pastoralist. It was this aspect of the Amorite ethnic identity that probably set it apart from other contemporary ethnicities.

This reconstruction of the Amorite identity as representing an ideal pastoralist life-way has parallels elsewhere. In a study on Maasai identity in East Africa, Galaty (1982) found that the Maasai defined their own and others' identities on three levels nested within each other. At the most universal, mythological level, Maasai distinguished between themselves, being 'people of cattle', and hunters and agriculturalists. Myths explained why hunters and agriculturalists were non-Maasai by definition. At the intermediate level, representing a division of the Maasai identity itself, there were true Maasai or pastoralists, Maasai hunters, and Maasai agriculturalists. Finally, the third level represented a division of labour within true pastoralist groups. At this level, there were true pastoralists or Maasai, as opposed to the class of craftsmen, and the class of diviners. Even though at each level the strict meaning of Maasai was pastoralist, the Maasai identity could at times include other persons and groups as well. Even at the mythological level, some non-Maasai could be classified as Maasai with respect to outsiders. There thus existed clear distinctions between how the Maasai identity was defined, and how it was applied to different groups or persons. Nevertheless, the pastoralist ideal of Maasai identity was present at each level of identification (Galaty 1982).

Similar distinctions between ancestry and practiced subsistence strategies also played an important role in more recent Near Eastern agriculturalist/pastoralist communities. In the Balikh Valley, Syria, there was at least from the nineteenth century AD onward a clear distinction between 'Arab, *hadar*, and *badu* (Lewis 1988: 686). 'Arab was used for people who were semi-nomadic at that time or had been in the past, and who were of tribal descent. *Hadar* was used for people who had always been sedentary, and who could not claim a tribal descent. *Badu* indicated people who had been fully nomadic camel-pastoralists. Even though these distinctions had become virtually meaningless from an economic perspective by the time of Lewis' ethnographic research, the maintenance of these distinctions indicated the persistence of the ideals that were associated with each identity.

In this sense, then, Amorites may perhaps be best understood as 'people of sheep and goat', thereby referring to an aspect of Amorite identity through which all Amorites identified themselves without actually always being involved in it. The discrepancy between ideal and reality is perhaps best reflected in the apparent inconsistency that can be found in the literary composition of *The marriage of Martu*, noted by Porter (2007: 108). On the one hand, this composition described the god Martu as living in a tent, eating raw meat and so forth (see Section 7.2.2), whereas on the other hand it was said that he lived in the city of Inab. It is clear, then, that in this text the Amorite ideal is distorted as part of the story being told, but that the composition may have nevertheless reflected the complexities of the interaction between Amorites and others, and among Amorites themselves. King Zimri-Lim of Mari called himself 'king of Mari and the land of the *hana*', thereby further emphasizing the Amorite pastoralist ideal.

If it is accepted that people could adopt the Amorite ethnic identity, there must have been clear advantages to this choice. Given the fact that the strong association with pastoralism was a defining and integral aspect of Amorite identity, it is likely that any advantages resulted from precisely this association. The strong association with, but not exclusive focus on, pastoralism makes it likely that the Amorite identity functioned as a social expression of an exchange network between pastoralist and agriculturalist groups. The occurrence of duplicate place names across Mesopotamia may also point to a consciously maintained shared identity. The shared Amorite identity positively influenced the emergence and maintenance of intra-ethnic exchange networks, giving Amorites an adaptive advantage that allowed them to respond adequately to any crisis situation they might encounter.

7.4 Conclusions

The present chapter has reviewed textual evidence for the presence of Amorites in Mesopotamia. Amorites were mentioned in historical records from *c.* 2500 BC onwards, with a strong emphasis on the Ur III period, as well as the early second millennium BC. This evidence indicates that the perception of Amorites by those who wrote about them changed over time. Whereas the mid-third millennium texts are difficult to assess, the Ur III texts clearly show a discrepancy between the perception of Amorites in literary sources and administrative documents. The literary sources provide a stereotypical view of Amorites as being uncivilized and hostile toward the Ur III state, whereas the administrative texts indicate that Amorites were well-integrated into Ur III society. The early-second millennium material shows that Amorites could be found in all levels of society, and that many rulers across greater Mesopotamia were proud of their Amorite heritage. The Mari archives furthermore show that Amorite states were not organized on an agriculturalist-pastoralist dichotomy, as has been previously emphasized, but that these states were based on tribal identities that encompassed both economies. This evidence suggests that the incorporation of both the pastoral as well as the agricultural economy may have been the defining aspect of Amorite identity. The most important conclusion must be that the Amorite ethnicity emerged as an identity that can be interpreted as signifying a social network extending over large distances. The spatial distribution of the literary sources points to a strong focus on lower Mesopotamia, as well as the Syrian Euphrates and Khabur basins. Thus, the Amorite identity not only transcended boundaries between subsistence systems, but environmental boundaries as well. These characteristics gave Amorites a behavioural flexibility when responding to environmental stress. The development of the Amorite identity with respect to environmental change, and with respect to the hypotheses that were formulated in Section 3.5.3, will be discussed in the following chapter (Section 8.3).

8 Social responses to environmental change

8.1 Introduction

The palaeoenvironmental data presented in Section 2.3 suggest that northern Mesopotamia experienced increasing aridity from *c.* 2500 BC onwards. Although the effects may have been quite localized at that time, the data suggest that aridity increased toward the end of the third millennium BC. For that period, all local northern Mesopotamian palaeoclimate records, that is, Kazane Höyük, Göbekli Tepe, Wadi Avedji, Wadi Jaghjagh, and Tell Leilan, indicated a shift towards greater aridity (Fig. 2.7F–J). These records also suggest that this shift was not a short-lived event, but initiated a period with a structurally drier climate than existed during the early third millennium. Furthermore, osteological and botanical remains suggest that humans and plants may have experienced nutritional stress and water shortage, supporting the hypothesis that the late third millennium BC arid transition indeed affected human communities and led to more droughts. This climate scenario, then, is the background against which the changes in social structures that were reconstructed in Chapters 5 through 7 should be evaluated. The reconstructed developments for each survey region will be analyzed and discussed in light of the expected interaction patterns. Then, the textual evidence presented in Chapter 7 will be evaluated and compared with the development of the pastoral economy that was reconstructed in Chapter 6. Finally, this chapter hopes to place the reconstructions in a wider regional context, and to assess the degree to which responses across northern Mesopotamia were similar, and how local responses related to other developments.

8.2 The development of sedentary-sedentary relations

In Section 3.5.1, it was argued that a decrease in the carrying capacity within a region would lead to increased competition between groups relying on the same resources. For groups relying on subsistence farming, this competition would likely focus on good quality arable land and on water. Increasing competition over these resources would likely result in the development of social mechanisms regulating access to and use rights of these resources. Based on the theoretical framework outlined in Section 3.5, a convex rank-size distribution is especially expected during periods of resource stress. Chapter 5 described the selection of three case-studies in which this hypothesis was tested. These case-studies were the region covered by the Euphrates-Birecik Dam Survey (B-EDS) in Turkey, the Syrian part of the Balikh Valley that was covered by the Balikh Survey (BS), and the region covered by the North Jazira Survey (NJS) in Iraq. Diachronic developments in site size, as well as rank-size distributions, were calculated as proxies of inter-site competition. Furthermore, regional population trends were calculated to assess the degree to which northern Mesopotamia was subject to de-urbanization and de-population during the late third–early second millennium BC.

The B-EDS region was characterized by a low degree of urbanization and a low population density throughout the third and early second millennium BC. Population generally increased, with a minor decrease between 2100 and 1700 BC. Average settlement size steadily increased throughout this period. Above-average population growth is recorded for the Late EBA/Early MBA period, and the end of the Late MBA (Fig. 5.10A). The B-EDS rank-size plots were extremely primate and no significant diachronic change could be observed due to the overlapping error ranges (Fig. 5.11). The BS demographic curve showed alternating periods of population growth and decline during the third and early second millennium BC. The settled population grew during periods VIab, VIIa, and VIIb, and declined during periods VIcd and VIIc (Fig. 5.10B). At the same time, average settlement population peaked around 2000 BC at the VIcd–VIIa transition, only to go into a steep decline toward 1600 BC. Whereas the population growth rate during period VIab fell within the range of natural growth, the period VIIa–b growth rate exceeded the natural growth rate (Fig. 5.10B). The rank-size analysis showed a probably convex size distribution for the VIIa–VIIb and VIIb–VIIc transitions, but the large confidence zones for especially the 2400 BC and 2000 BC plots impede a diachronic reconstruction (Fig. 5.12). Finally, the NJS showed a steep population growth during the early third millennium, followed by a period of high demographic stability that ended in a marked population decline from 1750 BC onwards (Fig. 5.10C). Average settlement size paralleled this development. Population growth exceeded natural growth during the early third millennium, but remained within the natural range for the entire second half of the third millennium, as well as the first quarter of the second millennium BC (Fig. 5.10C). The NJS plots were slightly primate, with a possibly primo-convex distribution observed for 3000 BC. However, the large confidence zone for each reconstructed rank-size plot suggests that these distributions must be treated with caution and that conclusions drawn from them are not very reliable (Fig. 5.13).

From these data, two interesting observations can be made with respect to the scenario of environmental change: all regions experienced population growth at some point in time during the late third–early second millennium BC when aridity increased; and a convex distribution is only observed in the BS region at the VIIa–VIIb and VIIb–VIIc transitions at c. 1800 and 1700 BC, respectively, whereas in the B-EDS and NJS regions, there is no evidence at all for a convex size distribution. These observations directly allow the conclusion to be drawn that no direct correlation existed between increasing aridity and population decline in settled communities. This conclusion is supported by the evidence from other surveys (Section 5.4.1). In each survey region, the number of sites and the aggregate site size increased and declined considerably during the late third–early second millennium BC, but no apparent pattern and certainly no complete synchronicity can be directly inferred from these developments. Furthermore, the absence of convex settlement distributions, except for a short period in the Balikh Valley, suggests that climate change may have had a relatively minor impact on the organization of the settlement systems that were analyzed in this study. The following sections will analyze the developments in each survey region in more detail, as well as discuss the role of climate change in the reconstructed developments.

8.2.1 The Upper Euphrates: the emergence of Carchemish

The number of settlements in the B-EDS region is low throughout the third and early second millennium BC. The possible slight change toward a more primate rank-size distribution could reflect the growing importance of Carchemish toward the late second millennium, as is possibly also reflected in the site's increased size. Clearly, these settlement developments fall outside the expected convex settlement pattern resulting from inter-site competition and growing resource stress.

It was already suggested that the area covered by the B-EDS probably represented the northern half of the Carchemish sector, whereas the southern half is located to the south of the modern Turkish–Syrian border (Section 5.3.2). This would suggest, then, that the primate nature of the B-EDS rank-size plots resulted from the fact that the total settlement system had not been correctly identified. Nevertheless, it should be realized that the areas north and south of Carchemish were very distinct in terms of environment and settlement patterns. The area north of Carchemish, that is, the B-EDS region, was characterized by relatively small settlements that were located close together, whereas the area south of Carchemish had a much more dispersed settlement system (Peltenburg 2007a: 17–18). Secondly, Carchemish is located in the transitional zone between the stable dry-farming wheat-based agricultural production system of the northern part of the valley, and the barley-based system in the southern part of the valley (see Fig. 2.5). Finally, Carchemish lies in the transitional zone between the dense oak woodlands of the Taurus proper, and the woodland steppe that characterized much of the Upper Syrian Euphrates (Peltenburg 2007a: fig. 1.3).

Given the fact that the B-EDS area falls well within the zone of dry-farming, it may be questioned whether the region ever experienced a sufficiently severe drought as to permanently affect the groups living in the region. This situation would explain why the expected settlement pattern did not emerge in this region as a response to increasing aridity. Nevertheless, a correlation between climate change and the settlement development in the B-EDS region may be tentatively suggested. Large sites in this region may have played an important role in north–south interaction. The location of this region between different environmental zones would suggest that Carchemish functioned as a gateway city between these regions (see Burghardt 1971).

Summarizing, the development of the settlement system in the B-EDS region seems to have been relatively unaffected by climate change during the late third millennium. The emergence of Carchemish as an important gateway city between the northern and southern environmental and cultural zones may potentially be linked to increasing aridity, as communities in the more marginal regions sought to establish social relations with the north.

8.2.2 The Balikh Valley: cycles of competition

Close inspection of the data from the BS region suggests that population growth, rather than climate change may have been the driving force behind increasing resource competition. During the third and early second millennium BC, the Balikh Valley experienced two periods of population growth, resulting in peak populations at 2400 and 1700 BC. Wilkinson suggested that the recent water discharge of the Balikh, before the introduction of large-scale motorized irrigation, was potentially sufficient to irrigate agricultural fields sustaining a population of 7000–30000 persons (1998: 80–1).[99] It can safely be assumed that the late third–early second millennium discharge was at most the same as this amount. The reconstructed population between 3000 and 1600 BC continuously exceeds the lower end of this range. The population peak of 2400 BC, and particularly that of 1700 BC, approached the upper limit of this range, suggesting that agricultural production may have been subject to stress and periodic failure. It is exactly during these two periods of maximum population that the strongest evidence for competition between communities can be found.[100]

Towards 2400 BC, the inter-community social tension building up in the Balikh Valley was reflected in the construction of city walls documented for the middle and late third millennium BC. Hammam et-Turkman (BS 175) was walled around 2350 BC (Thissen 1989: 197). Around the same time, city walls were built at Jidle (BS 275) and possibly Sahlan (BS 247) (Mallowan 1946: 134, 138), Kazane Höyük in the Harran area to the north of the BS region (Rosen 1997: 397), and the city wall of Tell Bi'a (BS 1) was renewed (Miglus and Strommenger 2002: 21). These building activities reflect a real concern for safety and occurred at a time when pressure on resources was growing. In this system, no site was able to assert control over the entire region to extract resources for its own benefit, a system that may have functioned for some large third millennium centres elsewhere in northern Mesopotamia (Wilkinson 1994). One reason why this type of settlement hierarchy did not appear in the Balikh Valley may have been that not land but water was the restricting resource.

During the early second millennium, there is direct evidence for competition over water in the letter to Shamshi-Adad from the inhabitants of Tuttul/Tell Bi'a. They complained that the city of Zalpah, located upstream along the Balikh and possibly to be identified with Hammam et-Turkman (van Loon and Meijer 1988: xxvi), cut off the waters of the Balikh, thereby depriving Tuttul of vital irrigation water (Dossin 1974; Villard 1987). Given the fact that the immediate surroundings of Tell Bi'a are far more easily irrigated by using water from the Balikh than the Euphrates (Schirmer 1987: 71), this complaint seems fully justified if there actually was water shortage. Tell Bi'a may not have been the only settlement to suffer from reduced river discharge, and the resulting competition over resources may have led to settlement nucleation and the formation of distinct territories, as evidenced in the convex rank-size distributions for the period 1800–1700 BC.

---

99 It remains unclear whether irrigation was actually practised in the Balikh Valley during the third and early second millennium BC. Clearly identifiable ancient irrigation canals in the valley dated to the Hellenistic period and later (Wilkinson 1998: 69). Botanical samples from Hammam et-Turkman period VI and VII contexts contained too few weed species indicative of a wet environment, and by inference, of irrigation, to posit the practice of irrigation (van Zeist et al. 1988: 713).

100 The present reconstruction of the Balikh Valley demographic developments supersedes, and slightly differs from, the analysis presented in Wossink forthcoming. The differences mainly derive from the inclusion of the results of the rank-size analysis, necessitating a modification of the scenario of human-environment interaction that was offered earlier.

This scenario differs considerably from Curvers' (1991: 215) interpretation of the settlement system in the Balikh Valley during the Early and Middle Bronze Age. Curvers' model was primarily driven by a cycle of waxing and waning southern Mesopotamian influence on northern city-states. According to Curvers, the mid-third millennium BC population peak, which he called 'process of urbanization', resulted from the 'catalytic effect of Uruk expansion towards Syria and Anatolia', and was associated with the emergence of a ranked society in the Balikh Valley. During the late third millennium BC, the complexity of this society increased due to influences from southern Mesopotamia. At the end of the third millennium, this system collapsed due to internal social and political stress. The second 'process of urbanization' was initiated by local urban rulers who had a strong pastoral background and who pursued a policy of (re)settlement and sedentarization. These rulers sought to establish their power base by associating themselves with powerful foreign dynasties. Eventually, after 1700 BC, this system collapsed because these foreign powers withdrew their support from the local rulers in the Balikh Valley. Curvers emphasized the influence of external, primarily southern Mesopotamian, political entities, that is, the Uruk phenomenon, Ebla, the Akkadian and Ur III states, and Mari., This study however focuses on local processes as the primary causes of the observed socio-cultural changes. Local human-environment interaction are emphasized over the still debated impact of far-off political entities, and emphasizes political, cultural, and economic continuity in the Balikh Valley. In this way, there is no need to hypothesize significant southern Mesopotamian influence on the Jezirah, for which archaeological and textual evidence remains ambiguous (see Section 2.5).

Summarizing, it can be argued that in this scenario, both the mid-third millennium BC as well as the early second millennium BC cycles of increasing competition were primarily triggered by population growth, resulting in higher population pressure on available resources. The onset of drier climatic conditions during the second half of the third millennium, resulting in reduced river discharge and less reliable rainfall, may have strained agricultural production and led to a stronger reliance on irrigation, but was otherwise not a driving force behind the settlement developments that were reconstructed for the Balikh Valley for the period 3000–1500 BC.

8.2.3 The northeastern Jezirah: centralization and exchange

The reconstructed settlement data from the NJS region are difficult to interpret. No clear convex settlement distribution can be discerned due to the large error margin of the NJS rank-size plot. In fact, all plots demonstrate a slight tendency toward a primate or primo-convex distribution. It has been suggested that the NJS region was most likely to represent an area that did not coincide with an entire settlement system, as a result of which the rank-size plots would be flawed (Section 5.3). The proximity of large sites outside the survey area would suggest on the one hand that the NJS region may harbour parts of other settlement systems, and on the other hand that the system that can be associated with Tell al-Hawa was not entirely included in the NJS region.

Despite these uncertainties, a convincing case has been made for the existence of a mid–late third millennium hierarchical settlement system centred on Tell al-Hawa (Wilkinson 1994; Wilkinson and Tucker 1995). The model was based on the survey evidence from the NJS, consisting not only of the settlement record, but also of the hollow ways and extensive sherd scatters around major sites. According to this model, the hierarchical system was centred on a large central site. This site's population was so large that in a dry year with low agricultural production, its own catchment area, hypothesized to be an area around the site with a radius of up to 5 km, could no longer sustain the population, and surplus had to be imported from surrounding sites. Through the extraction of agricultural surplus from surrounding sites, which remained below their site catchment's carrying capacity, the central site was able to grow beyond its own catchment area's carrying capacity. However, whereas the satellites around Tell al-Hawa were abandoned, some settlements around the second-rank sites remained occupied, suggesting a slightly different land-use system for the central sites and the second-rank sites. This late third millennium system probably emerged out of a Late Chalcolithic settlement landscape characterized by small independent villages, through an early third millennium system characterized by central sites surrounded by satellites within a 5 km zone (Wilkinson 1994; Wilkinson and Tucker 1995).

This hierarchical settlement system relied on strong interaction between the central site and the secondary settlements. Grain deliveries must have been recorded, and there must have been regular traffic between the settlements within the system, especially in times of local labour shortages during harvest (Wilkinson 1994: 503). Some evidence for these contacts came from the area of Titrish Höyük along the Turkish Euphrates, where a hierarchical settlement system similar to that of Tell al-Hawa was argued to have existed. At Titrish Höyük, a particular type of large storage jar was abundantly present at the central site, but largely absent at the secondary sites, and this difference was speculatively argued to represent a flow of grain toward the central site. This pattern could also be identified in the NJS area, but not as strong as in the area of Titrish Höyük (Wilkinson 1994: 494–5). The slight primate tendency observed in the rank-size plots of the NJS region could potentially also point to a vertically integrated settlement system, although this would require a better understanding of the limits of the Tell al-Hawa settlement system as a whole (see Section 5.3.2).

Apart from the emergence of large, centralized sites, the model also aimed to explain the evidence for agricultural intensification. Shortening of fallow and applying animal manure to the fields would have been the primary methods of intensification. Whereas fallow-shortening cannot be proven from archaeological remains, there is circumstantial evidence for animal manuring. Many mid–late third millennium BC urban centres in the Upper Khabur area and the northeastern Jezirah are surrounded by hollow ways, supposed to have come into being through continued traffic by persons and animals (van Liere and Lauffray 1954; Wilkinson 1993; Ur 2003; Rutishauser 2008).[101] Around many of these sites, extensive sherd scatters can be found that extend as far away from the site as many hollow ways. Wilkinson suggested that these sherd scatters represent the material remains of settlement refuse being applied as manure to agricultural fields (1989), and that the hollow ways may have indicated the limit of the zone of intensive agricultural production. Given that the sherd scatters occur around sites that reached their largest size during the mid–late third millennium, it seems reasonable to assume that manuring was adopted as a method to achieve higher agricultural production.[102] Calculations of mean annual production per hectare, combined with reconstructed population levels for these large sites, have suggested that these populations could indeed only be sustained under an agricultural regime combining short or no fallow periods with manuring and/or irrigation (Wilkinson 1994, 2004; Altaweel 2008).

Looking more closely at this model of land-use around urban sites, it becomes apparent that the model combines two developments that are not necessarily related. First, the model proposes how a central site can attain a size beyond that which can be sustained by its immediate catchment area through extracting surplus from neighbouring sites, and secondly, the model proposes how the overall population of the entire settlement system can grow through agricultural intensification. The first development is entirely concerned with the organization of the settlement system, whereas the second development is concerned with the correlation between agricultural intensification and population growth within the entire settlement system. This means that agricultural intensification will occur once a certain population threshold is surpassed, irrespective of how this population is dispersed across the landscape.

---

101 Although this explanation of radial lines around northern Mesopotamian sites is widely accepted, other hypotheses exist. It has been suggested that the radial lines functioned as water harvesting networks, although this explanation does not necessarily rule out the possibility that they functioned as roads at some point in time (McLellan and Porter 1995; McLellan et al. 2000). Weiss doubted the ancient nature of the hollow ways and suggested that they were of much more recent date (Ristvet 2005: 26; Weiss 2007).

102 Ethnographic research indicates that manure derived from human settlements usually comes from animal pens and stables where few other activities are carried out, and that where manure is collected from open-air spaces, it is cleaned before being transported to the fields (Horne 1994: 53). This procedure would also reduce weight. An alternative explanation of sherd scatters is that they derived from an efficient type of irrigation called pitcher-irrigation, which was traditionally used in India and Iran (Barrow 1987: 240). Pitcher irrigation would leave an archaeological residue consisting of sherds spread across the landscape. If this were the case, one would expect the repertoire of pot shapes to be more limited, i.e. biased toward closed forms, than when these scatters would result from communal site refuse collected for manuring purposes. Additionally, because pitchers must have water-permeable walls, it is possible to test whether sherds could have resulted from pitcher irrigation even in the absence of good rims and other feature sherds.

Given that the total amount of settled area remained roughly the same, it is theoretically possible that the Tell al-Hawa region operated as a closed system free of any external influences or population movements (Wilkinson and Tucker 1995: 51). This is also suggested by the present analysis, which found that the NJS region was characterized by a relatively stable population growth falling entirely within the expected natural growth rates (see Section 5.3.1). The transition from a Late Chalcolithic 'resilient risk-reduction subsistence system' toward a mid–late third millennium 'brittle product-maximization system' (Wilkinson and Tucker 1995: 85) can therefore be fully explained in terms of natural population growth and associated agricultural intensification (see Boserup 1965).

The formation of a site hierarchy during the mid to late third millennium BC, and its continued existence during the early second millennium BC remains, however, unexplained. There must have been clear advantages for either the central site to assert its control over its neighbours, or for secondary sites to relinquish their surplus production to another settlement. Especially during periods of resource stress, second-rank settlements may have actively resisted the extraction of their agricultural surplus by the central site in the settlement system. In other words, why did a hierarchical settlement system arise, rather than a system of more independent and equal-sized settlements, which would be expected under conditions of increased resource stress? Wilkinson recently suggested that the emergence of anomalously large sites could have been stimulated by the formation of inter-regional exchange networks (2000b: 12; 2004: 186). A complex road network spread across northern Mesopotamia during the third and second millennium BC (Lebeau 2000b; Ur 2003; Marro 2004b, a), and many sites along these roads attained a size that cannot be explained solely by the subsistence potential of these sites' catchment areas. These networks would have been focused on the exchange of high-value commodities, as well as exchange between sedentary and (semi-)nomadic population components. The location of certain sites along important roads may have provided these sites with an advantage in resource access over neighbouring sites. Eventually, this advantage may have developed into unequal relations between settlements, whereby staples and high-value commodities moved in different directions, thereby creating a highly flexible economy combining agricultural, pastoral, and high-value commodities (Wilkinson 2000b: 14–16).

During the early second millennium BC, the number of settlements increased, and average settlement size declined, although Tell al-Hawa retained its status as the largest site in the region. The 2000 BC rank-size plot is slightly primate (Fig. 5.13C), whereas the 1500 BC plot shows a slight tendency towards a primo-convex distribution (Fig. 5.13D), although the large error-margin inhibits any further comments being made on these plots. Again, as with the late third millennium settlement pattern, the observed distributions for the early second millennium BC are not easily explained by the expected patterns for inter-settlement social relations. One possibility is that Tell al-Hawa retained the key function in the inter-regional exchange network that it acquired in the third millennium, and could therefore continue to assert its influence over neighbouring communities, thereby neglecting the effects of stronger inter-site competition. The close proximity of the NJS to the large sites along the Tigris makes it likely that the area profited from the Old Assyrian trade network extending into Anatolia (Wilkinson and Tucker 1995: 55).

Above, it was already suggested that large central sites not only developed in response to inter-regional exchange, but also played a role in exchange between sedentary and (semi-)nomadic communities. Although this topic will be more fully explored in the following section, some preliminary comments are necessary to explain the continued role of Tell al-Hawa as the dominant site in the NJS region during the early second millennium BC. It was argued that the extensive urbanization that characterized large parts of the Upper Khabur, as well as the NJS region, necessitated a relocation of specialized pastoral production toward non-urban areas (Section 6.5.1). This process, in turn, may have stimulated the emergence of central sites in the exchange between (semi-)nomadic pastoralist communities and sedentary agriculturalist communities. It may therefore be tentatively suggested that it was exactly this function as gateway city (see Burghardt 1971) between urbanized areas and more marginal areas frequented by (semi-)nomadic communities that allowed Tell al-Hawa, and sites like Tell Brak and Mari to continue to exist as important settlements dominating their environments. Thus, when aridity set in during the late third millennium, these settlements capitalized on their functions as gateway cities which they had acquired earlier that millennium.

Summarizing, natural population growth and the formation of exchange networks were the determining factors in shaping the settlement patterns that were observed in the NJS region. The initial formation of long-distance exchange networks during the mid–late third millennium, propelled by a demand for high-value and/or pastoral commodities, may have led to the prominence of Tell al-Hawa within the larger settlement system. It was probably exactly this central and perhaps monopolizing position with respect to exchange with (semi-)nomadic communities that allowed Tell al-Hawa to continue dominating its surroundings when late third–early second millennium BC increased aridity necessitated a stronger focus on agriculturalist-pastoralist interaction. In this sense, the onset of drier conditions reinforced rather than changed a social constellation that was already highly geared towards maintaining an urbanized society in a relatively marginal environment.

8.3 The development of sedentary-(semi-)nomadic relations

It has been argued that groups relying on different subsistence strategies would develop cooperative mechanisms in order to increase each group's chance of survival and access to contested and therefore scarce resources (Section 3.5.2). This expectation is based on the concept that specialization is on the one hand more productive than diversification, but on the other hand makes a group more vulnerable to environmental stress. Chapters 6 and 7 reviewed the evidence for the existence of a pastoral economy in northern Mesopotamia, as well as the evidence for the existence of a social identity of these pastoralists. The present section will discuss to what degree the reconstructed developments conform to the expected patterns.

Chapter 6 presented a scenario for the development of the pastoral economy in northern Mesopotamia during the third and early second millennium BC. It was suggested that a complex and fully developed pastoral economy had existed in Mesopotamia since the fourth millennium BC. During the mid-third millennium, urbanization in the Upper Khabur led to a relocation of pastoral production to more marginal areas to the south and to the west of the main urbanized areas of the central and eastern Upper Khabur, that is, the Middle Khabur and Jebel Abd al-Aziz areas, and the area around Tell Beydar, respectively. Eventually, the small-scale pastoral production along the Middle Khabur was supplemented by specialized production focused on the area of the Jebel Abd al-Aziz. This specialized distant village-based or transhumant pastoral production system ended somewhere during the late third millennium and was probably replaced by a system whereby specialized pastoral production was practised by more mobile, (semi-)nomadic communities. Early second millennium bone assemblages from northern Mesopotamia are scarce, but the textual evidence from Mari testifies to the importance of the pastoral economy to society overall, and to the high degree of mobility of pastoralist communities.

Chapter 7 discussed the evidence for the presence of Amorites in greater Mesopotamia, and aimed to identify the nature of Amorite identity, as well as its diachronic development. It was demonstrated that Amorite can be considered an ethnic identity during the early second millennium, and that this ethnic connotation also existed during earlier periods. This identity was expressed in a vocabulary focussing on a pastoralist ideal, even though being pastoralist was not essential to being Amorite. Textual references to Amorites show that they were present in greater Mesopotamia from at least the middle third millennium BC onwards, and that they operated on all levels of society. During the Ur III period, a dichotomy existed between public records displaying Amorites as non-human enemies of the state, and administrative records indicating that Amorites held all manner of positions in society. During the early second millennium, states emerged across Mesopotamia, led by rulers who emphasized their Amorite heritage. The evidence from Mari has shown that these states were based on tribal identities including both agriculturalist and pastoralist communities.

If these developments and the evidence for climate change at the end of the third millennium are examined more closely, it can be observed that there is a considerable degree of synchronicity between important changes in the pastoral economy, the Amorite identity, and climate change. With respect to the expected interaction between sedentary and (semi-)nomadic communities, the following scenario is suggested.

The date of the genesis of the Amorite identity cannot be fixed with certainty on the basis of the presently available evidence. Textual sources indicate that the Amorite ethnicity was recognized by the mid-third millennium BC, but it may have been present long before that time. Its increased importance towards the end of the third millennium may be associated with the trend towards specialization in the pastoralist economy that occurred during the mid-third millennium BC. This trend may have stimulated the emergence of an ethnic identity through which exchanges between different economic groups were carried out. This development was most pronounced in, although not exclusive to, in the wider Khabur area, making this region a potential candidate for an Amorite homeland, or part of it. The Amorite identity was adopted by individuals and groups who found the identity's potential for pastoralist-agriculturalist interaction advantageous. This advantage may have derived from the increased sensitivity to environmental stress of both the pastoral as well as the agricultural economy, which was a direct consequence of the development toward stronger specialization. However, as long as this economic system functioned properly, the Amorite identity may have either constituted a minority within Mesopotamian society at large, or Amorites may not have felt the need to emphasize their identity and to invoke ethnic solidarity. The second interpretation is favoured here because it does not necessarily exclude the possibility that Amorites indeed represented a minority, and because it accords well with the idea that ethnicity is a situational identity that is usually only emphasized under specific conditions. The relative scarcity of references to Amorites before the Ur III period should therefore not be understood as an absence of Amorites, but as an indication that the Amorite identity was not emphasized by its carriers, and that scribes found it unimportant to record this aspect of identity. This interpretation, then, suggests that Amorites may have constituted a significant ethnic group within parts of Mesopotamia well before the Ur III period.

The window into Mesopotamian history and society that is provided by the Ur III textual records clearly shows an entirely different picture. At that time, the Amorite ethnicity had certainly become an important aspect of identity, and many Amorites lived throughout greater Mesopotamia. Rather than arguing that the sudden increase in Amorite names and references during the late third/early second millennium BC resulted from population displacements, it is here suggested that this increase reflected the flexible and situational nature of ethnicity as identity. As argued earlier, ethnicity is an identity that can be adapted to very specific needs and situations. The growing importance of the Amorite identity from the Ur III period onward must therefore be explained in terms of the advantages associated with becoming Amorite.

As argued in Section 7.3.3, the Amorite identity constituted a social network between pastoralist and agriculturalist communities that was primarily defined through an idiom of pastoralist ideals. The growth of this network is visible in the Ur III records and resulted in the complex early second millennium BC social landscape of numerous Amorite kingdoms throughout greater Mesopotamia. It coincided with climate change and increasing aridity during the late third–early second millennium BC onward. Increasing environmental stress may have strained the agricultural and pastoral economies, and prompted closer cooperation and contact between the two economies in order to create a more diverse and reliable resource base. At the same time, the higher mobility of pastoralist communities that is suggested by the early second millennium textual evidence impeded regular contacts and exchanges, thereby reinforcing the need for a social network that strengthened ties between agriculturalist and pastoralist communities. It is in this sense, then, that the growing importance of the Amorite identity should be interpreted. Through this identity, Amorites could create an extensive exchange network that was organized along constructed or perceived kinship lines and based on ethnic solidarity and more generalized reciprocity than would be possible among ethnically differentiated groups. The Amorite identity and its associated values and the social network it embodied represented a behavioural alternative that could be implemented under certain circumstances (see Section 3.4). Eventually, this adaptive advantage not only allowed

Amorite groups to be more successful than their competitors, but would also generate an ethnic shift toward becoming Amorite, eventually resulting in the patchwork of Amorite kingdoms that can be observed for the early second millennium BC.

The result of this process of becoming Amorite is most clearly seen in the Mari archives that allow the most detailed view of the organization of an Amorite state. As was argued in Section 7.2.3, this archive has shown the importance of tribal affiliation, and it has recently been shown that many conflicts were not fought over the nomadic pastoralist-sedentary agriculturalist or tribe-state divide, but over tribal boundaries. This interpretation is in line with the interpretation offered here, namely that the Amorite identity distinguished itself from other identities primarily through its ability to combine agriculturalist and pastoralist production modes. The unique view afforded by the Mari archives possibly allows us to propose that these tribes functioned as factions within the Amorite ethnicity, in the sense proposed by Brumfiel (1994). It could be tentatively suggested that such factions emerged out of a desire by individual city-states to attach themselves to particular (semi-)nomadic groups, or vice versa, thereby creating stronger social and reciprocal obligations. The strict association of settlements with groups such as the Sim'alites or the Yaminites (see Section 5.4.2) suggests that factionalism may have been significant in the Mari kingdom, as is also suggested by the frequent conflicts between these tribal confederacies (see Section 7.2.3). Ristvet recently combined textual, geological, and archaeological evidence to locate pasture areas in the Upper Khabur (2008), and it might be tentatively suggested that distinct areas were closely associated with individual cities.

Because the Amorite identity is associated with specialized pastoralism, it may also be associated with urbanism. Throughout history, increased demographic stress seems to have been one of several processes that can induce specialization of pastoral production. The Amorite identity could be understood as the social reflection of this causal relationship between urbanism and specialized pastoralism, and the social element that binds these components together (cf. Porter 2004: 73-4). The increased importance of the Amorite identity would, then, not only reflect social interaction between agricultural and pastoral communities in general, but it would also be a reflection of an urban culture that tried to maintain its identity and existence in the face of climate change. Additionally, the Amorite identity may eventually have eased transitions between these components by bridging social and ethnic boundaries and differences in social status. The identity thus accommodated the demographic fluidity that can be observed in more recent Near Eastern societies, as described, for example, for the Biqa Valley (Marfoe 1979). The spread of the Amorite identity may therefore have not only signified a reorganization of economic relations between different social units, but possibly also a re-evaluation of each unit's social status.

Summarizing, it has been argued that the Amorite ethnic identity represented a social network between pastoralist and agriculturalist communities that emerged as a response to increasing aridity and associated environmental stress. The explicitly kinship-based ethnic identity allowed Amorite groups to participate in an exchange network based on ethnic solidarity and generalized reciprocity. The success of these groups eventually resulted in an ethnic shift whereby other groups sought to associate themselves with this network by becoming Amorite. The emergence of Amorite kingdoms and dispersal of Amorite names across greater Mesopotamia should therefore not be understood as evidence for moving people, but as evidence for a moving identity.[103]

---

103 The Amorite ethnicity was not the only ethnic identity to appear in greater Mesopotamia at the end of the third millennium. The records from Ebla do not include Hurrian names, and these seem to be absent in the Beydar archives as well. However, certainly from the time of Naram-Sin, Hurrians were present in the Upper Khabur area. Whereas the Amorite identity was more strongly focused on northern Mesopotamia and the southern alluvium, the Hurrian identity probably extended northwards into the Caucasus (Steinkeller 1998). It is interesting to note that the boundary between the Hurrian and Amorite identity must have been located in northern Mesopotamia, an area that was extremely sensitive to the climatic fluctuations that led to the increasing importance of the Amorite identity, and that both identities extended into distinct environmental zones. It could very tentatively be suggested that this ethnic boundary may have shifted north and south as a result of groups aligning themselves with either Hurrian or Amorite neighbours, depending on the adaptive advantage inherent in each alternative. This situation would reflect the nature of ethnic change described by Barth for the border region of Afghanistan and Pakistan (1969b).

8.4 Conclusions

The previous analysis of diachronic settlement developments in three regions of northern Mesopotamia has resulted in three different scenarios of changing human-environment interaction during the late third–early second millennium BC. In each region, natural population growth and the effects of mid-third millennium urbanization were found to be much more important factors in determining human-environment interaction than increasing aridity during the late third–early second millennium BC. Where social responses to climate change could be detected, as was the case for the B-EDS and the NJS, these responses were found to be primarily reinforcing trends that had already been set in motion as a result of mid-third millennium urbanization. Furthermore, the expected pattern of inter-site competition, driven by resource stress, was only encountered in the BS region, and there it was primarily correlated to population growth, with increasing aridity only acting as a reinforcing mechanism. In the NJS region, the prior existence of an exchange network focused on Tell al-Hawa probably ensured the continuing dominance of that site and prevented the emergence of inter-site competition.

The analysis of pastoralist-agriculturalist interaction has shown that the development of the Amorite identity can be explained by the theoretical framework presented in Section 3.5. The Amorite identity functioned as a social network that enabled pastoralist and agriculturalist groups to maintain stable exchange relations. These relations provided Amorite groups with mutual access to diverse resources, thereby stabilizing the groups' respective economies. In this sense, the adoption of the Amorite identity throughout the Near East during the late third–early second millennium may have been fundamental to the creation of the resilient dimorphic economy that was the hallmark of the Mari state.

Concluding, the present study has shown that climate change impacted upon human societies in marginal areas of the Near East, but also that careful analysis of local environmental, cultural, and economic conditions is required in order to distinguish climate-induced changes from other developments. It has been shown that a good understanding of prevailing socio-cultural conditions is required in order to understand why a human group does, or does not respond to climate change as expected. More importantly, this study has made clear that local climate change scenarios, as for example presented in the Leilan Climate Change Model (see Section 1.3), can rarely if ever be extrapolated to other regions or periods without major adjustments. This observation also suggests that although lessons may be drawn from the past, the past does not dictate the future. Finally, the case-studies analyzed here, and the diversity of identified responses, show above all the extreme resiliency of human communities when faced with climate change.

8.5 Final comments: looking back and looking into the future

This study has shown that effective research into responses to environmental change requires (*a*) a good chronological control of both palaeoclimatic as well as culture-historical data, (*b*) a clear understanding of the effects of environmental change on human communities, (*c*) proof that past human groups would disappear if they failed to adapt to environmental change, and (*d*) a theoretical framework that helps to model expected responses to environmental change in order to detect environmentally triggered socio-cultural developments in the multitude of developments that make up cultural change (see Chapters 1 and 3). By evaluating the degree to which the present study has succeeded in fulfilling these requirements, it is possible to highlight areas of potential further research to increase understanding of human-environment interaction in northern Mesopotamia during the late third–early second millennium BC.

Chronological control of both palaeoclimate proxy records as well as archaeological and historical chronologies remains problematic. The lack of local well-dated palaeoclimate proxy records is unlikely to be resolved in the near future. A relatively high precision is now being achieved for the Soreq Cave and Lake Van proxy records. Nevertheless, the Soreq Cave records rely on U-series dates, which were taken at intervals and from which an average sedimentation rate was calculated for the intervals. This means that dating unique climate events rather than long-term environmental trends remains problematic and associated with a considerable degree of uncertainty.

Archaeological and cultural chronologies in northern Mesopotamia generally suffer from a lack of high-quality radiocarbon dates. The continuing debates on the Early Jezirah (EJ) periodization and the earliest date for Khabur ware, the definition of characteristic pottery shapes for each EJ sub-period and each northern Mesopotamian culture area, and the correlation between material culture and historical chronology demonstrate that an independent high-precision radiocarbon chronology for Mesopotamia is becoming increasingly vital. Until the establishment of such a chronology, the drawing of correlations between unique climate events and unique historical events should be avoided, and it is perhaps better to try to reconcile long-term processes, for which both the palaeoclimate and archaeological data are currently better suited.

The reconstruction of the actual effects of environmental change is impeded by the lack high-precision palaeoclimate proxy records for northern Mesopotamia. The local northern Mesopotamian records all point to a significant shift towards aridity somewhere during the third millennium BC (see Section 2.3.5). However, the available records usually only indicate a relative shift and cannot be used to calculate changes in annual rainfall or river discharges. Furthermore, the detected shifts were only precisely dated by comparison with more precise, but more distant proxy records, which may have been unaffected by some of the more regional northern Mesopotamian climatic fluctuations. Although recent climate modelling using better dated but more distant proxy records is starting to fill this gap, the quality and resolution of these models is still relatively low and their validity remains to be tested. This lack of high-precision palaeoclimate data reinforces the need to test whether climate change was actually sufficiently severe to threaten human life-ways. This study has collected some evidence that this might have been the case (Section 2.3.8), but there is a need for site-specific as well as regional studies dealing with differences across northern Mesopotamia.

Finally, there is a potentially large diversity in responses to environmental change. For this reason, it is necessary to consider the full range of possible responses, instead of focusing on a single response and present an argument based on its occurrence or absence. As this study has shown, reliance on settlement data alone provides a distorted view of northern Mesopotamian human-environment interaction. The reason for this distortion must be sought in the fact that at least a part of the total response structure took place beyond the realm of settled life, namely in those parts of northern Mesopotamia that were home to (semi-)nomadic communities. This complex interaction, and the changes therein, has only been detected through the adoption of a theoretical framework that explicitly aimed to address a wide range of responses, and predict which response is adopted under changing environmental conditions. In this sense, then, past climate changes perhaps challenge today's archaeologists as much as that these changes challenged the peoples of the past.

Bibliography

Abdi, K. 2003: The early development of pastoralism in the Central Zagros Mountains. *Journal of World Prehistory* 17/4, 395–448.

Abruzzi, W.S. 1982: Ecological theory and ethnic differentiation among human populations. *Current Anthropology* 23/1, 13–35.

Abu Assaf, A. 1978/9: Archäologische Geländebegehung in den Bezirken Homs, Tartus, Lataqiya und Raqqa. *Archiv für Orientforschung* 26, 176–7.

Adams, R.M. 1965: *Land behind Baghdad: a history of settlements on the Diyala Plains*. Chicago.

—. 2006: Shepherds at Umma in the Third Dynasty of Ur: interlocutors with a world beyond the scribal field of ordered vision. *Journal of the Economic and Social History of the Orient* 49/2, 133–69.

Akkermans, P.M.M.G. 1984: Archäologische Geländebegehung im Balih-Tal. *Archiv für Orientforschung* 31, 188–90.

—. 1993: *Villages in the steppe. Later Neolithic settlement and subsistence in the Balikh valley, northern Syria*. Ann Arbor.

Akkermans, P.M.M.G. and G.M. Schwartz 2003: *The archaeology of Syria. From complex hunter-gatherers to early urban societies (ca. 16,000–3000 BC)*. Cambridge World Archaeology. Cambridge.

Algaze, G., R. Breuninger, and J. Knudstad 1994: The Tigris-Euphrates archaeological reconnaissance project: final report on the Birecik and Carchemish Dam survey areas. *Anatolica* 20, 1–96.

Algaze, G., R. Breuninger, C. Lightfoot, and M. Rosenberg 1991: the Tigris-Euphrates archaeological reconnaissance project: a preliminary report of the 1989–1990 seasons. *Anatolica* 17, 175–240.

Altaweel, M. 2006: Excavations in Iraq: the Ray Jazirah Project, first report. *Iraq* 68, 155–81.

—. 2007: Excavations in Iraq: the Jazirah Salvage Project, second report. *Iraq* 69, 117–44.

—. 2008: Investigating agricultural sustainability and strategies in northern Mesopotamia: results produced using a socio-ecological modeling approach. *Journal of Archaeological Science* 35/4, 821–35.

Ammerman, A.J. 1981: Surveys and archaeological research. *Annual Review of Anthropology* 10, 63–88.

Anastasio, S., M. Lebeau, and M. Sauvage 2004: *Atlas of preclassical Upper Mesopotamia*. Subartu, 13. Turnhout.

Anbar, M. 1991: *Les tribus amurrites de Mari*. Orbis Biblicus et Orientalis, 108. Freiburg.

Archi, A. 1985: Mardu in the Ebla texts. *Orientalia* 54, 7–13.

Archi, A. and M.G. Biga 2003: A victory over Mari and the fall of Ebla. *Journal of Cuneiform Studies* 55, 1–44.

Arz, H.W., F. Lamy, and J. Patzold 2006: A pronounced dry event recorded around 4.2 ka in brine sediments from the Northern Red Sea. *Quaternary Research* 66/3, 432–41.

Aurenche, O. 1981: Essai de démographie archéologique. L'exemple des villages du Proche-Orient ancien. *Paléorient* 7/1, 93–105.

Aurenche, O. and P. Desfarges 1983: Travaux d'ethnoarchéologie en Syrie et Jordanie. Rapports préliminaires. *Syria* 60, 147–85.

Bahrani, Z. 2006: Race and ethnicity in Mesopotamian antiquity. *World Archaeology* 38/1, 48–59.

Baillie, M.G.L. 1998: Review of Dalfes, H. N., G. Kukla and H. Weiss: third millennium bc climate change and Old World collapse. *Journal of Archaeological Science* 25/2, 185–6.

Ball, W., D.J. Tucker, and T.J. Wilkinson 1989: The Tell al-Hawa project. Archaeological investigations in the north Jazira 1986–87. *Iraq* 51, 1–66.

Balossi, F., G.M. Di Nocera, and M. Frangipane 2007: The contribution of a small site to the study of settlement changes on the Turkish Middle Euphrates between the third and second millennium B.C: preliminary stratigraphic data from Zeytinli Bahçe Höyük. In C. Kuzucuoğlu and C. Marro (eds.): *Sociétés humaines et changement climatique à la fin du troisième millénaire: une crise a-t-elle eu lieu en Haute Mésopotamie?* Varia Anatolica, 19. Istanbul, 355–81.

Banning, E.B. 2002: *Archaeological survey.* Manuals in Archaeological Method, Theory, and Technique. New York.

Bar-Matthews, M., A. Ayalon, M. Gilmour, A. Matthews, and C.J. Hawkesworth 2003: Sea-land oxygen isotopic relationships from planktonic foraminifera and speleothems in the Eastern Mediterranean region and their implication for paleorainfall during interglacial intervals. *Geochimica et Cosmochimica Acta* 67/17, 3181–99.

Bar-Matthews, M., A. Ayalon, and A. Kaufman 1997: Late Quaternary paleoclimate in the Eastern Mediterranean region from stable isotope analysis of speleothems at Soreq Cave, Israel. *Quaternary Research* 47/2, 155–68.

Bar-Matthews, M., A. Ayalon, and A. Kaufman 1998: Middle to Late Holocene (6,500 yr. period) paleoclimate in the Eastern Mediterranean region from stable isotopic composition of speleothems from Soreq Cave, Israel. In A.S. Issar and N. Brown (eds.): *Water, environment and society in times of climatic change.* Water Science and Technology Library, 31. Dordrecht, 203–14.

Barfield, T.J. 1993: *The nomadic alternative.* Englewood Cliffs.

Barge, O. and B. Moulin 2008: The development of the Syrian steppe during the Early Bronze Age. In H. Kühne, R.M. Czichon, and F.J. Kreppner (eds.): *Proceedings of the 4th International Congress of the Archaeology of the Ancient Near East, 1.* Wiesbaden, 19–28.

Barnard, A. 1992: Social and spatial boundary maintenance among Southern African hunter-gatherers. In M.J. Casimir and A. Rao (eds.): *Mobility and territoriality. Social and spatial boundaries among foragers, fishers, pastoralists and peripatetics.* New York, 137–51.

Barrow, C. 1987: *Water resources and agricultural development in the tropics.* Longman Development Studies. Burnt Mill.

Barth, F. 1969a: Introduction. In F. Barth (ed.): *Ethnic groups and boundaries. The social organization of culture difference.* Boston, 9–38.

—. 1969b: Pathan identity and its maintenance. In F. Barth (ed.): *Ethnic groups and boundaries. The social organization of culture difference.* Boston, 117–34.

Bartl, K. 1994: *Frühislamische Besiedlung im Balih-Tal, Nordsyrien.* Berliner Beiträge zum Vorderen Orient, 15. Berlin.

Bates, D.G. and S.H. Lees 1977: The role of exchange in productive specialization. *American Anthropologist* 79/4, 824–41.

Battini-Villard, L. 1999: *L'espace domestique en Mésopotamie de la IIIe dynastie de'Ur à l'époque paléo-babylonienne.* BAR International Series, 767. Oxford.

Bauer, J., R.K. Englund, and M. Krebernik 1998: *Mesopotamien. Späturuk-Zeit und Frühdynastische Zeit.* Orbis Biblicus et Orientalis, 160/1. Freiburg.

Beaulieu, P.-A. 2005: The god Amurru as emblem of ethnic and cultural identity. In W.H. van Soldt, R. Kalvelagen, and D. Katz (eds.): *Ethnicity in ancient Mesopotamia: papers read at the 48th Rencontre Assyriologique Internationale, Leiden, 1–4 July 2002.* Uitgaven van het Nederlands Instituut voor het Nabije Oosten te Leiden, 102. Leiden, 31–46.

Bell, B. 1971: The dark ages in ancient history. I. The first dark age in Egypt. *American Journal of Archaeology* 75/1, 1–26.

Bentley, G.C. 1987: Ethnicity and practice. *Comparative Studies in Society and History* 29/1, 24–55.

Bernbeck, R. 1993: *Steppe als Kulturlandschaft: das 'Agig-Gebiet Ostsyriens vom Neolithikum bis zur islamischen Zeit.* Berliner Beiträge zum Vorderen Orient. Ausgrabungen, 1. Berlin.

Besonen, M. and M. Cremaschi n.d.: Geomorphological field survey report. Tell Leilan, June 2002. Available at: http://leilan.yale.edu/works/geo_report/index.html (Accessed 10 April, 2008).

Bilgen, A.N. 2001: Early Bronze Age vessels (grave goods?) from Harabebezikan Höyük. In N. Tuna, J. Öztürk, and J. Velibeyoğlu (eds.): *Salvage Project of the Archaeological Heritage of the Ilısu and Carchemish Dam Reservoirs. Activities in 1999*. Ankara, 437–52.

Binford, L.R. 2001: *Constructing frames of reference. An analytical method for archaeological theory building using hunter-gatherer and environmental data sets*. Berkeley.

Bintliff, J.L. 1997: Further considerations on the population of ancient Boeotia. In J.L. Bintliff (ed.): *Recent developments in the history and archaeology of central Greece: proceedings of the 6th International Boeotian conference*. BAR International Series, 666. Oxford, 231–52.

—. 2000: The concepts of 'site' and 'offsite' archaeology in surface artefact survey. In M. Pasquinucci and F. Trément (eds.): *Non-destructive techniques applied to landscape archaeology*. The Archaeology of Mediterranean Landscapes, 4. Oxford, 200–15.

Bintliff, J.L., M. Kuna, and N. Venclová (eds.) 2000: *The future of surface artefact survey in Europe*. Sheffield Archaeological Monographs, 13. Sheffield.

Black, J.A., G. Cunningham, J. Ebeling, E. Flückiger-Hawker, E. Robson, J. Taylor, and G. Zólyomi 1998–2006: The Electronic Text Corpus of Sumerian Literature. Available at: http://etcsl.orinst. ox.ac.uk/ (Accessed 19 November, 2008).

Boerma, J.A.K. 1988: Soils and environment of Tell Hammam et-Turkman. In M.N. van Loon (ed.): *Hammam et-Turkman I. Report on the University of Amsterdam's 1981–1984 excavations in Syria*. Uitgaven van het Nederlands Historisch-Archaeologisch Instituut te Istanbul, 63. Istanbul, 1–11.

Boessneck, J. 1988: Tierknochenfunde vom Tell Chuēra / Nordost-Syrien. In U. Moortgat-Correns: *Tell Chuēra in Nordost-Syrien. Vorläufige Berichte über die neunte und zehnte Grabungskampagne 1982 und 1983*. Tell Chuēra in Nordost-Syrien, 9. Berlin, 79–98.

Boone, J.L. 1992: Competition, conflict, and the development of social hierarchies. In E.A. Smith and B. Winterhalder (eds.): *Evolutionary ecology and human behavior*. Foundations of Human Behavior. New York, 301–37.

Boserup, E. 1965: *The conditions of agricultural growth. The economics of agrarian change under population pressure*. London.

Bottema, S. 1997: Third millennium climate in the Near East based upon pollen evidence. In H.N. Dalfes, G. Kukla, and H. Weiss (eds.): *Third millennium BC climate change and Old World collapse*. NATO ASI Series, I/49. Berlin, 489–515.

Braidwood, R.J. 1937: *Mounds in the plain of Antioch: an archeological survey*. Oriental Institute Publications, 48. Chicago.

Braun, D.P. and S. Plog 1982: Evolution of "tribal" social networks: theory and prehistoric North American evidence. *American Antiquity* 47/3, 504–25.

Brinkman, J.A. 1977 (1964): Appendix: Mesopotamian chronology of the historical period. In A.L. Oppenheim: *Ancient Mesopotamia. Portrait of a dead civilization*. Chicago, 335–48.

Brooks, N. 2006: Cultural responses to aridity in the Middle Holocene and increased social complexity. *Quaternary International* 151/1, 29–49.

Bruins, H.J. 2001: Near East: towards an integrated ^{14}C time foundation. *Radiocarbon* 43/3, 1147–54.

Bruins, H.J., J. van der Plicht, and A. Mazar 2003a: ^{14}C Dates from Tel Rehov: Iron-Age chronology, pharaohs, and Hebrew kings. *Science* 300/5617, 315–18.

Bruins, H.J., J.J. Akong'a, M.M.E.M. Rutten, and G.M. Kressel 2003b: *Drought planning and rainwater harvesting for arid-zone pastoralists: the Turkana and Maasai (Kenya) and the Negev Bedouin (Israel)*. NIRP Research for Policy Series, 17. Amsterdam.

Brumfiel, E.M. 1994: Factional competition and political development in the New World: an introduction. In E.M. Brumfiel and J.W. Fox (eds.): *Factional competition and political development in the New World*. New Directions in Archaeology. Cambridge, 3–13.

Bryson, R.A. and R.U. Bryson 1997: High resolution simulations of regional Holocene climate: North Africa and the Near East. In H.N. Dalfes, G. Kukla, and H. Weiss (eds.): *Third millennium BC climate change and Old World collapse.* NATO ASI Series, I/49. Berlin, 565–93.

Bucak, E., R. Dittmann, S.K. Huh, F. Jarecki, S. Kiltz, U. Röttger, and I. Vorontsov 2004: Short report on the excavations at Şavi Höyük in 2001. In N. Tuna, J. Öztürk, and J. Velibeyoğlu (eds.): *Salvage Project of the Archaeological Heritage of the Ilısu and Carchemish Dam Reservoirs. Activities in 2001.* Ankara, 174–8.

Buccellati, G. 1966: *The Amorites of the Ur III period.* Pubblicazioni del Seminario di Semitistica, Ricerche, 1. Naples.

—. 1990: 'River bank', 'high country', and 'pasture land': the growth of nomadism on the Middle Euphrates and the Khabur. In S.M. Eichler, M. Wäfler, and D. Warburton (eds.): *Tall al-Hamidiyah 2.* Orbis Biblicus et Orientalis, 6. Freiburg, 87–117.

—. 1992: Ebla and the Amorites. In C.H. Gordon and G.A. Rendsburg (eds.): *Eblaitica: essays on the Ebla archives and Eblaite language. Volume 3.* Publications of the Center for Ebla Research at New York University. Winona Lake, 83–104.

—. 1997: Amorites. In E.M. Meyers (ed.): *The Oxford encyclopedia of archaeology in the Near East. Volume 1.* New York, 107–11.

—. 2008: The origin of the tribe and of 'industrial' agropastoralism in Syro-Mesopotamia. In H. Barnard and W. Wendrich (eds.): *The archaeology of mobility. Old World and New World nomadism.* Cotsen Advanced Seminar Series, 4. Los Angeles, 141–59.

Buccellati, G. and M. Kelly-Buccellati 1999: Das archäologische Projekt Tall Mozan/Urkeš. *Mitteilungen der Deutschen Orient-Gesellschaft zu Berlin* 131, 7–16.

—. 2000: The royal palace of Urkesh. Report on the 12th season at Tell Mozan/Urkesh: excavations in area AA, June–October 1999. *Mitteilungen der Deutschen Orient-Gesellschaft zu Berlin* 132, 133–83.

Buitenhuis, H. 1988: *Archeozoölogisch onderzoek langs de Midden-Eufraat. Onderzoek van het faunamateriaal uit zes nederzettingen in Zuidoost-Turkije en Noord-Syrië daterend van ca. 10.000 BP tot 1400 AD.* Unpublished dissertation, Rijksuniversiteit Groningen.

—. forthcoming: Preliminary report on faunal remains from Tell Hammam et-Turkman, North Syria. In D.J.W. Meijer (ed.): *Hammam et-Turkman II.*

Bunnens, G. 2007: Site hierarchy in the Tishrin Dam area and third millennium geopolitics in Northern Syria. In E. Peltenburg (ed.): *Euphrates river valley settlement: the Carchemish sector in the third millennium BC.* Levant Supplementary Series, 5. Oxford, 43–54.

Burghardt, A.F. 1971: A hypothesis about gateway cities. *Annals of the Association of American Geographers* 61/2, 269–85.

Buringh, P. 1960: *Soils and soil conditions in Iraq.* Baghdad.

Butterlin, P. 2007: Mari, les šakkanakku et la crise de la fin du troisième millénaire. In C. Kuzucuoğlu and C. Marro (eds.): *Sociétés humaines et changement climatique à la fin du troisième millénaire: une crise a-t-elle eu lieu en Haute Mésopotamie?* Varia Anatolica, 19. Istanbul, 227–45.

Butzer, K.W. 1982: *Archaeology as human ecology: method and theory for a contextual approach.* Cambridge.

—. 1997: Sociopolitical discontinuity in the Near East C. 2200 B.C.E.: scenarios from Palestine and Egypt. In H.N. Dalfes, G. Kukla, and H. Weiss (eds.): *Third millennium BC climate change and Old World collapse.* NATO ASI Series, I/49. Berlin, 245–96.

Carneiro, R.L. 1970: A theory of the origin of the state. *Science* 169/3947, 733–38.

Carneiro, R.L. and D.F. Hilse 1966: On determining the probable rate of population growth during the Neolithic. *American Anthropologist* 68/1, 177–81.

Casana, J. 2007: Structural transformations in settlement systems of the northern Levant. *American Journal of Archaeology* 111/2, 195–221.

Cashdan, E. 1987: Trade and its origins on the Botletli River, Botswana. *Journal of Anthropological Research* 43/2, 121–38.

—. 2001a: Ethnic diversity and its environmental determinants: effects of climate, pathogens, and habitat diversity. *American Anthropologist* 103/4, 968–91.

—. 2001b: Ethnocentrism and xenophobia: a cross-cultural study. *Current Anthropology* 42/5, 760–65.

Casimir, M.J. 1992: The dimensions of territoriality: an introduction. In M.J. Casimir and A. Rao (eds.): *Mobility and territoriality. Social and spatial boundaries among foragers, fishers, pastoralists and peripatetics.* New York, 1–26.

Castel, C. 2007: L'abandon d'Al-Rawda (Syrie) à la fin du troisième millénaire: premières tentatives d'explication. In C. Kuzucuoğlu and C. Marro (eds.): *Sociétés humaines et changement climatique à la fin du troisième millénaire: une crise a-t-elle eu lieu en Haute Mésopotamie?* Varia Anatolica, 19. Istanbul, 159–78.

Castel, C. and E. Peltenburg 2007: Urbanism on the margins: third millennium BC Al-Rawda in the arid zone of Syria. *Antiquity* 81/313, 601–16.

Cauvin, J. 1970: Mission 1969 en Djezireh (Syrie). *Bulletin de la Société Préhistorique Française* 67, 286–7.

Chamberlain, A. 2006: *Demography in archaeology.* Cambridge Manuals in Archaeology. Cambridge.

Chapman, J. 1999: Archaeological proxy-data for demographic reconstructions: facts, factoids or fiction? In J. Bintliff and K. Sbonias (eds.): *Reconstructing past population trends in Mediterranean Europe (3000 BC–AD 1800).* The Archaeology of Mediterranean Landscapes, 1. Oxford, 65–76.

Charpin, D. 1987: Šubat-Enlil et le pays d'Apum. *MARI* 5, 129–40.

—. 1993: La Syrie à l'époque de Mari: des invasions amorites à la chute de Mari. In S. Cluzan, E. Delpont, and J. Mouliérac (eds.): *Syrie. Mémoire et civilisation.* Paris, 144–9.

—. 2003: La «toponymie en mirroir» dans le Proche-Orient amorrite. *Revue d'Assyriologie et d'Archéologie orientale* 97, 3–34.

—. 2004: Nomades et sédentaires dans l'armée de Mari du temps de Yahdun-Lîm. In C. Nicolle (ed.): *Nomades et sédentaires dans le Proche Orient ancien.* Amurru, 3. Paris, 83–94.

Charpin, D. and J.-M. Durand 1986: «Fils de Sim'al»: les origines tribales des rois de Mari. *Revue d'Assyriologie* 80, 141–83.

Charpin, D., D.O. Edzard, and M. Stol 2004: *Mesopotamien. Die altbabylonische Zeit. Annäherungen 4.* Orbis biblicus et orientalis, 160/4. Freiburg.

Charpin, D. and N. Ziegler 2003: *Florilegium marianum V. Mari et le proche-orient à l'époque amorrite: essai d'histoire politique.* Mémoires de N.A.B.U., 6. Paris.

Cherry, J.F. 1979: Four problems in Cycladic prehistory. In J.L. Davis and J.F. Cherry (eds.): *Papers in Cycladic prehistory.* UCLAMon, 14. Los Angeles, 22–47.

Clason, A.T. and H. Buitenhuis 1978: A preliminary report on the faunal remains of Nahr el Homr, Hadidi and Ta'as in the Tabqa Dam region in Syria. *Journal of Archaeological Science* 5/1, 75–83.

—. 1998: Patterns in animal food resources in the Bronze Age in the Orient. In H. Buitenhuis, L. Bartosiewicz, and A.M. Choyke (eds.): *Archaeozoology of the Near East III. Proceedings of the third international symposium on the archaeozoology of southwestern Asia and adjacent areas.* ARC-Publication, 18. Groningen, 233–42.

Coombes, P. and K. Barber 2005: Environmental determinism in Holocene research: causality or coincidence? *Area* 37/3, 303–11.

Cooper, J.S. 1983: *The curse of Agade.* Johns Hopkins Near Eastern Studies. Baltimore.

Cooper, L. 2006: *Early urbanism on the Syrian Euphrates.* New York.

Copeland, L. 1979: Observations on the prehistory of the Balikh Valley, Syria, during the 7th to 4th millennia B.C. *Paléorient* 5, 251–75.

—. 1982: Prehistoric tells in the lower Balikh valley, Syria: report on the survey of 1978. *Annales Archéologiques Arabes Syriennes* 32, 251–71.

Córdoba, J.M. 1988: Prospección en el valle de río Balih (Syria). Informe provisional. *Aula Orientalis* 6/2, 149–88.

Costanza, R., L. Graumlich, W. Steffen, C. Crumley, J. Dearing, K. Hibbard, R. Leemans, C. Redman, and D. Schimel 2007: Sustainability or collapse: what can we learn from integrating the history of humans and the rest of nature? *AMBIO: a Journal of the Human Environment* 36/7, 522–7.

Courty, M.-A. 1994: Le cadre paléogéographique des occupations humaines dans le bassin du Haut-Khabur (Syrie du nord-est). Premiers resultats. *Paléorient* 20/1, 21–59.

—. 2001: Evidence at Tell Brak for the Late EDIII/Early Akkadian air blast event (4 kyr BP). In D. Oates, J. Oates, and H. McDonald (eds.): *Excavations at Tell Brak, 2. Nagar in the third millennium BC.* Cambridge, 367–72.

Courty, M.-A. and H. Weiss 1997: The scenario of environmental degradation in the Tell Leilan region, NE Syria, during the late third millennium abrupt climate change. In H.N. Dalfes, G. Kukla, and H. Weiss (eds.): *Third millennium BC climate change and Old World collapse.* NATO ASI Series, I/49. Berlin, 107–47.

Cowgill, G.L. 1975: On causes and consequences of ancient and modern population changes. *American Anthropologist* 77/3, 505–25.

Cribb, R. 1991: *Nomads in archaeology.* New Studies in Archaeology. Cambridge.

Crumley, C.L. 1994: Historical ecology. A multidimensional ecological orientation. In C.L. Crumley (ed.): *Historical ecology. Cultural knowledge and changing landscapes.* School of American Research Advanced Seminar Series. Santa Fe, 1–16.

Cullen, H.M., P.B. deMenocal, S. Hemming, G. Hemming, F.H. Brown, T. Guilderson, and F. Sirocko 2000: Climate change and the collapse of the Akkadian empire: evidence from the deep sea. *Geology* 28/4, 379–82.

Curvers, H.H. 1991: *Bronze Age society in the Balikh Drainage (Syria).* Unpublished dissertation, Universiteit van Amsterdam.

d'Hont, O. 2004: Entre sédentarité et nomadisme: éléments pour une définition de ces deux termes pris dans l'histoire du peuplement de la moyenne vallée de l'Euphrate depuis l'avènement de l'Islam. In C. Nicolle (ed.): *Nomades et sédentaires dans le Proche Orient ancien.* Amurru, 3. Paris, 13–24.

Danti, M.D. 1997: Regional surveys and excavations. In R.L. Zettler: *Subsistence and settlement in a marginal environment: Tell es-Sweyhat, 1989–1995 preliminary report.* MASCA Research Papers in Science and Archaeology, 14. Philadelphia, 85–94.

—. 2006–7: The expansion of settlement into the Balikh-Euphrates uplands. Available at: http://www.jezireh.org/sweyhat_regionalsurvey.html (Accessed 7 January, 2008).

de Roche, C.D. 1983: Population estimates from settlement area and number of residences. *Journal of Field Archaeology* 10/2, 187–92.

Deckers, K. and S. Riehl 2007: Fluvial environmental contexts for archaeological sites in the Upper Khabur basin (northeastern Syria). *Quaternary Research* 67/3, 337–48.

deMenocal, P.B. 2001: Cultural responses to climate change during the late Holocene. *Science* 292, 667–73.

Deveci, A. and Y. Mergen 1999: Zeytin Bahçeli Höyük 1998: preliminary report. In N. Tuna and J. Öztürk (eds.): *Salvage Project of the Archaeological Heritage of the Ilısu and Carchemish Dam Reservoirs. Activities in 1998* Ankara, 113–18.

Dewar, R.E. 1991: Incorporating variation in occupation span into settlement-pattern analysis. *American Antiquity* 56/4, 604–20.

—. 1994: Contending with contemporaneity: a reply to Kintigh. *American Antiquity* 59/1, 149–52.

Diamond, J. 2005: *Collapse: How societies choose to fail or survive.* London.

Dittmann, R., C. Grewe, S.K. Huh, and C. Schmidt 2001: Bericht über einen Survey des Şavi Höyük im Gebiet des Kargemiş-Stausees, Südost Turkei 1999. In R.M. Boehmer and J. Maran (eds.): *Lux orientis: Archäologie zwischen Asien und Europa. Festschrift für Harald Hauptmann zum 65. Geburtstag.* Internationale Archäologie Studia honoraria, 12. Rahden, 97–111.

Dittmann, R., S.K. Huh, T. Mitschang, E. Müller, U. Röttger, C. Schmidt, and D. Wicke 2002: report on the first campaign of excavations at Şavi Höyük in 2000. In N. Tuna, J. Öztürk, and J. Velibeyoğlu (eds.): *Salvage Project of the Archaeological Heritage of the Ilısu and Carchemish Dam Reservoirs. Activities in 2000*. Ankara, 233–40.

Dohmann-Pfälzner, H. and P. Pfälzner 2001: Ausgrabungen der Deutschen Orient-Gesellschaft in der zentralen Oberstadt van Tall Mozan/Urkeš. Bericht über die in Kooperation mit dem IIMAS durchgeführte Kampagne 2000. *Mitteilungen der Deutschen Orient-Gesellschaft zu Berlin* 133, 97–139.

Dossin, G. 1974: Le site de Tuttul-sur-Balîh. *Revue d'Assyriologie* 68, 25–34.

Drennan, R.D. and C.E. Peterson 2004: Comparing archaeological settlement systems with rank-size graphs: a measure of shape and statistical confidence. *Journal of Archaeological Science* 31/5, 533–49.

Durand, J.-M. 1985: La situation historique des *šakkanakku*: Nouvelle approche. *MARI* 4, 147–72.

—. 1992: Unité et diversités au Proche-Orient à l'époque amorrite. In D. Charpin and F. Joannès (eds.): *La circulation des biens, des personnes et des idées dan le Proche-Orient ancien. Actes de la XXXVIIIe Rencontre Assyriologique Internationale (Paris, 8–10 juillet 1991)*. Paris, 97–128.

—. 1997: *Les documents épistolaires du palais de Mari. Tome I*. Littératures Anciennes du Proche-Orient, 16. Paris.

—. 1998: *Les documents épistolaires du palais de Mari. Tome II*. Littératures Anciennes du Proche-Orient, 17. Paris.

—. 2004a: Mari und die Assyrer. In J.-W. Meyer and W. Sommerfeld (eds.): *2000 v. Chr.: politische, wirtschaftliche und kulturelle Entwicklung im Zeichen einer Jahrtausendwende*. Colloquien der Deutschen Orient-Gesellschaft, 3. Saarbrücken, 371–92.

—. 2004b: Peuplement et sociétés à l'époque amorrite. (I) Les clans bensim'alites. In C. Nicolle (ed.): *Nomades et sédentaires dans le Proche Orient ancien*. Amurru, 3. Paris, 111–97.

Dyson-Hudson, R. and N. Dyson-Hudson 1980: Nomadic pastoralism. *Annual Review of Anthropology* 9, 15–61.

Dyson-Hudson, R. and E.A. Smith 1978: Human territoriality: an ecological reassessment. *American Anthropologist* 80/1, 21–41.

Eastwood, W.J., N. Roberts, H.F. Lamb, and J.C. Tibby 1999: Holocene environmental change in southwest Turkey: a palaeoecological record of lake and catchment-related changes. *Quaternary Science Reviews* 18/4–5, 671–95.

Échallier, J.-C. and F. Braemer 1995: Nature et fonctions des «desert kites»: données et hypothèses nouvelles. *Paléorient* 21/1, 35–63.

Edzard, D.O. 1997: *Gudea and his dynasty*. The Royal Inscriptions of Mesopotamia, Early Periods, 3/1. Toronto.

Eichler, S.M., V. Haas, D. Steudler, M. Wäfler, and D. Warburton 1985: *Tall al-Hamidiya 1. Vorbericht 1984*. Orbis Biblicus et Orientalis, 4. Freiburg.

Eidem, J. 2000: Northern Jezira in the 18th century BC. Aspects of geo-political patterns. In O. Rouault and M. Wäfler (eds.): *La Djéziré et l'Euphrate syriens de la protohistoire à la fin du IIe millénaire av. J.-C.: tendances dans l'interprétation historique des données nouvelles*. Subartu, 7. Turnhout, 255–64.

Eidem, J., I. Finkel, and M. Bonechi 2001: The third-millennium inscriptions. In D. Oates, J. Oates, and H. McDonald (eds.): *Excavations at Tell Brak, 2. Nagar in the third millennium BC*. Cambridge, 99–120.

Eidem, J. and D. Warburton 1996: In the land of Nagar: survey around Tell Brak. *Iraq* 58, 51–64.

Einwag, B. 1993: Vorbericht über die archäologische Geländebegehung in der Westgazira. *Damaszener Mitteilungen* 7, 23–43.

—. 1998: *Die Keramik aus dem Bereich des Palastes A in Tall Bi'a/Tuttul und das Problem der frühen Mittleren Bronzezeit*. Münchener Vorderasiatische Studien, 19. München.

Ember, C.R. and M. Ember 1992: Resource unpredictability, mistrust, and war: a cross-cultural study. *Journal of Conflict Resolution* 36/2, 242–62.

Emberling, G. 1997: Ethnicity in complex societies: archaeological perspectives. *Journal of Archaeological Research* 5/4, 295–344.

Enzel, Y., R. Bookman, D. Sharon, H. Gvirtzman, U. Dayan, B. Ziv, and M. Stein 2003: Late Holocene climates of the Near East deduced from Dead Sea level variations and modern regional winter rainfall. *Quaternary Research* 60, 263–73.

Ergenzinger, P.J., W. Frey, H. Kühne, and H. Kurschner 1988: The reconstruction of environment, irrigation and development of settlement on the Habur in north-east Syria. In J.L. Bintliff, D.A. Davidson, and E.G. Grant (eds.): *Conceptual issues in environmental archaeology*. Edinburgh, 108–28.

Evershed, R.P., S. Payne, A.G. Sherratt, M.S. Copley, J. Coolidge, D. Urem-Kotsu, K. Kotsakis, M. Ozdogan, A.E. Ozdogan, O. Nieuwenhuyse, P.M.M.G. Akkermans, D. Bailey, R.-R. Andeescu, S. Campbell, S. Farid, I. Hodder, N. Yalman, M. Ozbasaran, E. Bicakci, Y. Garfinkel, T. Levy, and M.M. Burton 2008: Earliest date for milk use in the Near East and southeastern Europe linked to cattle herding. *Nature*, doi: 10.1038/nature07180.

Faivre, X. n.d.: *Histoire du peuplement du "Triangle du Habur" au Bronze moyen et au début du Bronze récent: étude céramologique d'après les fouilles de Tell Mohammed Diyab (1987–1996) et la prospection dans le bassin occidental du Haut Habur (1989–1997)*. Unpublished doctoral thesis, Université Charles de Gaulle-Lille.

Falconer, S.E. and S.E. Savage 1995: Heartlands and hinterlands: Alternative trajectories of early urbanization in Mesopotamia and the southern Levant. *American Antiquity* 60/1, 37–58.

Falsone, G. and P. Sconzo 2007: The 'champagne-cup' period at Carchemish. A review of the Early Bronze Age levels on the Acropolis Mound and the problem of the Inner Town. In E. Peltenburg (ed.): *Euphrates river valley settlement: the Carchemish sector in the third millennium BC*. Levant Supplementary Series, 5. Oxford, 73–93.

Fielden, K. 1978/9: Archäologische Geländebegehung im Gebiet des Unteren Ġaġġaġ. *Archiv für Orientforschung* 26, 172.

Finkelstein, J.J. 1966: The genealogy of the Hammurapi dynasty. *Journal of Cuneiform Studies* 20, 95–118.

Flannery, K.V. 1976: Sampling by intensive surface collection. In K.V. Flannery (ed.): *The early Mesoamerican village*. San Diego, 51–62.

Fleming, D.E. 2004: *Democracy's ancient ancestors: Mari and early collective governance*. Cambridge.

Forbes, H. 1989: Of grandfathers and grand theories: the hierarchised ordering of responses to hazard in a Greek rural community. In P. Halstead and J. O'Shea (eds.): *Bad year economics: Cultural responses to risk and uncertainty*. New Directions in Archaeology. Cambridge, 87–97.

Fortin, M. 2000: Économie et société dans la moyenne vallée du Khabour durant la période de Ninevite 5. In O. Rouault and M. Wäfler (eds.): *La Djéziré et l'Euphrate syriens de la protohistoire à la fin du IIe millénaire av. J.-C.: tendances dans l'interprétation historique des données nouvelles*. Subartu, 7. Turnhout, 111–36.

Frangipane, M., C. Alvaro, F. Balossi, and G. Siracusano 2002: The 2000 campaign at Zeytinlibahçe Höyük. In N. Tuna, J. Öztürk, and J. Velibeyoğlu (eds.): *Salvage Project of the Archaeological Heritage of the Ilısu and Carchemish Dam Reservoirs. Activities in 2000*. Ankara, 83–99.

Frangipane, M., F. Balossi, G.M. Di Nocera, A. Palmieri, and G. Siracusano 2004: The 2001 excavation campaign at Zeytinlibahçe Höyük: preliminariy results. In N. Tuna, J. Öztürk, and J. Velibeyoğlu (eds.): *Salvage Project of the Archaeological Heritage of the Ilısu and Carchemish Dam Reservoirs. Activities in 2001*. Ankara, 35–56.

Frangipane, M. and E. Bucak 2001: Excavations and research at Zeytinlibahçe Höyük, 1999. In N. Tuna, J. Öztürk, and J. Velibeyoğlu (eds.): *Salvage Project of the Archaeological Heritage of the Ilısu and Carchemish Dam Reservoirs. Activities in 1999*. Ankara, 109–131.

Frayne, D. 1990: *Old Babylonian period (2003–1595 BC)*. The Royal Inscriptions of Mesopotamia, Early Periods, 4. Toronto.

—. 1993a: *Sargonic and Gutian periods (2334–2113 BC)*. The Royal Inscriptions of Mesopotamia, Early Periods, 2. Toronto.

—. 1993b: *Ur III period (2112–2004 BC)*. The Royal Inscriptions of Mesopotamia, Early Periods, 3/2. Toronto.

—. 2008: *Presargonic period (2700–2350 BC)*. The Royal Inscriptions of Mesopotamia, Early Periods, 1. Toronto.

Frumkin, A. 2009: Stable isotopes of a subfossil Tamarix tree from the Dead Sea region, Israel, and their implications for the Intermediate Bronze Age environmental crisis. *Quaternary Research* 71/3, 319–28.

Frumkin, A., G. Kadan, Y. Enzel, and Y. Eyal 2001: Radiocarbon chronology of the Holocene Dead Sea: attempting a regional correlation. *Radiocarbon* 43/3, 1179–89.

Fuensanta, J.G. 2007: The Tilbes Project (Birecik Dam, Turkish Euphrates): the Early Bronze evidence. In E. Peltenburg (ed.): *Euphrates river valley settlement: The Carchemish sector in the third millennium BC.* Levant Supplementary Series, 5. Oxford, 142–51.

Fuensanta, J.G., E. Bucak, E. Crivelli, P. Charvat, and R. Moya 2005: The research of the Tilbes Project, 2004. *Kazı Sonuçları Toplantısı* 27.

Fuensanta, J.G., E. Bucak, E. Crivelli, H. Karabulut, H. Sauren, P. Charvat, and R. Moya 2006: The Tilbes project research in 2005: Surtepe Höyük excavations. *Kazı Sonuçları Toplantısı* 28, 457–70.

Fuensanta, J.G., P. Charvat, E. Bucak, R. Moya, and M.A. Jimenez 2002a: Tilbes and Tilvez Höyük salvage excavation report. *Kazı Sonuçları Toplantısı* 24, 369–76.

Fuensanta, J.G., P. Charvat, E. Bucak, M.A. Jimenez, P. Kvetina, and F. Velinski 2002b: 2001 Surtepe Höyük salvage excavation report. *Kazı Sonuçları Toplantısı* 24, 105–12.

Fuensanta, J.G., M.S. Rothman, and E. Bucak 1998: 1997 Salvage excavations at Tilbeş Höyük. *Kazı Sonuçları Toplantısı* 20, 207–18.

Fuensanta, J.G., M.S. Rothman, P. Charvat, and E. Bucak 2001: Tilbeş Höyük salvage project excavation. *Kazı Sonuçları Toplantısı* 23, 131–44.

Galaty, J.G. 1982: Being "Maasai"; being "People-of-Cattle": ethnic shifters in East Africa. *American Ethnologist* 9/1, 1–20.

Gallet, Y., M. Le Goff, A. Genevey, J.-C. Margueron, and P. Matthiae 2008: Geomagnetic field intensity behavior in the Middle East between ~3000 BC and ~1500 BC. *Geophysical Research Letters* 35/L02307, doi: 10.1029/2007GL031991.

Galvin, K.F. 1987: Forms of finance and forms of production: the evolution of specialized livestock production in the ancient Near East. In E.M. Brumfiel and T.K. Earle (eds.): *Specialization, exchange, and complex societies.* New Directions in Archaeology. Cambridge, 119–29.

Gasche, H., J.A. Armstrong, S.W. Cole, and V.G. Gurzadyan 1998: *Dating the fall of Babylon. A reappraisal of second-millennium chronology (A joint Ghent–Chicago–Harvard project).* Mesopotamian History and Environment Memoirs, 4. Ghent.

Gelb, I.J. 1980: *Computer-aided analysis of Amorite.* Assyriological Studies, 21. Chicago.

Goodfriend, G.A. 1999: Terrestrial stable isotope records of Late Quaternary paleoclimates in the eastern Mediterranean region. *Quaternary Science Reviews* 18/4–5, 501–13.

Göyünç, N. and W. Hütteroth 1997: *Land an der Grenze. Osmanische Verwaltung im heutigen türkisch–syrisch–irakischen Grenzgebiet im 16. Jahrhundert.* Istanbul.

Gragson, T. 1998: Potential versus actual vegetation: human behavior in a landscape medium. In W. Balée and C.L. Crumley (eds.): *Advances in historical ecology.* Historical Ecology Series. New York, 213–31.

Green, M.W. 1980: Animal husbandry at Uruk in the archaic period. *Journal of Near Eastern Studies* 39/1, 1–35.

Gremmen, W.H.E. and S. Bottema 1991: Palynological investigations in the Syrian Gazira. In H. Kühne (ed.): *Die rezente Umwelt von Tall Seh Hamad und Daten zur Umweltrekonstruktion der Assyrischen Stadt Dur-Katlimmu.* Berichte der Ausgrabung Tall Seh Hamad, Dur-Katlimmu, 1. Berlin, 105–16.

Gronenborn, D. 2005: Klimaforschung und Archäologie. In D. Gronenborn (ed.): *Klimaveränderung und Kulturwandel in neolithischen Gesellschaften Mitteleuropas, 6700–2200 v. Chr.* RGZM Tagungen, 1. Mainz, 1–16.

Guichard, M. 2002: Le Šubartum occidental à l'avènement de Zimrî-Lîm. In D. Charpin and J.-M. Durand (eds.): *Florilegium marianum VI. Recueil d'études à la mémoire d'André Parrot.* Mémoires de N.A.B.U., 7. Paris, 119–68.

Halstead, P. and J. O'Shea 1989a: Introduction: cultural responses to risk and uncertainty. In P. Halstead and J. O'Shea (eds.): *Bad year economics: cultural responses to risk and uncertainty.* New Directions in Archaeology. Cambridge, 1–7.

— (eds.) 1989b: *Bad year economics: cultural responses to risk and uncertainty.* New Directions in Archaeology. Cambridge.

Hassan, F.A. 1981: *Demographic archaeology.* Studies in Archaeology. New York.

Hawkins, J.D. 1980: Karkamis. In D.O. Edzard (ed.): *Reallexikon der Assyriologie und Vorderasiatische Archäologie, V.* Berlin, 426–46.

Hempelmann, R. 2008: Kharab Sayyar: the foundation of the Early Bronze Age settlement. In J.M. Córdoba, M. Molist, M.C. Pérez, I. Rubio, and S. Martínez (eds.): *Proceedings of the 5th International Congress on the Archaeology of the Ancient Near East. Madrid, April 3–8 2006. Vol. II.* Madrid, 153–64.

Henry, A.G. and D.R. Piperno 2008: Using plant microfossils from dental calculus to recover human diet: a case study from Tell al-Raqa'i, Syria. *Journal of Archaeological Science* 35/7, 1943–50.

Herveux, L. 2004: Étude archéobotanique préliminaire de tell al-Rawda, site de la fin du Bronze ancien en Syrie intérieure. *Akkadica* 125/1, 79–91.

Hobbs, J.J. 1990: *Bedouin life in the Egyptian wilderness.* Cairo.

Hole, F. 1991: Middle Khabur settlement and agriculture in the Ninevite V period. *Bulletin of the Canadian Society for Mesopotamian Studies* 21, 17–30.

—. 1997a: Evidence for mid-Holocene environmental change in the Western Khabur Drainage, Northeastern Syria. In H.N. Dalfes, G. Kukla, and H. Weiss (eds.): *Third millennium BC climate change and Old World collapse.* NATO ASI series, I/49. Berlin, 39–66.

—. 1997b: Palaeoenvironment and human society in the Jezireh of northern Mesopotamia 20,000–6,000 B.P. *Paléorient* 23/2, 39–49.

—. 1999: Economic implications of possible storage structures at Tell Ziyadeh, NE Syria. *Journal of Field Archaeology* 26/3, 267–83.

—. 2009: Pastoral mobility as an adaptation. In J. Szuchman (ed.): *Nomads, tribes, and the state in the ancient Near East.* Oriental Institute Seminars, 5. Chicago, 261–83.

Hole, F. and B.F. Zaitchik 2007: Policies, plans, practice, and prospects: Irrigation in northeastern Syria. *Land Degradation & Development* 18/2, 133–52.

Horne, L. 1993: Occupational and locational instability in arid land settlement. In C.M. Cameron and S.A. Tomka (eds.): *Abandonment of settlements and regions: ethnoarchaeological and archaeological approaches.* New Directions in Archaeology. Cambridge, 43–53.

—. 1994: *Village spaces. Settlement and society in northeastern Iran.* Smithsonian Series in Archaeological Inquiry. Washington.

Horowitz, D.L. 1985: *Ethnic groups in conflict.* Berkeley.

IJzereef, G.F. 2001: Animal remains. In M.N. Van Loon (ed.): *Selenkahiye. Final report on the University of Chicago and University of Amsterdam excavations in the Tabqa Reservoir, northern Syria, 1967–1975.* Uitgaven van het Nederlands Historisch-Archaeologisch Instituut te Istanbul, 91. Istanbul, 569–85.

IPCC 2007: *Climate change 2007: Synthesis report. Contribution of Working Groups I, II and III to the Fourth Assessment Report of the Intergovernmental Panel on Climate Change.* Geneva.

Issar, A.S. and M. Zohar 2004: *Climate change. Environment and civilization in the Middle East.* Berlin.

Jacobsen, T. and R.M. Adams 1958: Salt and silt in ancient Mesopotamian agriculture. Progressive changes in soil salinity and sedimentation contributed to the breakup of past civilizations. *Science* 128, 1251–58.

Johnson, A.W. and T.K. Earle 2000: *The evolution of human societies. From foraging group to agrarian state. Second edition.* Stanford.

Johnson, G.A. 1977: Aspects of regional analysis in archaeology. *Annual Review of Anthropology* 6, 479–508.

—. 1980: Rank-size convexity and system integration: a view from archaeology. *Economic Geography* 56/3, 234–47.

—. 1981: Monitoring complex system integration and boundary phenomena with settlement size data. In S. Van der Leeuw (ed.): *Archaeological approaches to the study of complexity.* Cingula, 6. Amsterdam, 144–88.

Jones, S. 1997: *The archaeology of ethnicity. Constructing identities in the past and present.* London.

Kamp, K.A. and N. Yoffee 1980: Ethnicity in ancient Western Asia during the early second millennium B.C.: archaeological assessments and ethnoarchaeological prospectives. *Bulletin of the American Schools of Oriental Research* 237, 85–104.

Kaptijn, E. 2009: *Life on the watershed. Reconstructing subsistence in a steppe region using archaeological survey: a diachronic perspective on habitation in the Jordan Valley.* Dissertation, Leiden University.

Kaptijn, E., L. Petit, E. Grootveld, F. Hourani, G. van der Kooij, and O. al-Ghul 2005: Dayr 'Alla regional project: Settling the steppe. First campaign 2004. *Annual of the Department of Antiquities of Jordan* 49, 89–99.

Keenan, D.J. 2002: Why early-historical radiocarbon dates downwind from the Mediterranean are too early. *Radiocarbon* 44/1, 225–37.

Kelly-Buccellati, M. 2002: Ein hurritischer Gang in die Unterwelt. *Mitteilungen der Deutschen Orient-Gesellschaft zu Berlin* 134, 131–48.

Kelly, R.L. 1995: *The foraging spectrum. Diversity in hunter-gatherer lifeways.* Washington.

Kenyon, K.M. 1966: *Amorites and Canaanites.* Schweich Lectures on Biblical Archaeology. London.

Khazanov, A.M. 2009: Specific characteristics of Chalcolithic and Bronze Age pastoralism in the Near East. In J. Szuchman (ed.): *Nomads, tribes, and the state in the ancient Near East.* Oriental Institute Seminars, 5. Chicago, 119–27.

Khoury, P.S. and J. Kostiner (eds.) 1990a: *Tribes and state formation in the Middle East.* Berkeley.

—. 1990b: Introduction: Tribes and the complexities of state formation in the Middle East. In P.S. Khoury and J. Kostiner (eds.): *Tribes and state formation in the Middle East.* Berkeley, 1–22.

Kintigh, K.W. 1994: Contending with contemporaneity in settlement pattern studies. *American Antiquity* 59/1, 143–8.

Klengel, H. 1972: *Zwischen Zelt und Palast. Die Begegnung von Nomaden und Seßhaften im alten Vorderasien.* Vienna.

Klengel, H. 1992: *Syria: 3000 to 300 B.C.: a handbook of political history.* Berlin.

Köhler-Rollefson, I. 1992: A model for the development of nomadic pastoralism on the Transjordanian Plateau. In O. Bar-Yosef and A.M. Khazanov (eds.): *Pastoralism in the Levant. Archaeological materials in anthropological perspectives.* Monographs in World Archaeology, 10. Madison, 11–18.

—. 1996: The one-humped camel in Asia: Origin, utilization and mechanisms of dispersal. In D.R. Harris (ed.): *The origins and spread of agriculture and pastoralism in Eurasia.* London, 282–94.

Köhler-Rollefson, I. and G.O. Rollefson 1990: The impact of Neolithic subsistence strategies on the environment: the case of 'Ain Ghazal, Jordan. In S. Bottema, G. Entjes-Nieborg, and W. van Zeist (eds.): *Man's role in the shaping of the Eastern Mediterranean landscape.* Rotterdam, 1–14.

Kohler, T.A. and C.R. Van West 1996: The calculus of self-interest in the development of cooperation: sociopolitical development and risk among the Northern Anasazi. In J.A. Tainter and B.B. Tainter (eds.): *Evolving complexity and environmental risk in the prehistoric Southwest*. Santa Fe Institute Studies in the Sciences of Complexity Proceedings, 24. Reading, 169–96.

Kohlmeyer, K. 1981: "Wovon man nicht sprechen kann" Grenzen der Interpretation von bei Oberflächenbegehungen gewonnenen archäologischen Informationen. *Mitteilungen der Deutschen Orient-Gesellschaft zu Berlin* 113, 53–79.

—. 1984: Euphrat-Survey. Die mit Mitteln der Gerda Henkel Stiftung durchgeführte archäologische Geländebegehung im syrischen Euphrattal. *Mitteilungen der Deutschen Orient-Gesellschaft zu Berlin* 116, 95–118.

Koliński, R. and J. Piątkowska-Małecka 2008: Animals in the steppe: patterns of animal husbandry as a reflection of changing environmental conditions in the Khabur Triangle. In H. Kühne, R.M. Czichon, and F.J. Kreppner (eds.): *Proceedings of the 4th International Congress of the Archaeology of the Ancient Near East, 1*. Wiesbaden, 115–27.

Kouchoukos, N.T. 1998: *Landscape and social change in late prehistoric Mesopotamia*. Dissertation, Yale University.

Kozbe, G. and M.S. Rothman 2005: Chronology and function at Yarım Höyük, part II. *Anatolica* 31/111–44.

Kramer, C. 1982: *Village ethnoarchaeology: rural Iran in archaeological perspective*. Studies in Archaeology. New York.

Krom, M.D., J.D. Stanley, R.A. Cliff, and J.C. Woodward 2002: Nile River sediment fluctuations over the past 7000 yr and their key role in sapropel development. *Geology* 30/1, 71–4.

Kuniholm, P.I., B. Kromer, S.W. Manning, M. Newton, C.E. Latini, and M.J. Bruce 1996: Anatolian tree rings and the absolute chronology of the eastern Mediterranean, 2220–718 BC. *Nature* 381/6585, 780–3.

Kupper, J.-R. 1957: *Les nomades en Mésopotamie au temps des rois de Mari*. Bibliothèque de la Faculté de Philosophie et Lettres de l'Université de Liège, 142. Paris.

Kuzucuoğlu, C. 2007a: Climatic and environmental trends during the third millennium B.C. in Upper Mesopotamia. In C. Kuzucuoğlu and C. Marro (eds.): *Sociétés humaines et changement climatique à la fin du troisième millénaire: une crise a-t-elle eu lieu en Haute Mésopotamie?* Varia Anatolica, 19. Istanbul, 459–80.

—. 2007b: Integrating environmental matters in cultural trends. In C. Kuzucuoğlu and C. Marro (eds.): *Sociétés humaines et changement climatique à la fin du troisième millénaire: une crise a-t-elle eu lieu en Haute Mésopotamie?* Varia Anatolica, 19. Istanbul, 21–33.

Landmann, G., A. Reimer, G. Lemcke, and S. Kempe 1996: Dating Late Glacial abrupt climate changes in 14,570 yr long continuous varve record of Lake Van, Turkey. *Palaeogeography, Palaeoclimatology, Palaeoecology* 122, 107–18.

Lawler, A. 2006: North versus South, Mesopotamian style. *Science* 312, 1458–63.

le Strange, G. 1905: *The lands of the eastern Caliphate. Mesopotamia, Persia, and Central Asia from the Moslem conquest to the time of Timur*. Cambridge Geographical Series. Cambridge.

Lebeau, M. 2000a: Stratified archaeological evidence and compared periodizations in the Syrian Jezirah during the third millennium B.C. In C. Marro and H. Hauptmann (eds.): *Chronologies des pays du Caucase et de l'Euphrate aux IVe–IIIe millénaires*. Varia Anatolica, 11. Istanbul, 167–92.

—. 2000b: Les voies de communication en Haute Mésopotamie au IIIe millénaire avant notre ère. In O. Rouault and M. Wäfler (eds.): *La Djéziré et l'Euphrate syriens de la protohistoire à la fin du IIe millénaire av. J.-C.: tendances dans l'interprétation historique des données nouvelles*. Subartu, 7. Turnhout, 157–62.

Lemcke, G. and M. Sturm 1997: $\delta^{18}O$ and trace element measurements as proxy for the reconstruction of climate changes at Lake Van (Turkey): preliminary results. In H.N. Dalfes, G. Kukla, and H. Weiss (eds.): *Third millennium BC climate change and Old World collapse.* NATO ASI Series, I/49. Berlin, 653–78.

Lévêque, A. n.d.: *Atlas de Syrie. Agriculture.* Beyrouth.

Lewis, N.N. 1987: *Nomads and settlers in Syria and Jordan, 1800–1980.* Cambridge Middle East Library. Cambridge.

—. 1988: The Balikh Valley and its people. In M.N. van Loon (ed.): *Hammam et-Turkman I. Report on the University of Amsterdam's 1981–84 excavations in Syria.* Uitgaven van het Nederlands Historisch-Archaeologisch Instituut te Istanbul, 63. Istanbul, 683–695.

Lönnqvist, M. 2006: Archaeological surveys of the Jebel Bishri. The preliminary report of the Finnish mission to Syria, 2000–2004. *Kaskal* 3, 203–40.

—. 2008: Were nomadic Amorites on the move? Migration, invasion and gradual infiltration as mechanisms for cultural transitions. In H. Kühne, R.M. Czichon, and F.J. Kreppner (eds.): *Proceedings of the 4th International Congress of the Archaeology of the Ancient Near East, 2.* Wiesbaden, 195–215.

Luke, J.T. 1965: *Pastoralism and politics in the Mari period: a re-examination of the character and political significance of the major West Semitic tribal groups on the Middle Euphrates, ca. 1828–1758 B.C.* American Schools of Oriental Research Dissertation Series, 3. Cambridge (MA).

Lyon, J. 2000: Middle Assyrian expansion and settlement development in the Syrian Jazira: the view from the Balikh Valley. In R.M. Jas (ed.): *Rainfall and agriculture in northern Mesopotamia.* Uitgaven van het Nederlands Historisch-Archaeologisch Instituut te Istanbul, 88. Istanbul, 89–126.

Lyonnet, B. 1996: La prospection archéologique de la partie occidentale du Haut Khabur (Syrie du nord-est): méthodes, résultats et questions autour de l'occupation aux IIIe et IIe millénaires av. n. è. *Amurru* 1, 363–76.

—. 1998: Le peuplement de la Djéziré occidentale au début du 3e millénaire, villes circulaires et pastoralisme: questions et hypothèses. In M. Lebeau (ed.): *About Subartu: Studies devoted to Upper Mesopotamia.* Subartu, 4/1. Turnhout, 179–93.

—. (ed.) 2000: *Prospection archéologique du Haut-Khabur occidental (Syrie du N.E.) Volume I.* Bibliothèque Archéologique et Historique, 155. Beyrouth.

—. 2004: Le nomadisme et l'archéologie: problèmes d'identification. Le cas de la partie occidentale de la Djéziré aux 3ème et début du 2ème millénaire avant notre ère. In C. Nicolle (ed.): *Nomades et sédentaires dans le Proche Orient ancien.* Amurru, 3. Paris, 25–49.

Madella, M. and D.Q. Fuller 2006: Palaeoecology and the Harappan civilisation of South Asia: A reconsideration. *Quaternary Science Reviews* 25/11–12, 1283–1301.

Magny, M., B. Vanniere, G. Zanchetta, E. Fouache, G. Touchais, L. Petrika, C. Coussot, A.-V. Walter-Simonnet, and F. Arnaud 2009: Possible complexity of the climatic event around 4300–3800 cal. BP in the central and western Mediterranean. *The Holocene* 19/6, 823–33.

Maisels, C.K. 1990: *The emergence of civilization. From hunting and gathering to agriculture, cities, and the state in the Near East.* London.

Mallowan, M.E.L. 1936: The excavations at Tall Chagar Bazar, and an archaeological survey of the Habur region, 1934–5. *Iraq* 3/1, 1–86.

—. 1946: Excavations in the Balikh Valley, 1938. *Iraq* 8, 111–62.

Manning, S.W., M. Barbetti, B. Kromer, P.I. Kuniholm, I. Levin, M.W. Newton, and P.J. Reimer 2002: No systematic early bias to Mediterranean ^{14}C ages: radiocarbon measurements from tree-ring and air samples provide tight limits to age offsets. *Radiocarbon* 44/3, 739–54.

Manning, S.W., B. Kromer, P.I. Kuniholm, and M.W. Newton 2001: Anatolian tree rings and a new chronology for the East Mediterranean Bronze-Iron Ages. *Science* 294/5551, 2532–5.

Manning, S.W., C.B. Ramsey, W. Kutschera, T. Higham, B. Kromer, P. Steier, and E.M. Wild 2006: Chronology for the Aegean Late Bronze Age 1700–1400 B.C. *Science* 312/5773, 565–9.

Marchesi, G. 2006: *LUMMA in the onomasticon and literature of ancient Mesopotamia.* History of the Ancient Near East/Studies, 10. Padova.

Marfoe, L. 1979: The integrative transformation: patterns of sociopolitical organization in southern Syria. *Bulletin of the American Schools of Oriental Research* 234, 1–42.

Margueron, J.-C. 1991: Mari, l'Euphrate, et le Khabur au milieu du IIIe millénaire. *Bulletin of the Canadian Society for Mesopotamian Studies* 21, 79–100.

Marro, C. 2004a: Itinéraires et voies de circulation du Caucase à l'Euphrate: le rôle des nomades dans le système d'échanges et l'économie protohistorique des IVème–IIIème millénaires avant notre ère. In C. Nicolle (ed.): *Nomades et sédentaires dans le Proche Orient ancien.* Amurru, 3. Paris, 51–62.

—. 2004b: Upper Mesopotamia and the Caucasus: an essay on the evolution of routes and road networks from the Old Assyrian kingdom to the Ottoman empire. In A. Sagona (ed.): *A view from the highlands: Archaeological studies in honour of Charles Burney.* Ancient Near Eastern studies. Supplement, 12. Leuven, 91–120.

—. 2007a: Upper-Mesopotamia and the late third millennium crisis hypothesis: state of the art and issues at stake. In C. Kuzucuoğlu and C. Marro (eds.): *Sociétés humaines et changement climatique à la fin du troisième millénaire: une crise a-t-elle eu lieu en Haute Mésopotamie?* Varia Anatolica, 19. Istanbul, 13–20.

—. 2007b: Continuity and change in the Birecik Valley at the end of the third millennium B.C: the archaeological evidence from Horum Höyük. In C. Kuzucuoğlu and C. Marro (eds.): *Sociétés humaines et changement climatique à la fin du troisième millénaire: une crise a-t-elle eu lieu en Haute Mésopotamie?* Varia Anatolica, 19. Istanbul, 383–401.

Marro, C., A. Tibet, and F. Bulgan 2000: Fouilles de sauvetage de Horum Höyük (province de Gaziantep): Quatrième rapport préliminaire. *Anatolia Antiqua* 8, 257–78.

Marro, C., A. Tibet, and R. Ergeç 1997: Fouilles de sauvetage de Horum Höyük (province de Gaziantep): Premier rapport préliminaire. *Anatolia Antiqua* 5, 371–91.

—. 1998: Fouilles de sauvetage de Horum Höyük (province de Gaziantep): Troisième rapport préliminaire. *Anatolia Antiqua* 7, 285–307.

Matthews, V.H. 1978: *Pastoral nomadism in the Mari Kingdom (ca. 1830–1760 B.C.).* American Schools of Oriental Research Dissertation Series, 3. Cambridge, Mass.

Matthews, W. 2003: Microstratigraphic sequences: Indications of uses and concepts of space. In R. Matthews (ed.): *Excavations at Tell Brak, 4. Exploring an upper Mesopotamian regional centre, 1994–1996.* Cambridge, 377–88.

Mayewski, P.A., E.E. Rohling, J.C. Stager, W. Karlén, K.A. Maasch, L.D. Meeker, E.A. Meyerson, F. Gasse, S. van Kreveld, K. Holmgren, J. Lee-Thorp, G. Rosqvist, F. Rack, M. Staubwasser, R.R. Schneider, and E.J. Steig 2004: Holocene climate variability. *Quaternary Research* 62/3, 243–55.

McCorriston, J. 1992: The Halaf environment and human activities in the Khabur Drainage, Syria. *Journal of Field Archaeology* 19/3, 315–33.

—. 1997: The fiber revolution. Textile extensification, alienation, and social stratification in ancient Mesopotamia. *Current Anthropology* 38/4, 517–49.

McIntosh, R.J., J.A. Tainter, and S.K. McIntosh 2000a: Climate, history, and human action. In R.J. McIntosh, J.A. Tainter, and S.K. McIntosh (eds.): *The way the wind blows: climate, history, and human action.* Historical Ecology Series. New York, 1–42.

— (eds.) 2000b: *The way the wind blows: climate, history, and human action.* Historical Ecology Series. New York.

McLellan, T.L., R. Grayson, and C. Oglesby 2000: Bronze Age water harvesting in north Syria. In O. Rouault and M. Wäfler (eds.): *La Djéziré et l'Euphrate syriens de la protohistoire à la fin du IIe millénaire av. J.-C.: tendances dans l'interprétation historique des données nouvelles.* Subartu, 7. Turnhout, 137–55.

McLellan, T.L. and A. Porter 1995: Jawa and North Syria. *Studies in the History and Archaeology of Jordan* 5, 49–65.

Meijer, D.J.W. 1986: *A survey in Northeastern Syria*. Uitgaven van het Nederlands Historisch-Archaeologisch Instituut te Istanbul, 58. Istanbul.

—. 1989: Hammam al-Turkman on the Balikh: First results of the fifth campaign, 1988. *Akkadica* 64–65, 1–12.

—. 1990: An archaeological survey: some assumptions and ideas. In S.M. Eichler, M. Wäfler, and D. Warburton (eds.): *Tall al-Hamidiyah 2*. Orbis Biblicus et Orientalis, 6. Freiburg, 31–45.

—. 1996: Tell Hammam al-Turkman: preliminary report on the seventh campaign, May–July 1995. *Anatolica* 22, 181–93.

—. 2000: Ecology and archaeology: perceptions and questions. In L. Milano, S. de Martino, F.M. Fales, and G.B. Lanfranchi (eds.): *Landscapes: territories, frontiers and horizons in the ancient Near East*. History of the Ancient Near East, 3. Padova, 203–12.

—. 2001: Architecture and stratigraphy. In M.N. Van Loon (ed.): *Selenkahiye. Final report on the University of Chicago and University of Amsterdam excavations in the Tabqa Reservoir, northern Syria, 1967–1975*. Uitgaven van het Nederlands Historisch-Archaeologisch Instituut te Istanbul, 91. Istanbul, 25–125.

—. 2007: The area of the Balikh between ca. 2500 and 1700 BC. In P. Matthiae, F. Pinnock, L. Nigro, and L. Peyronel (eds.): *Proceedings of the international colloquium: From relative chronology to absolute chronology: The second millennium BC in Syria–Palestine (Rome, 29th November–1st December 2001)*. Contributi del Centro Linceo Interdisciplinare «Beniamino Segre», 117. Rome, 313–26.

—. forthcoming: Nomadism, pastoralism, and town and country: about the roaming elements in the Syrian Middle Bronze Age. In P.R. de Miroschedji (ed.): *Proceedings of 3ICAANE (Paris, 2002)*.

Meyer, J.-W. 1997: Djebelet el-Beda: eine Stätte der Ahnenverehrung? *Altorientalische Forschungen* 24/2, 294–309.

—. 2006: Zur Frage der Urbanisierung von Tell Chuera. In P. Butterlin, M. Lebeau, J.-Y. Monchambert, J.-L. Montero Fenollós, and B. Muller (eds.): *Les espaces syro-mésopotamiens. Dimensions de l'expérience humaine au Proche-Orient ancien. Volume d'hommage offert à Jean-Claude Margueron*. Subartu, 17. Turnhout, 179–89.

Michalowski, P. 1976: *The royal correspondence of Ur*. Dissertation, Yale University.

—. 2005: Literary works from the court of king Ishbi-Erra of Isin. In Y. Sefati, P. Artzi, C. Cohen, B.L. Eichler, and V.A. Hurowitz (eds.): *"An experienced scribe who neglects nothing": ancient Near Eastern studies in honor of Jacob Klein*. Bethesda, 199–211.

Miglus, P.A. and E. Strommenger 2002: *Stadtbefestigungen, Häuser und Tempel*. Ausgrabungen in Tall Bi'a/Tuttul, 8. Saarbrücken.

Migowski, C., M. Stein, S. Prasad, J.F.W. Negendank, and A. Agnon 2006: Holocene climate variability and cultural evolution in the Near East from the Dead Sea sedimentary record. *Quaternary Research* 66/3, 421–31.

Millet Albà, A. 2004: La localisation des terroirs benjaminites du royaume de Mari. In C. Nicolle (ed.): *Nomades et sédentaires dans le Proche Orient ancien*. Amurru, 3. Paris, 225–34.

Minc, L.D. and K.P. Smith 1989: The spirit of survival: cultural responses to resource variability in North Alaska. In P. Halstead and J. O'Shea (eds.): *Bad year economics: cultural responses to risk and uncertainty*. New Directions in Archaeology. Cambridge, 8–39.

Minnis, P.E. 1996: Notes on economic uncertainty and human behavior in the prehistoric North American Southwest. In J.A. Tainter and B.B. Tainter (eds.): *Evolving complexity and environmental risk in the prehistoric Southwest*. Santa Fe Institute Studies in the Sciences of Complexity Proceedings, 24. Reading, 57–78.

Moore, A.M.T., G. Hillman, C., and A.J. Legge (eds.) 2000: *Village on the Euphrates. From foraging to farming at Abu Hureyra*. Oxford.

Moortgat-Correns, U. 1972: *Die Bildwerke von Djebelet el Bēdā in ihrer räuumlichen und zeitlichen Umwelt.* Berlin.

Mulders, M.A. 1969: *The arid soils of the Balikh Basin (Syria).* Unpublished dissertation, Universiteit van Utrecht.

Neely, J.A. and H.T. Wright 1994: *Early settlement and irrigation on the Deh Luran Plain. Village and early state societies in Southwestern Iran.* University of Michigan Museum of Anthropology Technical Report, 26. Ann Arbor.

Nettle, D. 1996: Language diversity in West Africa: an ecological approach. *Journal of Anthropological Archaeology* 15/4, 403–38.

—. 1998: Explaining global patterns of language diversity. *Journal of Anthropological Archaeology* 17/4, 354–74.

Neumann, F., C. Schölzel, T. Litt, A. Hense, and M. Stein 2006: Holocene vegetation and climate history of the northern Golan heights (Near East). *Vegetation History and Archaeobotany* 16, 329–46.

Nicolle, C. 2006: *Tell Mohammed Diyab 3. Travaux de 1992–2000 sur les buttes A et B.* Paris.

Oates, D. and J. Oates 2001: Archaeological reconstruction and historical commentary. In D. Oates, J. Oates, and H. McDonald (eds.): *Excavations at Tell Brak, 2. Nagar in the third millennium BC.* Cambridge, 379–96.

Oates, D., J. Oates, and H. McDonald (eds.) 2001: *Excavations at Tell Brak, 2. Nagar in the third millennium BC.* Cambridge.

Oates, J. 2004: Archaeology in Mesopotamia: digging deeper at Tell Brak (Albert Reckitt Archaeological Lecture, British Academy, 2004). Available at: http://www.mcdonald.cam.ac.uk/projects/brak/index.htm (Accessed 7 November, 2008).

Ökse, A.T. 2005: Gre Virike: A ritual centre for Early Bronze Age rural communities on the Middle Euphrates. *Antiquity* 79/306, http://antiquity.ac.uk/ProjGall/okse/index.html (Accessed 17 March, 2009).

—. 2006: Gre Virike (period I) Early Bronze Age ritual facilities on the Middle Euphrates River. *Anatolica* 32, 1–27.

—. 2007: A 'high' terrace at Gre Virike to the north of Carchemish: power of local rulers as founders? In E. Peltenburg (ed.): *Euphrates river valley settlement: the Carchemish sector in the third millennium BC.* Levant Supplementary Series, 5. Oxford, 94–104.

Ökse, A.T., A. Engin, V.M. Tekinalp, H.U. Dağ, and A. Görüş 2001: Research at Mezraa Höyük, 1999. In N. Tuna, J. Öztürk, and J. Velibeyoğlu (eds.): *Salvage Project of the Archaeological Heritage of the Ilısu and Carchemish Dam Reservoirs. Activities in 1999.* Ankara, 213–32.

Ökse, A.T. and V.M. Tekinalp 1999: Mezraa Höyük: research in 1998. In N. Tuna and J. Öztürk (eds.): *Salvage Project of the Archaeological Heritage of the Ilısu and Carchemish Dam Reservoirs. Activities in 1998* Ankara, 199–214.

Osborne, R. 2004: Demography and survey. In S.E. Alcock and J.F. Cherry (eds.): *Side-by-side survey: comparative regional studies in the Mediterranean world.* Oxford, 163–172.

Otto, A. 2006: *Alltag und Gesellschaft zur Spätbronzezeit: eine Fallstudie aus Tall Bazi (Syrien).* Subartu, 19. Turnhout.

Parkinson, W.A. 2002: Introduction: archaeology and tribal societies. In W.A. Parkinson (ed.): *The archaeology of tribal societies.* Archaeological Series, 15. Ann Arbor, 1–12.

Pearson, J.A., H. Buitenhuis, R.E.M. Hedges, L. Martin, N. Russell, and K.C. Twiss 2007: New light on early caprine herding strategies from isotope analysis: a case study from Neolithic Anatolia. *Journal of Archaeological Science* 34/12, 2170–9.

Peltenburg, E. 2000: From nucleation to dispersal. Late third millennium BC settlement pattern transformation in the Near East and Aegean. In O. Rouault and M. Wäfler (eds.): *La Djéziré et l'Euphrate syriens de la protohistoire à la fin du IIe millénaire av. J.-C.: tendances dans l'interprétation historique des données nouvelles.* Subartu, 7. Turnhout, 183–206.

—. 2007a: New perspectives on the Carchemish sector of the Middle Euphrates River valley in the 3rd millennium BC. In E. Peltenburg (ed.): *Euphrates river valley settlement: the Carchemish sector in the third millennium BC*. Levant Supplementary Series, 5. Oxford, 1–24.

—. 2007b: Diverse settlement pattern changes in the middle Euphrates Valley in the later third millennium B.C: the contribution of Jerablus Tahtani. In C. Kuzucuoğlu and C. Marro (eds.): *Sociétés humaines et changement climatique à la fin du troisième millénaire: une crise a-t-elle eu lieu en Haute Mésopotamie?* Varia Anatolica, 19. Istanbul, 247–66.

— (ed.) 2007c: *Euphrates river valley settlement: the Carchemish sector in the third millennium BC*. Levant Supplementary Series, 5. Oxford.

—. forthcoming: The emergence of Carchemish as a major polity: contributions from the Land of Carchemish Project (Syria), 2006. In: *Proceedings of 6ICAANE (Rome, 2008)*.

Peters, J., D. Helmer, A. von den Driesch, and M. Saña Segui 1999: Early animal husbandry in the Northern Levant. *Paléorient* 25/2, 27–48.

Peterson, C.E. and R.D. Drennan 2005: Communities, settlements, sites, and surveys: regional-scale analysis of prehistoric human interaction. *American Antiquity* 70/1, 5–30.

Petit, L.P. forthcoming: *Settlement dynamics in the Middle Jordan Valley during Iron II*

Pfälzner, P. 1998: Eine Modifikation der Periodisierung Nordmesopotamiens im 3. Jtsd. v. Chr. *Mitteilungen der Deutschen Orient-Gesellschaft zu Berlin* 130, 69–71.

—. 2001: *Haus und Haushalt: Wohnformen des dritten Jahrtausends vor Christus in Nordmesopotamien*. Damaszener Forschungen, 9. Mainz am Rhein.

Pfälzner, P. and A. Wissing 2004: Urbanismus in der Unterstadt von Urkeš. Ergebnisse einer geomagnetischen Prospektion und eines archäologischen Surveys in der südostlichen Unterstadt von Tall Mozan im Sommer 2002. *Mitteilungen der Deutschen Orient-Gesellschaft zu Berlin* 136, 41–86.

Plog, S. and J.L. Hantman 1990: Chronology construction and the study of prehistoric culture change. *Journal of Field Archaeology* 17/4, 439–56.

Poidebard, A. 1934: *La trace de Rome dans le désert de Syrie: le Limes de Trajan à la conquête arabe. Recherches aériennes (1925–1932)*. Paris.

Pollock, S. 1999: *Ancient Mesopotamia: the Eden that never was*. Case Studies in Early Societies. Cambridge.

Porter, A. 2002: The dynamics of death: ancestors, pastoralism, and the origins of a third-millennium city in Syria. *Bulletin of the American Schools of Oriental Research* 325, 1–36.

—. 2004: The urban nomad: countering old clichés. In C. Nicolle (ed.): *Nomades et sédentaires dans le Proche Orient ancien*. Amurru, 3. Paris, 69–74.

—. 2007: You say potato, I say... typology, chronology and the origins of the Amorites. In C. Kuzucuoğlu and C. Marro (eds.): *Sociétés humaines et changement climatique à la fin du troisième millénaire: une crise a-t-elle eu lieu en Haute Mésopotamie?* Varia Anatolica, 19. Istanbul, 69–115.

—. 2009: Beyond dimorphism: ideologies and materialities of kinship as time-space distanciation. In J. Szuchman (ed.): *Nomads, tribes, and the state in the ancient Near East*. Oriental Institute Seminars, 5. Chicago, 201–25.

Postgate, N. 1994: How many Sumerians per hectare? – Probing the anatomy of an early city. *Cambridge Archaeological Journal* 4/1, 47–65.

Powell, M.A. 1990: Maße und Gewichte. In D.O. Edzard (ed.): *Reallexikon der Assyriologie und Vorderasiatische Archäologie, VII*. Berlin.

Pruß, A. 2004: Remarks on the chronological periods. In S. Anastasio, M. Lebeau, and M. Sauvage: *Atlas of preclassical Upper Mesopotamia*. Subartu, 13. Turnhout, 7–21.

Pustovoytov, K., K. Schmidt, and H. Taubald 2007: Evidence for Holocene environmental changes in the northern Fertile Crescent provided by pedogenic carbonate coatings. *Quaternary Research* 67/3, 315–27.

Rappaport, R.A. 1979: *Ecology, meaning, and religion*. Berkeley.

Rautman, A.E. 1996: Risk, reciprocity, and the operation of social networks. In J.A. Tainter and B.B. Tainter (eds.): *Evolving complexity and environmental risk in the prehistoric Southwest.* Santa Fe Institute Studies in the Sciences of Complexity Proceedings, 24. Reading, 197–222.

Read, D.W. and S.A. LeBlanc 2003: Population growth, carrying capacity, and conflict. *Current Anthropology* 44/1, 59–85.

Reade, J. 2001: Assyrian king-lists, the royal tombs of Ur, and Indus origins. *Journal of Near Eastern Studies* 60/1, 1–19.

Redding, R.W. 1993: Subsistence security as a selective pressure favoring increasing cultural complexity. *Bulletin on Sumerian Agriculture* 7, 77–98.

Redman, C.L. 1982: Archaeological survey and the study of Mesopotamian urban systems. *Journal of Field Archaeology* 9/3, 375–82.

Redman, C.L. and P.J. Watson 1970: Systematic, intensive surface collection. *American Antiquity* 35/3, 279–91.

Reichel, C.D. 2004: Appendix B: site gazetteer. In T.J. Wilkinson: *On the margin of the Euphrates. Settlement and land use at Tell es-Sweyhat and in the upper Lake Assad area, Syria.* Oriental Institute Publications, 124. Chicago, 335–48.

Reifenberg, A. 1952: The soils of Syria and the Lebanon. *Journal of Soil Science* 3/1, 68–88.

Renfrew, C. 1972: *The emergence of civilisation: the Cyclades and the Aegean in the third millennium B.C.* Studies in Prehistory. London.

Renfrew, C. and P. Bahn 1996 (1991): *Archaeology. Theories, methods and practice.* London.

Riehl, S. 2008: Climate and agriculture in the ancient Near East: a synthesis of the archaeobotanical and stable carbon isotope evidence. *Vegetation History and Archaeobotany* 17/Supplement 1, 43–51.

—. 2009: Archaeobotanical evidence for the interrelationship of agricultural decision-making and climate change in the ancient Near East. *Quaternary International* 197/1–2, 93–114.

Riehl, S., R.A. Bryson, and K. Pustovoytov 2008: Changing growing conditions for crops during the Near Eastern Bronze Age (3000–1200 BC): The stable carbon isotope evidence. *Journal of Archaeological Science* 35/4, 1011–22.

Ristvet, L. 2005: *Settlement, economy, and society in the Tell Leilan Region, Syria, 3000–1000 BC.* Unpublished dissertation, University of Cambridge.

—. 2008: Legal and archaeological territories of the second millennium BC in northern Mesopotamia. *Antiquity* 82/317, 585–99.

Ristvet, L. and H. Weiss 2005: The Habur region in the late third and early second millennium BC. In W. Orthmann (ed.): *The history and archaeology of Syria*, 1. Saarbrucken.

Roberts, B.K. 1996: *Landscapes of settlement. Prehistory to the present.* London.

Röllig, W. and H. Kühne 1977/8: The Lower Habur. A preliminary report on a survey conducted by the Tubinger Atlas des Vorderen Orients in 1975. *Annales Archéologiques Arabes Syriennes* 27–28, 115–40.

—. 1983: The Lower Habur second preliminary report on a survey in 1977. *Annales Archéologiques Arabes Syriennes* 33/2, 187–99.

Rosen, A.M. 1995: The social response to environmental change in Early Bronze Age Canaan. *Journal of Anthropological Archaeology* 14, 26–44.

—. 1997: The geoarchaeology of Holocene environments and land use at Kazane Hoyuk, S.E. Turkey. *Geoarchaeology* 12/4, 395–416.

—. 2007: *Civilizing climate. Social responses to climate change in the ancient Near East.* Lanham.

Rosen, A.M. and S.A. Rosen 2001: Determinist or not determinist?: climate, environment, and archaeological explanation in the Levant. In S.R. Wolff (ed.): *Studies in the archaeology of Israel and neighboring lands: in memory of Douglas L. Esse.* ASOR Books, 5. Chicago, 535–49.

Rosen, S.A. 2003: Early multi-resource nomadism: excavations at the Camel Site in the central Negev. *Antiquity* 77/298, 749–61.

—. 2008: Desert pastoral nomadism in the *longue durée*. A case study from the Negev and the Southern Levantine deserts. In H. Barnard and W. Wendrich (eds.): *The archaeology of mobility. Old World and New World nomadism.* Cotsen Advanced Seminar Series, 4. Los Angeles, 115–40.

Rothman, M.S., R. Ergeç, N.F. Miller, J.A. Weber, and G. Kozbe 1998: Yarim Höyük and the Uruk expansion (Part I). *Anatolica* 24, 65–99.

Rothman, M.S. and J.G. Fuensanta 2003: The archaeology of the Early Bronze I and II periods in southeastern Turkey and north Syria. In M. Özdoğan, H. Hauptmann, and N. Başgelen (eds.): *From village to cities. Early villages in the Near East. Studies presented to Ufuk Esin.* Istanbul, 583–622.

Rowe, A. 2002: Structural responses to resource stress: evidence for the evolution of social and economic associations from Jordanian pastoral society. In W. Wendrich and G. Van der Kooij (eds.): *Moving matters. Ethnoarchaeology in the Near East. Proceedings of the international seminar held at Cairo, 7–10 december 1998.* CNWS Publications, 111. Leiden, 29–43.

Rowton, M.B. 1973a: Urban autonomy in a nomadic environment. *Journal of Near Eastern Studies* 32/1/2, 201–15.

—. 1973b: Autonomy and nomadism in Western Asia. *Orientalia* 42, 247–258.

—. 1974: Enclosed nomadism. *Journal of the Economic and Social History of the Orient* 17, 1–30.

—. 1980: Pastoralism and the periphery in evolutionary perspective. In M.-T. Barrelet (ed.): *L'archéologie de l'Iraq du début de l'époque neólithique à 333 avant notre ère: perspectives et limites de l'interprétation anthropologique des documents.* Colloques Internationaux du Centre National de la Recherche Scientifique, 580. Paris.

Russell, K.W. 1988: *After Eden. The behavioral ecology of early food production in the Near East and North Africa.* BAR International Series, 391. Oxford.

Rutishauser, S. 2008: Hollow ways in der Fernerkundung. In T. Hofmeier and O. Kaelin (eds.): *Stückwerk. Beiträge zur Kulturgeschichte des Alten Orients.* Berlin, 215–40.

Sachau, E. 1883: *Reise in Syrien und Mesopotamien.* Leipzig.

Sahlins, M. 1972: *Stone age economics.* New York.

Said, R. 1993: *The river Nile. Geology, hydrology and utilization.* Oxford.

Sallaberger, W. 1996: Grain accounts: personnel lists and expenditure documents. In F. Ismail, W. Sallaberger, P. Talon, and K. van Lerberghe (eds.): *Administrative documents from Tell Beydar (seasons 1993–1995).* Subartu, 2. Turnhout, 81–106.

—. 1998: Ein Synchronismus der Urkunden von Tell Beydar mit Mari und Ebla? In M. Lebeau (ed.): *About Subartu: studies devoted to Upper Mesopotamia.* Subartu, 4/2. Turnhout, 23–39.

—. 2004a: Relative Chronologie von der späten frühdynastischen bis zur altbabylonischen Zeit. In J.-W. Meyer and W. Sommerfeld (eds.): *2000 v. Chr.: politische, wirtschaftliche und kulturelle Entwicklung im Zeichen einer Jahrtausendwende.* Colloquien der Deutschen Orient-Gesellschaft, 3. Saarbrücken, 15–43.

—. 2004b: A note on the sheep and goat flocks. Introduction to texts 141–167. In L. Milano, W. Sallaberger, P. Talon, and K. van Lerberghe (eds.): *Third millennium cuneiform texts from Tell Beydar (seasons 1996–2002).* Subartu, 12. Turnhout, 13–21.

—. 2007: From urban culture to nomadism: A history of Upper Mesopotamia in the late third millennium. In C. Kuzucuoğlu and C. Marro (eds.): *Sociétés humaines et changement climatique à la fin du troisième millénaire: une crise a-t-elle eu lieu en Haute Mésopotamie?* Varia Anatolica, 19. Istanbul, 417–56.

Sallaberger, W. and P. Talon 1996: Transliterated texts. In F. Ismail, W. Sallaberger, P. Talon, and K. van Lerberghe (eds.): *Administrative documents from Tell Beydar (seasons 1993–1995).* Subartu, 2. Turnhout, 127–74.

Salzman, P.C. 1978a: Ideology and change in Middle Eastern tribal societies. *Man* 13/4, 618–37.

—. 1978b: Does complementary opposition exist? *American Anthropologist* 80/1, 53–70.

Savage, S.H. 1997: Assessing departures from log-normality in the rank-size rule. *Journal of Archaeological Science* 24/3, 233–44.

Schacht, R.M. 1984: The contemporaneity problem. *American Antiquity* 49/4, 678–95.

Schiffer, M.B. 1987: *Formation processes in the archaeological record.* Albuquerque.

Schirmer, W. 1987: Landschaftsgeschichte um Tall Bi'a am syrischen Euphrat. *Mitteilungen der Deutschen Orient-Gesellschaft zu Berlin* 117, 57–71.

Schwartz, G.M. 1994: Rural economic specialization and early urbanization in the Khabur Valley, Syria. In G.M. Schwartz and S.E. Falconer (eds.): *Archaeological views from the countryside: village communities in early complex societies.* Washington, 19–36.

—. 2007: Taking the long view on collapse: A Syrian perspective. In C. Kuzucuoğlu and C. Marro (eds.): *Sociétés humaines et changement climatique à la fin du troisième millénaire: une crise a-t-elle eu lieu en Haute Mésopotamie?* Varia Anatolica, 19. Istanbul, 45–67.

Schwartz, G.M. and H.H. Curvers 1992: Tell al-Raqa'i 1989 and 1990: Further investigations at a small rural site of early urban northern Mesopotamia. *American Journal of Archaeology* 96/3, 397–419.

Schwartz, G.M., H.H. Curvers, F.A. Gerritsen, J.A. MacCormack, N.F. Miller, and J.A. Weber 2000: Excavation and survey in the Jabbul Plain, Western Syria: the Umm el-Marra project 1996–1997. *American Journal of Archaeology* 104/3, 419–62.

Sertok, K. 2007: Fruit stands and the definition of a cultural area around Carchemish. In E. Peltenburg (ed.): *Euphrates river valley settlement: the Carchemish sector in the third millennium BC.* Levant Supplementary Series, 5. Oxford, 238–49.

Sertok, K. and R. Ergeç 1999a: A new Early Bronze Age cemetery: excavations near the Birecik Dam, southeastern Turkey. *Anatolica* 25, 87–107.

—. 1999b: Seraga Höyük: Research in 1998. In N. Tuna and J. Öztürk (eds.): *Salvage Project of the Archaeological Heritage of the Ilısu and Carchemish Dam Reservoirs. Activities in 1998* Ankara, 168–74.

Sertok, K. and F. Kulakoğlu 2001: Results of the 1999 season excavations at Şaraga Höyük. In N. Tuna, J. Öztürk, and J. Velibeyoğlu (eds.): *Salvage Project of the Archaeological Heritage of the Ilısu and Carchemish Dam Reservoirs. Activities in 1999.* Ankara, 475–486.

—. 2002: Şaraga Höyük 2000. In N. Tuna, J. Öztürk, and J. Velibeyoğlu (eds.): *Salvage Project of the Archaeological Heritage of the Ilısu and Carchemish Dam Reservoirs. Activities in 2000.* Ankara, 370–81.

Sertok, K., F. Kulakoğlu, and F. Squadrone 2004: Şaraga Höyük 2001. In N. Tuna, J. Öztürk, and J. Velibeyoğlu (eds.): *Salvage Project of the Archaeological Heritage of the Ilısu and Carchemish Dam Reservoirs. Activities in 2001.* Ankara, 308–15.

—. 2007: Living along and together with the Euphrates. The effects of the Euphrates on a long-life settlement as Şaraga Höyük. In C. Kuzucuoğlu and C. Marro (eds.): *Sociétés humaines et changement climatique à la fin du troisième millénaire: une crise a-t-elle eu lieu en Haute Mésopotamie?* Varia Anatolica, 19. Istanbul, 341–53.

Sherratt, A. 1981: Plough and pastoralism: aspects of the secondary products revolution. In I. Hodder, G. Isaac, and N. Hammond (eds.): *Pattern of the past: studies in honour of David Clark.* Cambridge, 261–305.

Sołtysiak, A. forthcoming: Human remains from Tell Barri, area G. In R. Pierobon-Benoit (ed.): *Tell Barri/Kahat 3.*

Sommerfeld, W. 2000: Narām-Sîn, die "Große Revolte" und MAR.TU^{ki}. In J. Marzahn and H. Neumann (eds.): *Assyriologica et Semitica. Festschrift für Joachim Oelsner anläßlich seines 65. Geburtstages am 18. Februar 1997.* Alter Orient und Altes Testament, 252. Münster, 419–36.

Staubwasser, M. and H. Weiss 2006: Holocene climate and cultural evolution in late prehistoric-early historic West Asia. *Quaternary Research* 66/3, 372–87.

Stein, G.J. 1987: Regional economic integration in early state societies: third millennium B.C. pastoral production in Gritille, southeast Turkey. *Paléorient* 13/2, 101–11.

Stein, G.J., R. Bernbeck, C. Coursey, A. McMahon, N.F. Miller, A. Misir, J. Nicola, H. Pittman, S. Pollock, and H.T. Wright 1996: Uruk colonies and Anatolian communities: an interim report on the 1992–1993 excavations at Hacinebi, Turkey. *American Journal of Archaeology* 100/2, 205–60.

Stein, G.J. and P. Wattenmaker 1990: The 1987 Tell Leilan regional survey: preliminary report. In N.F. Miller (ed.): *Economy and settlement in the Near East: analyses of ancient sites and materials*. MASCA Research Papers in Science and Archaeology. Supplement, 7. Philadephia, 8–18.

Steinkeller, P. 1992: Early Semitic literature and third millennium seals with mythological motifs. In P. Fronzaroli (ed.): *Literature and literary language at Ebla*. Quaderni di Semitistica, 18. Firenze, 243–83.

—. 1998: The historical background of Urkesh and the Hurrian beginnings in northern Mesopotamia. In G. Buccellati and M. Kelly-Buccellati (eds.): *Urkesh and the Hurrians. Studies in honor of Lloyd Cotsen* Bibliotheca Mesopotamica, 26. Malibu, 75–98.

Stevens, L.R., E. Ito, A. Schwalb, and H.E. Wright Jr. 2006: Timing of atmospheric precipitation in the Zagros Mountains inferred from a multi-proxy record from Lake Mirabad, Iran. *Quaternary Research* 66/3, 494–500.

Stevens, L.R., H.E. Wright Jr., and E. Ito 2001: Proposed changes in seasonality of climate during the Lateglacial and Holocene at Lake Zeribar, Iran. *The Holocene* 11/6, 747–55.

Stol, M. 1995: Old Babylonian cattle. *Bulletin on Sumerian Agriculture* 8, 173–213.

Stone, E.C. 1981: Texts, architecture and ethnographic analogy: patterns of residence in Old Babylonian Nippur. *Iraq* 43, 19–33.

—. 1987: *Nippur neighborhoods*. Studies in Ancient Oriental Civilization, 44. Chicago.

Stone, G.D. 1994: Agricultural intensification and perimetrics: ethnoarchaeological evidence from Nigeria. *Current Anthropology* 35/3, 317–24.

Stone, T. 2003: Social identity and ethnic interaction in the Western Pueblos of the American Southwest. *Journal of Archaeological Method and Theory* 10/1, 31–67.

Stordeur, D. 1993: Sédentaires et nomades du PPNB final dans le désert de Palmyre (Syrie). *Paléorient* 19/1, 187–204.

Stothers, R.B. 1999: Volcanic dry fogs, climate cooling, and plague pandemics in Europe and the Middle East. *Climatic Change* 42, 713–23.

Streck, M.P. 2000: *Das amurritische Onomastikon der altbabylonischen Zeit I: die Amurriter, die onomastische Forschung, Orthographie und Phonologie, Nominalphonologie*. Alter Orient und Altes Testament, 271/1. Münster.

—. 2002: Zwischen Weide, Dorf und Stadt: sozio-ökonomische Strukturen des amurritischen Nomadismus am Mittleren Euphrat. *Baghdader Mitteilungen* 33, 155–209.

—. 2004: Die Amurriter der altbaylonischen Zeit im Spiegel des Onomastikons. Eine ethno-linguistische Evaluierung. In J.-W. Meyer and W. Sommerfeld (eds.): *2000 v. Chr.: politische, wirtschaftliche und kulturelle Entwicklung im Zeichen einer Jahrtausendwend*. Colloquien der Deutschen Orient-Gesellschaft, 3. Saarbrücken, 313–55.

Strommenger, E. 1981: Die archäologischen Forschungen in Tall Bi'a 1980. *Mitteilungen der Deutschen Orient-Gesellschaft zu Berlin* 113, 23–34.

Sumner, W.M. 1979: Estimating population by analogy: an example. In C. Kramer (ed.): *Ethnoarchaeology: Implications of ethnography for archaeology*. New York, 164–74.

—. 1990: An archaeological estimate of population trends since 6000 BC in the Kur River Basin, Fars Province, Iran. In M. Taddei (ed.): *South Asian archaeology 1987: proceedings of the ninth International Conference of the Association of South Asian Archaeologists in Western Europe, held in the Fondazione Giorgio Cini, Island of San Giorgio Maggiore, Venice*. Serie orientale Roma, 66. Rome, 3–16.

Sykes, M. 1915: *The Caliphs' last heritage. A short history of the Turkish empire*. London.

Szarzynska, K. 2002: *Sheep husbandry and production of wool, garments and cloths in archaic Sumer*. Warsaw.

Tainter, J.A. 1988: *The collapse of complex societies*. New Studies in Archaeology. Cambridge.

Tainter, J.A. and B.B. Tainter (eds.) 1996: *Evolving complexity and environmental risk in the prehistoric Southwest*. Santa Fe Institute Studies in the Sciences of Complexity Proceedings, 24. Reading.

Talon, P. 1996: Personal names. In F. Ismail, W. Sallaberger, P. Talon, and K. van Lerberghe (eds.): *Administrative documents from Tell Beydar (seasons 1993–1995)*. Subartu, 2. Turnhout, 75–80.

Taşkıran, H. 2002: The Palaeolithic survey in the Carchemish Dam reservoir region: 2000 season. In N. Tuna, J. Öztürk, and J. Velibeyoğlu (eds.): *Salvage Project of the Archaeological Heritage of the Ilısu and Carchemish Dam Reservoirs. Activities in 2000*. Ankara, 413–29.

Thalen, D.C.P. 1979: *Ecology and utilization of desert shrub rangelands in Iraq*. The Hague.

Thissen, L. 1989: An Early Bronze III pottery region between the Middle Euphrates and Habur: new evidence from Tell Hammam et-Turkman. In O.M.C. Haex, H.H. Curvers, and P.M.M.G. Akkermans (eds.): *To the Euphrates and beyond. Archaeological studies in honour of Maurits N. van Loon*. Rotterdam, 195–211.

Ur, J.A. 2003: CORONA satellite photography and ancient road networks: a northern Mesopotamian case study. *Antiquity* 77/295, 102–15.

—. 2004: *Urbanism and society in the third millennium Upper Khabur Basin*. Unpublished dissertation, University of Chicago.

van de Mieroop, M. 1997: *The ancient Mesopotamian city*. Oxford.

—. 2007: *A history of the ancient Near East, ca. 3000–323 BC. Second edition*. Blackwell History of the Ancient World. Malden.

van der Kooij, G. 2001: The vicissitudes of life at Dayr 'Allā during the first millennium BC, seen in a wider context. In G. Bisheh (ed.): *Studies in the history and archaeology of Jordan VII: Jordan by the millenia*. Amman, 295–303.

van der Steen, E. 2009: Tribal societies in the nineteenth century: a model. In J. Szuchman (ed.): *Nomads, tribes, and the state in the ancient Near East*. Oriental Institute Seminars, 5. Chicago, 105–17.

van Driel, G. 2005: Ethnicity, how to cope with the subject. In W.H. van Soldt, R. Kalvelagen, and D. Katz (eds.): *Ethnicity in ancient Mesopotamia: papers read at the 48th Rencontre Assyriologique Internationale, Leiden, 1–4 July 2002*. Uitgaven van het Nederlands Instituut voor het Nabije Oosten te Leiden, 102. Leiden, 1–10.

van Liere, W.J. 1963: Capitals and citadels of Bronze-Iron Age Syria in their relationship to land and water. *Les Annales Archéologiques de Syrie* 13, 109–22.

van Liere, W.J. and J. Lauffray 1954: Nouvelle prospection archeologique dans la Haute Jezireh Syrienne. *Annales Archéologiques Arabes Syriennes* 4/5, 129–48.

van Loon, M.N. (ed.) 1988: *Hammam et-Turkman I. Report on the University of Amsterdam's 1981–84 excavations in Syria*. Uitgaven van het Nederlands Historisch-Archaeologisch Instituut te Istanbul, 63. Istanbul.

van Loon, M.N. and D.J.W. Meijer 1988: Foreword. In M.N. van Loon (ed.): *Hammam et-Turkman I. Report on the University of Amsterdam's 1981–1984 excavations in Syria*. Uitgaven van het Nederlands Historisch-Archaeologisch Instituut te Istanbul, 63. Istanbul, xxv–xxix.

van Neer, W. and B. Decupere 2000: Faunal remains from Tell Beydar (seasons 1992–1997). In K. van Lerberghe and G. Voet (eds.): *Tell Beydar: Environmental and technical studies*. Subartu, 6. Turnhout, 69–115.

van Zeist, W. and S. Bottema 1977: Palynological investigations in western Iran. *Palaeohistoria* 19, 19–85.

van Zeist, W., W. Waterbolk-van Rooijen, and S. Bottema 1988: Some notes on the plant husbandry of Tell Hammam et-Turkman. In M.N. van Loon (ed.): *Hammam et-Turkman I. Report on the University of Amsterdam's 1981–84 excavations in Syria*. Uitgaven van het Nederlands Historisch-Archaeologisch Instituut te Istanbul, 63. Istanbul, 705–15.

van Zeist, W. and H. Woldring 1978: A postglacial pollen diagram from Lake Van in east Anatolia. *Review of Palaeobotany and Palynology* 26, 249–76.

Vila, E. 1995: Analyse de la faune des secteurs nord et sud du Steinbau 1. In W. Orthmann, R. Hempelmann, H. Klein, C. Kühne, M. Novak, A. Pruß, E. Vila, H.-M. Weicken, and A. Wener: *Ausgrabungen in Tell Chuēra in Nordost-Syrien I: Vorbericht über die Grabungskampagnen 1986 bis 1992*. Vorderasiatische Forschungen der Max Freiherr von Oppenheim-Stiftung, 2. Saarbrücken, 267–79.

Villard, P. 1987: Un conflit d'autorités à propos des eaux du Balih. *MARI* 5, 591–96.

von den Driesch, A. 1993: Faunal remains from Habuba Kabira in Syria. In H. Buitenhuis and A.T. Clason (eds.): *Archaeozoology of the Near East. Proceedings of the first international symposium on the archaeozoology of southwestern Asia and adjacent areas*. Leiden, 52–9.

Waetzoldt, H. 1996: Privathäuser. Ihre Grösse, Einrichtung und die Zahl der Bewohner. In K.R. Veenhof (ed.): *Houses and households in Ancient Mesopotamia*. Uitgaven van het Nederlands Historisch-Archaeologisch Instituut te Istanbul, 78. Istanbul, 145–52.

Wäfler, M. 2001: *Tall al-Hamidiya 3. Zur historischen Geographie von Idamaras zur Zeit der Archive von Mari (2) und Subat-enlil/Sehna*. Orbis Biblicus et Orientalis, 21. Freiburg.

Wattenmaker, P. 1987: Town and village economies in an early state society. *Paléorient* 13/2, 113–22.

—. 1998: *Household and state in Upper Mesopotamia: specialized economy and the social uses of goods in an early complex society*. Smithsonian Series in Archaeological Inquiry. Washington.

Weber, J.A. 1997: Faunal remains from Tell es-Sweyhat and Hajji Ibrahim. In R.L. Zettler: *Subsistence and settlement in a marginal environment: Tell es-Sweyhat, 1989–1995 preliminary report*. MASCA Research Papers in Science and Archaeology, 14. Philadelphia, 133–67.

Weiss, H. 1977: Periodization, population and early state formation in Khuzistan. In L.D. Levine and T.C. Young Jr. (eds.): *Mountains and lowlands: essays in the archaeology of greater Mesopotamia*. Bibliotheca Mesopotamica, 7. Malibu, 347–70.

—. 1986: The origins of Tell Leilan and the conquest of space in third millennium Mesopotamia. In H. Weiss (ed.): *The origins of cities in dry-farming Syria and Mesopotamia in the third millennium B.C.* Guilford, 71–108.

—. 1997: Leilan, Tell. In E.M. Meyers (ed.): *The Oxford encyclopedia of archaeology in the Near East. Volume 3*. New York, 341–7.

—. 2000: Beyond the Younger Dryas. Collapse as adaptation to abrupt climate change in ancient West Asia and the Eastern Mediterranean. In G. Bawden and R.M. Reycraft (eds.): *Environmental disaster and the archaeology of human response*. Anthropological Papers, 7. Albuquerque, 75–98.

—. 2007: Tell Leilan Project. Available at: http://leilan.yale.edu/index.html (Accessed 13 February, 2008).

Weiss, H. and R.S. Bradley 2001: What drives societal collapse? *Science* 291, 609–610.

Weiss, H., M.-A. Courty, W. Wetterstrom, F. Guichard, L.M. Senior, R. Meadow, and A. Curnow 1993: The genesis and collapse of third millennium north Mesopotamian civilization. *Science* 261, 995–1004.

Weiss, H., F. deLillis, D. deMoulins, J. Eidem, T. Guilderson, U. Kasten, T. Larsen, L. Mori, L. Ristvet, E. Rova, and W. Wetterstrom 2002–3: Revising the contours of history at Tell Leilan. *Annales Archéologiques Arabes Syriennes* 45–46, 59–74.

Wendrich, W. and H. Barnard 2008: The archaeology of mobility: definitions and research approaches. In H. Barnard and W. Wendrich (eds.): *The archaeology of mobility. Old World and New World nomadism*. Cotsen Advanced Seminar Series, 4. Los Angeles, 1–21.

Whallon, R. 2006: Social networks and information: non-"utilitarian" mobility among hunter-gatherers. *Journal of Anthropological Archaeology* 25/2, 259–70.

Whiting, R.M. 1995: Amorite tribes and nations of second-millennium Western Asia. In J.M. Sasson (ed.): *Civilizations of the Ancient Near East*. New York, 1231–42.

Wick, L., G. Lemcke, and M. Sturm 2003: Evidence of Lateglacial and Holocene climatic change and human impact in eastern Anatolia: high-resolution pollen, charcoal, isotopic and geochemical records from the laminated sediments of Lake Van, Turkey. *The Holocene* 13/5, 665–75.

Wilcke, C. 1969: Zur Geschichte der Amurriter in der Ur-III-Zeit. *Welt des Orients* 5, 1–31.

Wilkinson, T.J. 1989: Extensive sherd scatters and land-use intensity: some recent results. *Journal of Field Archaeology* 16/1, 31–46.

—. 1990: The development of settlement in the north Jazira between the 7th and 1st millennia B.C. *Iraq* 52, 49–62.

—. 1993: Linear hollows in the Jazira, Upper Mesopotamia. *Antiquity* 67, 548–62.

—. 1994: The structure and dynamics of dry-farming states in Upper Mesopotamia. *Current Anthropology* 35/5, 483–520.

—. 1997: Environmental fluctuations, agricultural production and collapse: a view from Bronze Age Upper Mesopotamia. In H.N. Dalfes, G. Kukla, and H. Weiss (eds.): *Third millennium BC climate change and Old World collapse*. NATO ASI series, I/49. Berlin, 67–106.

—. 1998: Water and human settlement in the Balikh Valley, Syria: investigations from 1992–1995. *Journal of Field Archaeology* 25/1, 63–87.

—. 1999: Demographic trends from archaeological survey: case studies from the Levant and the Near East. In J. Bintliff and K. Sbonias (eds.): *Reconstructing past population trends in Mediterranean Europe (3000 BC–AD 1800)*. The Archaeology of Mediterranean Landscapes, 1. Oxford, 45–64.

—. 2000a: Regional approaches to Mesopotamian archaeology: The contribution of archaeological surveys. *Journal of Archaeological Research* 8/3, 219–67.

—. 2000b: Settlement and land use in the zone of uncertainty in Upper Mesopotamia. In R.M. Jas (ed.): *Rainfall and agriculture in northern Mesopotamia*. Uitgaven van het Nederlands Historisch-Archaeologisch Instituut te Istanbul, 88. Istanbul, 1–35.

—. 2003: *Archaeological landscapes of the Near East*. Tucson.

—. 2004: *On the margin of the Euphrates. Settlement and land use at Tell es-Sweyhat and in the upper Lake Assad area, Syria*. Oriental Institute Publications, 124. Chicago.

—. n.d.: *Sabi Abyad: the geoarchaeology of a complex landscape*. Manuscript.

Wilkinson, T.J., J.H. Christiansen, J.A. Ur, M. Widell, and M. Altaweel 2007a: Urbanization within a dynamic environment: modeling Bronze Age communities in Upper Mesopotamia. *American Anthropologist* 109/1, 52–68.

Wilkinson, T.J., E. Peltenburg, A. McCarthy, E.B. Wilkinson, and M. Brown 2007b: Archaeology in the Land of Carchemish: landscape surveys in the area of Jerablus Tahtani, 2006. *Levant* 39, 213–47.

Wilkinson, T.J. and D.J. Tucker 1995: *Settlement development in the North Jazira, Iraq: a study of the archaeological landscape*. Iraq Archaeological Reports, 3. Baghdad.

Wilkinson, T.J., J.A. Ur, and J. Casana 2004: From nucleation to dispersal: trends in settlement pattern in the Northern Fertile Crescent. In S.E. Alcock and J.F. Cherry (eds.): *Side-by-side survey: comparative regional studies in the Mediterranean world*. Oxford, 189–205.

Winterhalder, B. 1980: Environmental analysis in human evolution and adaptation research. *Human Ecology* 8/2, 135–70.

—. 1994: Concepts in historical ecology. The view from evolutionary ecology. In C.L. Crumley (ed.): *Historical ecology. Cultural knowledge and changing landscapes*. School of American Research Advanced Seminar Series. Santa Fe, 17–41.

Winterhalder, B., F. Lu, and B. Tucker 1999: Risk-sensitive adaptive tactics: models and evidence from subsistence studies in biology and anthropology. *Journal of Archaeological Research* 7/4, 301–48.

Wirth, E. 1971: *Syrien. Eine geographische Landeskunde*. Wissenschaftliche Länderkunden, 4/5. Darmstadt.

Wolfart, R. 1967: *Geologie von Syrien und dem Libanon*. Beiträge zur Regionalen Geologie der Erde, 6. Berlin.

Wossink, A. forthcoming: Climate, history, and demography: a case-study from the Balikh Valley, Syria. In: *Regards croisés sur l'étude archéologique des paysages anciens. Nouvelles recherches dans le bassin méditerranéen, en Asie Centrale et au Proche et au Moyen Orient*. Lyon.

Yalçıklı, D. and V.M. Tekinalp 2002: Mezraa Höyük excavations, 2000. In N. Tuna, J. Öztürk, and J. Velibeyoğlu (eds.): *Salvage Project of the Archaeological Heritage of the Ilısu and Carchemish Dam Reservoirs. Activities in 2000*. Ankara, 188–210.

Yardimci, N. 1993: Excavations, surveys and restoration works at Harran. In M. Frangipane, H. Hauptmann, M. Liverani, P. Matthiae, and M.J. Mellink (eds.): *Between the rivers and over the mountains: Archaeologica anatolica et mesopotamica Alba Palmieri dedicata*. Rome, 437–49.

Yoffee, N. 1988: The collapse of ancient Mesopotamian states and civilization. In G.L. Cowgill and N. Yoffee (eds.): *The collapse of ancient states and civilizations*. Tucson, 44–68.

—. 1995: Political economy in Early Mesopotamian states. *Annual Review of Anthropology* 24, 281–311.

Zeder, M.A. 1994: Of kings and shepherds: specialized animal economy in Ur III Mesopotamia. In G.J. Stein and M.S. Rothman (eds.): *Chiefdoms and early states in the Near East: the organizational dynamics of complexity*. Monographs in World Archaeology, 18. Madison, 175–91.

—. 1998: Environment, economy and subsistence in northern Mesopotamia. In M. Fortin and O. Aurenche (eds.): *Espace naturel, espace habité en Syrie du Nord (10e–2e millénaires av. J.-C.)*. Bulletin of the Canadian Society for Mesopotamian Studies, 33. Quebec, 55–67.

Zettler, R.L. 2003: Reconstructing the world of ancient Mesopotamia: divided beginnings and holistic history. *Journal of the Economic and Social History of the Orient* 46/1, 3–45.

Zhang, D.D., P. Brecke, H.F. Lee, Y.-Q. He, and J. Zhang 2007: Global climate change, war, and population decline in recent human history. *Proceedings of the National Academy of Sciences* 104/49, 19214–19.

Samenvatting

De uitdaging van klimaatverandering. Competitie en samenwerking tussen landbouw- en pastoralistische gemeenschappen in noord-Mesopotamië (c. 3000–1600 v. Chr.)

Klimaatverandering heeft door de geschiedenis heen altijd een belangrijke invloed gehad op menselijk gedrag. Deze studie probeert inzicht te verschaffen in de complexe verbanden tussen dat gedrag en klimaatverandering. De nadruk ligt daarbij op het effect van klimaatverandering op de sociale interactie tussen menselijke gemeenschappen van verschillend karakter. Deze studie betoogt dat deze interactie zich langs een continuüm zal ontwikkelen met competitie aan de ene kant, en samenwerking aan de andere kant. De keuze voor een strategie wordt bepaald door de mate waarin een strategie voordeel biedt wanneer een gemeenschap met klimaatverandering geconfronteerd wordt. Dit model wordt toegepast op de samenlevingen van noord-Mesopotamië gedurende het derde en vroege tweede millennium voor Christus (3000–1600 v. Chr.), wanneer zich een verschuiving naar een droger klimaat voordoet. Binnen dit geografische, chronologische en klimatologische kader kijkt deze studie naar veranderingen in nederzettingspatronen als een maat voor competitie tussen sedentaire landbouwgemeenschappen, en naar de ontwikkeling van de Amoritische identiteit als een indicatie voor samenwerking tussen sedentaire en meer mobiele pastoralistische gemeenschappen.

Hoofdstuk 1 bespreekt de wetenschappelijke context en de vraagstelling. Dit onderzoek maakt deel uit van het multidisciplinaire project *Settling the steppe. The Archaeology of changing societies in Syro-Palestinian drylands during the Bronze and Iron Ages*. Dit project wil inzicht verschaffen in de variëteit en stabiliteit van menselijke bewoning in marginale gebieden in het Nabije Oosten. Binnen dit kader onderzoekt deze studie het effect van klimaatverandering op de organisatie en stabiliteit van sedentaire gemeenschappen in noord-Mesopotamië tussen 3000–1600 v. Chr. Er wordt geconstateerd dat veel studies naar menselijke reacties op klimaatverandering sterk klimaat-deterministisch zijn, zonder aandacht voor de mogelijke variëteit van menselijke reacties. De onderzoeksvraag is daarom als volgt geformuleerd: hoe ontwikkelden zich de sociale relaties tussen sedentaire groepen onderling, en tussen sedentaire en (semi-)nomadische groepen als reactie op klimaatverandering in noord-Mesopotamië gedurende het derde en vroege tweede millennium voor Christus?

Hoofdstuk 2 bespreekt de geografische, historische en klimatologische gegevenheden van noord-Mesopotamië. Het onderzoeksgebied beslaat delen van zuidoost-Turkije, noordoost-Syrië en noordwest-Irak. Dit gebied is grotendeels vlak en wordt doorsneden door twee grote rivieren, de Eufraat en de Tigris, en enkele kleinere. Noord-Mesopotamië kent een landklimaat met warme, droge zomers en gematigde, natte winters. De hoeveelheid jaarlijkse neerslag varieert van minder dan 200 mm. in het zuiden van het onderzoeksgebied tot meer dan 400 mm. in het noorden en noordoosten. Dit betekent dat regenlandbouw in een aanzienlijk gedeelte van het onderzoeksgebied met grote risico's verbonden is. Paleoklimatologisch onderzoek laat zien dat het klimaat van noord-Mesopotamië gedurende het vroege derde millennium voor Christus gunstiger was dan tegenwoordig. Studie van geologische profielen in noord-Mesopotamië heeft laten zien dat de overgang naar een droger klimaat in de tweede helft van het derde millennium voor Christus moet hebben plaatsgevonden. Klimaatreconstructies elders in het Nabije Oosten bevestigen dit beeld. Noord-Mesopotamië bestond gedurende deze periode uit een verzameling van min of meer onafhankelijke lokale (stad)staten. Grotere regionale staten konden zich niet lang handhaven, en ook externe machten, zoals Ebla, Akkad, en de Ur III staat, lijken vaak niet meer dan indirecte controle of invloed uitgeoefend te hebben.

Hoofdstuk 3 bespreekt het theoretische kader van deze studie. Op basis van antropologische literatuur over de reactie van (pre-)moderne menselijke samenlevingen op klimaatveranderingen stelt dit hoofdstuk een model voor dat aangeeft welke vormen van sociale interactie verwacht mogen worden als gevolg van klimaatveranderingen. Dit model verwacht enerzijds dat territorialiteit en competitie zullen toenemen tussen groepen die afhankelijk zijn van dezelfde natuurlijke bronnen, wanneer deze bronnen schaarser worden als gevolg van klimaatverandering. Deze verwachting zal getoetst worden aan de hand van archeologische gegevens over nederzettingssystemen en wordt verder uitgewerkt in Hoofdstukken 4 en 5. Het model verwacht anderzijds dat handel, uitwisseling en sociale cohesie zullen toenemen tussen groepen die afhankelijk zijn van verschillende natuurlijke bronnen. Deze verwachting zal getoetst worden aan de hand van paleozoölogische en historische gegevens en wordt verder uitgewerkt in Hoofdstukken 6 en 7.

Hoofdstuk 4 behandelt de analyse van nederzettingssystemen aan de hand van gegevens van archeologische surveys. Er wordt een model besproken om gelijktijdigheid van nederzettingen te simuleren, en archeologische en etnografische data worden gecombineerd om het aantal inwoners per nederzetting te berekenen. Dit hoofdstuk toont aan dat etnografische data niet zonder meer bruikbaar zijn voor archeologische *case-studies* en stelt een alternatief voor.

Hoofdstuk 5 bespreekt de resultaten van drie archeologische surveys die binnen het studiegebied vallen en vult deze resultaten aan met die van recente opgravingen. Deze surveys zijn de Birecik-Euphrates Dam Survey (B-EDS), de Balikh Survey (BS) en de North Jazira Survey (NJS). De methode zoals voorgesteld in Hoofdstuk 4 wordt toegepast op deze resultaten om tot een evenwichtige reconstructie te komen van het aantal gelijktijdige nederzettingen en het aantal mensen per regio. Deze reconstructie laat zien dat de demografische en nederzettingsontwikkelingen in ieder gebied zeer verschillend waren, en niet eenduidig verklaard kunnen worden. Daarnaast worden de resultaten van omliggende surveys behandeld, en worden historische teksten besproken die helpen om de resultaten van de analyse in context te plaatsen.

Hoofdstuk 6 behandelt het paleozoölogisch en historisch materiaal om de pastoralistische economie van noord-Mesopotamië gedurende het derde en vroege tweede millennium v. Chr. te reconstrueren. Deze analyse laat zien dat er gedurende deze periode een ontwikkeling plaatsvond naar specialisatie van schaap/geit pastoralisme, waarbij ook de mate van mobiliteit/nomadisme toenam. Het resultaat van deze ontwikkeling is duidelijk zichtbaar in het tekstmateriaal van Mari.

Hoofdstuk 7 behandelt de beschikbare historische bronnen over de Amorieten. De analyse van dit materiaal laat zien dat deze benaming een etnische identiteit aanduidt. In tegenstelling tot wat vroeger vaak verondersteld werd, waren de Amorieten reeds ver voor de Ur III periode in Mesopotamië aanwezig, en werd deze benaming niet enkel door nomadische pastoralisten gebruikt. In tegendeel, deze etnische identiteit functioneerde als een identiteit welke verschillende economische groepen met elkaar verbond en hen in staat stelde stevige sociale en economische banden met elkaar aan te gaan.

Hoofdstuk 8 brengt de data uit Hoofdstukken 4–7 samen en analyseert ze in het licht van de hypotheses uit Hoofdstuk 3. Er wordt geconstateerd dat de resultaten van de drie geanalyseerde archeologische surveys maar deels in overeenstemming zijn met de hypothese dat competitie tussen sedentaire landbouwgemeenschappen zal toenemen als reactie op klimaatverandering kan geconstateerd worden. In de B-EDS regio lijkt de groei van Carchemish weliswaar aan klimaatverandering gekoppeld te kunnen worden, maar wordt deze groei niet door competitie gestimuleerd, maar door Carchemish' gunstige positie voor de handel tussen verschillende landschapzones. In de BS regio laten de organisatie van het nederzettingsysteem, de constructie van stadsmuren en historische bronnen een toename in competitie tussen nederzettingen zien voor de periodes waarin de bevolking een maximale grootte bereikte. Hoewel deze reconstructie de hypothese bevestigt, lijkt klimaatverandering hier niet de primaire oorzaak te zijn geweest, maar slechts een secundaire factor die een proces versterkte dat al door normale bevolkingsgroei in gang was gezet. In de NJS regio is evenmin een directe correlatie zichtbaar tussen klimaatverandering en een toename in de competitie tussen sedentaire gemeenschappen. Het lijkt erop dat de dominante positie van Tell al-Hawa in het nederzettingsysteem, opgebouwd in het vroege derde millennium, het ontstaan van andere grote nederzettingen verhinderde. Er wordt tevens geconstateerd dat de gelijktijdige ontwikkeling naar gespecialiseerd (semi-)nomadisch pastoralisme, en de opkomst van de Amorieten gekoppeld kan worden aan gelijktijdige klimaatverandering en gezien kan worden als de vorming van een identiteit binnen welke verschillende economische groepen konden samenwerken. De opkomst van Amoritische staten moet dan ook niet gezien worden als een infiltratie van nomadische stammen, maar als de reflectie van een proces van Amoritisering waarbij heersers de Amoritische etniciteit adopteerden. De aanleiding hiertoe werd gevormd door de voordelen die de Amoritische identiteit aan haar dragers verleende; namelijk de toegang tot een handelsnetwerk tussen pastoralistische en landbouwgemeenschappen waardoor de effecten van klimaatverandering gemitigeerd konden worden.

Dankwoord

Op deze plaats wil ik graag collectief de collega's bedanken die een bijdrage aan de totstandkoming van dit proefschrift geleverd hebben middels het kritisch lezen van (delen van) dit proefschrift, discussies over de inhoud en richting van dit onderzoek, het toegang verlenen tot bronnen en artikelen waar ik zonder hen geen toegang toe gehad zou hebben en het verhelpen van allerhande computer- en bureaugerelateerde problemen. Daarnaast wil ik graag iedereen bedanken die op minder directe, maar niet minder belangrijke wijze aan dit proefschrift heeft bijgedragen door te laten zien dat het leven in een ivoren toren zo slecht nog niet is, mij deelgenoot te maken van de wondere wereld van de archeologie en anatomie van de Neanderthaler, de dagelijkse lunch tot een plezierig medium te maken voor discussies over de universiteit in het algemeen en de faculteit in het bijzonder en wat verder maar ter tafel kwam, de nodige afleiding van de wetenschap te verzorgen middels een biertje op zijn tijd, mijn paranimfen te willen zijn, en altijd interesse te tonen en de nodige morele steun te verlenen.

Curriculum vitae

Arne Wossink werd geboren op 5 mei 1978 te Hilversum. In 1996 behaalde hij zijn VWO-diploma aan de Minkema Scholengemeenschap te Woerden. Daarna begon hij aan de Universiteit Leiden met de studie Archeologie, die hij in 2003 afrondde met als specialisatie Archeologie van het Nabije Oosten. Aan het eind van zijn studie en tot korte tijd daarna was hij werkzaam bij een boekenantiquariaat in Leiden. In 2004 begon hij aan de Faculteit der Archeologie van de Universiteit Leiden aan het promotieonderzoek waarvan dit boek het resultaat is.